GAME OF MIRRORS:
CENTRE-PERIPHERY NATIONAL CONFLI

T0251122

To my wife, Maruja, and children, Arkaitz and Maia, with love

Game of Mirrors: Centre-Periphery National Conflicts

FRANCISCO LETAMENDIA
University of the Basque Country, Spain

Translated by
KAREN HATHERLEY

Routledge
Taylor & Francis Group

LONDON AND NEW YORK

Contents

PART II: NATIONALISMS AND VIOLENCE

Preface

Since the failure of communist systems, or of "actually existing socialism" in the words of Fernando Claudín, the old thesis of the end of ideologies seems to have triumphed. However, those ideologies destined to die seem to be, above all, theories of revolution. The end of the millennium is the triumph of the market and of its ideology; democratic liberalism.

Revolution is yesterday's problem; revolutionaries are often nostalgic romantics, sometimes fanatics of violence, but always on the margins. Political violence is today a criminal phenomenon. Western society and the entire world are governed by a unique system of values; the "ethical irrationality" of which Max Weber spoke is a relic without any future.

Nevertheless, historic and social reality is a great deal more different and complex than the model which proclaims the end of ideologies would suggest.

First, social revolutions are an exceptional event in the Western world. It is necessary to go back to the bourgeois revolutions of the 19th century to encounter real and lasting changes in the social system.

Second, the remaining and lasting social upheavals are of an entirely different kind. From the insurrection of the gueux in the Netherlands until the secession of Yugoslavia, from the American revolution until the war in Yugoslavia, the most frequent revolutionary changes have been quite different in nature: the endemic revolutionary problem of many Western countries has not been linked to the social question, but to the national question.

Third, the criminalisation of political violence waged against the democratic state does not correspond to factual reality. There is criteria to differentiate meaningful political violence from the terrorism of "desperadoes," the presence of real political support by a party which legitimates terrorism. All "social" terrorist groups which have been active in Western democracies may be referred to as social "desperadoes" without a trace of social support. However, the Corsican, Irish and Basque cases are entirely different: violence is legitimized by a significant minority of the population in Corsica, Ireland or the Basque Country and by the electoral strength of political parties such as Sinn Féin and Herri Batasuna.

Nationalism and the national question have represented a problem since the early years of the 19th century until the present day. Understanding these phenomena represents a challenge for political science; a challenge because the nation is not a natural phenomenon, rather it is the consequence of nationalism - nationalism created the nation, as Ernest Gellner said. It is a challenge also because nationalism is an ambiguous and ambivalent phenomenon.

All attempts made to reduce nationalism to one or several factors have not been credible (that was Stalin's mistake); it has multiple factors that are variable in space and time. Nationalism is a problem of beliefs and conscience linked to the historical action of nationalist groups. Stalin was a great Tsar, not an expert in social sciences.

The second difficulty derives from the distinction between the nationalism of the dominant and nationalism of the oppressed. Nationalist discourse is a combination of progressive, conservative and sui generis elements. After the second world war many theorists equated nationalism and fascism. Nevertheless, in the 1960s the political theorist Maurice Duverger argued the importance of distinguishing between two nationalisms: the imperial nationalism of fascists, colonialists and such like, on the one hand, and on the other, the revolutionary nationalism of Third World Countries. Today, the majority of political theorists admit that centre and periphery nationalisms are different and, therefore, adversaries. Only by using the centre-periphery model can the nationalist phenomenon be understood. Nevertheless, this is not the case with violence.

Letamendía's work is theoretical and, at the same time, comparative within a very wide spectrum, making use of a range of examples and facts.

The author does not limit himself to the Basque case, the reason behind his theoretical interest in these related phenomena. Paco Letamendía, "Ortzi," the defense lawyer in the Burgos trial, has first hand experience of Basque nationalism. He adds to it exemplary knowledge of the main manifestations of this phenomenon in the world. The national question and the question of violence are analysed in his book from the "critical decentred" position which Piaget has written about. Comparative politics is the best way to achieve both this position and clarity of perspective. Professor Francisco Letamendía's analyses are an essential contribution to this necessary reflection.

D.L. Seiler
Institut d'Études Politiques de Bordeaux.

Prologue

If national conflicts are absorbing for the lay-person, they have always been a multi-disciplinary and multi-faceted subject for the academic. I have defined for the purpose of study a perspective which is that of political science. I must acknowledge the intellectual debts for the book.

The first debt in terms of time is that owing to the supervisor of my thesis, René Gallissot, who, in the atmosphere of the University of Paris VIII and the French CNRS, helped me transform into theoretical reflection the questions generated by my previous experience as anti-Francoist lawyer and politician of the early post-Franco period.

Essentially, the first part of the book uses Rokkan's theory of the four cleavages that national and industrial revolutions have given rise to in Western societies, one of which places the centre and periphery in opposition to each. Daniel Seiler's (Professor of the Institute of Political Studies in Bordeaux with whom I am co-directing a project funded by the Secretary of Foreign Action of the Basque Country within the framework of Basque-Aquitaine co-operation) analysis of this conflict, with the aim of studying Western European political parties, is the central thread on which the first part of this book is based, and a decisive influence on the text. With respect to transformations in the centre-periphery conflict brought about by the belated decentralization of the welfare state and by the emergence of supra-state realities such as the European Union, I have taken the views of John Loughlin, Michael Keating and the team from the journal "Regional & Federal Politics" into account.

In the first part I highlight the reflected nature of peripheral nationalisms that mimetise the construction of national society and community by the center. In the second part I study the mimetic nature of the latter with respect to the legitimate violence of the state, which ends up, on occasion, counter-imitating these in a style which is very much a game of mirrors. The reference point for this part, which is not, in any event, Basque-centered, is research on radical Basque nationalism that I carried out for my doctoral thesis.

With reference to the study of specific nationalisms, I would like to highlight as particularly interesting, some of the conferences in which I have participated: "Nation-building and sub-cultures," Vaedbeck, Denmark, in 1993; "Os nacionalismos en Europe: pasado e presente," Santiago, Spain, in

1993; and "Peripheral regions and European integration," Belfast, Northern Ireland, in 1993. Regarding Eastern European nationalism, I must make mention of the following conferences: "National Movements and Social Practice, Comparative Analysis," Saint Petersburg, Russia, in 1992; "Restructuring identities and political spaces in Europe," Sieldce, Poland, in 1994; and "Ethno-politics on the threshold of the 21st century," Kazan, Tatarstan, in 1995. The works compiled in 1995 by Professor Berberoglu in the collective book, to which I contributed, *The National Question*, published in Philadelphia in 1995, has been particularly useful in the study of some Third World nationalisms.

This book brings together reflections on a subject which has interested me throughout my life. If the reader gets half the satisfaction which I have felt while writing it, then I will be pleased.

Francisco Letamendia
Algorta, Basque Country

Acknowledgments for the English Version

I would like to thank Karen Hatherley for her translation from Spanish which required much hard work and endless patience. Thank you also to Daniel Holder who helped on some of the translation and to Douglas Hamilton who subsequently revised the English text.

I would like also to thank Izaskun Elizondo, Leire Moure and Igor Ahedo, members of my research group in the University of Basque Country, for their help and hard work preparing the camera-ready copy of my book.

This book benefits from the research carried out within the framework of several projects on "Global and local, the European construction" financed by the Euskadi-Aquitaine Fund of the Basque Government.

I would like also to thank the Vice-Rector for Cultural Extension and University of the University of the Basque Country because of its financial help for the English translation of my book.

PART I
CENTRE AND PERIPHERY

1 The Centre (1): The Nation State

Nationalisms are not chance or deviant phenomena, but a social correlation of collective political action and social mobilisation deriving from a universal form of political organisation, the nation-state, which, in scarcely two centuries, has spread throughout the world. Nationalisms either operate through movements which drive the centre forward in order to carry out the national state revolution, or through a reaction to this process by specific ethnic, linguistic or religious groups (although in the majority of cases the language and religion form part of the attributes of the ethnic groups). The latter groups may counter-mobilise for many reasons: because they are obstructed or denied equal access to the processes of acculturation and socialisation; because an imbalance has evolved between their economic and/or cultural development and their political status; or because they are victims of discrimination or persecution due to their ethnic, religious or cultural identity. The peripheral reactive movements mimetise the process of creation carried out by the nation-state of the national community (*Gemeinschaft*) and, on occasion, of the national society (*Gesellschaft*). These reactions are, in their turn, liable to provoke new counter mobilisations of the centre in the form of centralist nationalisms. It is to the explanation of the very complex interactions which take place, in the era of the nation-state, between the centre and the periphery, by way of a game of mirrors, that this book is dedicated.

The European Genesis of the Nation State

Prior to the development of the nation-state, four forms of state organisation existed in Western civilisation, each one characterised by a particular type of relation between the centre and the periphery. They are: the plurality of city-states, in the Hellenic sphere; the universal community of the Roman Empire; the double power of the medieval Church-Empire; and the plurality of sovereign monarchic and territorial states of the Modern Age.

The political fragmentation of the Western part of the Roman Empire, brought about by barbarian invasions and the ruralisation of economic life and feudalism, was not an arbitrary process. Territorial principalities were forming which displayed a relative ethnic affinity.[1] The intermediate stage of the "kingdoms" began to emerge between the feudal fiefs and the imperial symbolic power, serving as a cohesive factor. Its authority was based on the one hand, on being a religious power, inherited from the barbarian chieftains and drawing inspiration from the biblical monarchies (and recognised as such by the unction of Christianity), and on the other, on being a superior political principle, in the sphere of the Greco-Roman tradition, which had situated itself above the feudal hierarchy (Legoff, 1984). In this way, a monarchical political formation, which foreshadowed the modern state, emerged in the late middle ages.

The absolutist system of the Modern Age was based on the construction of several political centres, the result of the transformation of a number of medieval kingdoms, which made up a system of differentiated, autonomous and centralised states engaged in reciprocal rivalry. The disregard of the pre-modern systems for the principle of territoriality, the lack of distinction between the private and public spheres which prevented the transfer of power from individual rulers to the superior state institution, disappeared in accordance with the depersonalisation of power. The modern state situated itself above society and guaranteed the continuity of order beyond the life of specific governments. However, the political centre also had to be distinguished from society, in other words, the separation between the political and social systems, between government and civil society, which liberalism will reinforce and preserve, had to take place.

The concentration of powers which were previously controlled, in a dispersed way, by municipal and ecclesiastical feudal rulers, appearing for the first time in the Glorious Revolution of 1688 in Britain - in which an elite formed by the gentry had given new expression to the liberal contribution of the previous Puritan Revolution - had occurred in some absolutist states of the 18th century. Enlightened monarchs such as Prussian King Frederick the Great, introduced institutions which would quickly become universal: a single language written and spoken throughout the territory; a universal and compulsory income tax for all citizens; and an unified army.

Other processes throughout the Modern Age prepared the way for the introduction of the nation-state. These were the conversions carried out by the Renaissance of the main "common" languages in the symbol of political unity, to which they owe their victory over Latin; mercantilism and the unification of currency; and the Protestant Reformation, which reinforced the idea that subjects of the same principality must form a single community (Vilar, 1980).

The move from absolutist state to nation-state culminated in the transformation of traditional society into a national modern society. Through "low" rationalisation the traditional structures were subjected to instrumental rationality (for example, the organisation of labour and commerce, the transport network, information and communication, institutions of private ownership and state bureaucracy). Through "high" rationalisation the traditional world views remained no longer principles of legitimation but became personal beliefs (Habermas, 1986). The nation-state transformed the subject into citizen and assembled in a new centre (political, economic and cultural) the forces which were previously dispersed.

The paradigmatic value of the French Revolution resides in the visibility of the traumatic break with traditional society (superior in this respect to the American Revolution, a revolt of white colonies against the English metropolis tax demands).

Several ideological strands of a specifically national character converge in the French Revolution:

1. the liberal concept of the nation which originated in England and Holland and was developed by Locke. It was individualist and representative in nature, based on freedom and property (which influenced the Girondists and Thermidorians);
2. the genealogical concept of the nation-fatherland of Montesquieu and Burke, popularised by the first Romanticism, according to which the nation is an organic body made of descendancy and blood and is therefore an expansion of the family, personified through a real person or natural leader;
3. the democratic concept of the nation, of Rousseauian origin, which argues that sovereignty is the collective based on "the will of the people", not on the individual citizen (a concept adopted by the Jacobins, especially by Robespierre) (Gallissot, 1976).

If the revolutionary period is taken in its entirety and the Napoleonic phase is included (1789-1815), the economic, political and cultural functions through which the new nation-state created the centre can be clearly distinguished. In this period fiscal privileges, provincial customs posts and guild regulations began to disappear. The Church became a national institution; the aim of education - in accordance with the teachings of Rousseau in *Thoughts on the Polish Government* - was to create patriots through the teaching of the national language. A single language became a cause of patriotic pride and an instrument for nationalisation. From the Revolution onwards teachers of the French language were sent out to all the regions where different languages were spoken. In 1794, Deputy Barrère claimed in his report to the Convention that dialects and extraneous languages spoken in France "perpetuate the prevalence of fanaticism and superstition and safeguard the rule of priests and nobility" (Shafer, 1964).

In the first phase, the armies, previously dependent on the king and the nobility, become "le peuple en armes" (the people armed); they were formed by national guards in which all citizens enlisted. This phase quickly opened the way to professional revolution (Perlmutter, 1977) in which the professional military - whose role model was Napoleon - prepared and organised the popular army. The officers were staunch defenders of the nation-state. With the nation-state the delineation of state territory by borders acquired a new meaning, being defended by might (but in the modern state there are only political, not natural, borders). And the concept of internal enemy emerged when national space closed against those who became "foreigners" within its borders.

The development of society accompanied that of community: the transfer of the group psychology to the nation culminated in an irreversible process, and the concept of the nation acquired an emotional charge which allowed it to demand enormous sacrifices, including that of one's own life, when national sovereignty had to be defended against "the enemy". The internal enemy, which included up to a certain hypothetical point, in the French Revolution, the monarchical tyrant, became an external and perfectly real enemy at a later date when the conventional troops undertook an external war. For this reason, when the Napoleonic legions conquered almost all of Europe to the exclusive advantage of imperial France, men of action and theory emerged in these conquered European states preaching at one and the same time national reunification, patriotism and the fight against the invader. With few exceptions, the flame of patriotism did not spread through loyalty to France, rather it was a reaction to French enemy.[11]

The Nation-State and the Formation of Society: Economic and Political Functions

The formation of the centre by the nation-state requires three types of functions:

1. economic - the creation of the market by the Industrial Revolution.
2. political - integration and mass participation.
3. cultural - socialisation and legitimating mechanisms.

The economic and political functions construct society; cultural functions construct community.

These processes, conflicting in nature, are based on relations of power, domination and subordination, inherent to the political sphere. Political power strives for the stability of power relationships which spring from the unequal distribution of resources within a specific society: economic resources (material goods and services); cultural resources (symbolic goods) and political resources (coercion) (Braud, 1985). The mode of production of economic goods in the early nation-states is capitalism. The mode of production of cultural goods is legitimacy. The mode of production of coercion came under the jurisdiction of the state in modern societies and expresses itself through the monopoly of legitimate violence. Domination consists in the stabilisation of the relations of power within each sphere; the specificity of political power is that of being a product of the conjunction of the three dominations.

In Western nation-states the conflicts which are found at the basis of these three modes of domination give rise to a fragmented and contradictory process of formation of the centre. They generate fracture lines, or cleavages, out of which emerge different forms of modern collective political action: social movements, including ethno-national ones, pressure groups and political parties. According to the Rokkan-Seiler approach (Seiler, 1980), the two revolutions, the national and industrial, which in their interrelation form the bourgeois Revolution, generated the four cleavages spanning Western societies, and from which arise the families of existing political parties. The Industrial Revolution brought about, on the functional axis, the owners-workers confrontation, divided by the unequal distribution of the resources produced. It is in this conflict that bourgeois and working-class political parties have their origins. On the territorial level the rural-urban confrontation developed, giving rise to agrarian political parties. The national revolution brought about, on a functional level, the church-state conflict, a dispute, particularly in catholic societies, over control of the education system, from which both anti-clerical and pro-religious political parties emerge. On its territorial level the centre-periphery conflict emerged where ethnic differences prevented or denied specific social groupings equal access to the process of nationalisation. This conflict, in any case, was fed by the tensions generated by all the other cleavages existing in modern societies.

The first generation of nation-states created a unified market, a space for the accumulation of capital and the movement of goods and services, including the labour force. The non-intervention stance in the internal capitalist logic[III] did not mean passivity; the state imposed, in opposition to the *Ancien Regime*, the free exchange of goods and services and the abolition of the local tolls, and eradicated the feudal concept of mortmain. This modernising project included support for new farming techniques, as against backward traditional methods, and the promotion of free industrial development, as against the restrictions of guilds. Mercantile capital was assigned a secondary role due to its preponderant rapine nature, and a new modern system of bank credit was introduced, as against moneylenders. It was, in short, the creation of a "national" unified market, in contrast to the old fragmentation of the past.

However, economic modernisation did not take place without previous accumulation, a ruthless historical process of creation of the proletariat by evicting peasants from their lands and artisans from their work places and tools. It demanded the generation of surplus in the rural world which will meet the food requirements of the increasing urban population. This process provided a mass base, those damned by capitalist development, for the legimitist movements, which aligned traditional elites with peasants and artisans threatened by proletarianisation, under the leadership of the clergy, organic intellectual of the *Ancien Regime* (Seiler, 1996).

The Industrial Revolution also created a hierarchy within the space of the state. Blind to the location of the accumulation of capital, it contemplated the appearance of one, or very few, economic centres, coinciding or otherwise with the political centre, surrounded by a constellation of peripheries which were subjected to relations of uneven development and forced to supply raw materials and food to the centre, or centres. This situation, whether or not it involved ethnic differences, created pockets of internal colonialism in specific regions (Andalusia, Estremadura and Galicia in Spain; Occitany and Brittany in France; Mezzogiorno in Italy; and the Old South in the United States). Where economic centres do not coincide with the political centre (as is the case in Catalonia and, until recently, the Basque Country in Spain; the Po Valley in Italy; Slovenia in the former Yugoslavia; and the Baltic Countries in the former Soviet Union) inverse colonialism may arise which could then lead to movements from the wealthy periphery demanding higher levels of self-government from the poor political centre. It may also be the case that poor peripheries ally with the centre against the wealthy peripheries.

The unification of the market is then a conflicting and, to a certain extent, illusory process. Although early liberalism conceived of nation-states as simple shells having no other role than that of being an insurmountable defence of the process of production and distribution, and denied the possibility of unequal exchange, the fact is that the tendency of capitalism toward the globalisation of the relations of production, consumption and distribution situates the centre-periphery conflict on a global scale, bringing about the dependence of "proletarian nations" on "bourgeois nations."

The primordial conflict generated by the Industrial Revolution between capitalists and the proletariat, which excluded the latter from the nation in the early phases of liberalism - the working class do not have a country, Marx claimed in 1848 - was an obstacle which had to be overcome politically. The nationalisation of workers through their political involvement, firmly established with the introduction of universal suffrage, took place in the West during the second half of the 19th century, often, in the most developed states, in complicity with imperialist expansion. In the case of the least developed states, it was based on their participation in a "sacred union" against foreign enemies or internal minorities.

The function of political integration began once the process of monopolisation of legitimate violence had been concluded by the military and police state apparatus. It required, on the one hand, the co-ordination of the various administrative, political, economic and cultural domains of the society; the weakness of any of these would mean a frail market, as was the case in Spain during the 19th century, one of many defects in its process of integration.

In this process the following stages can be distinguished: penetration of civil society, standardisation, mass participation and social redistribution (Rokkan, 1983). The first two stages take place in the liberal phase of nation-states, defined, except when the move from traditional to national society has been revolutionary, by weak popular participation, limited universal suffrage and a system of political representation dominated by the parties of "notables". The construction of the state is the task of a small core of competitive elites; the local echelon, lacking in autonomy, becomes the vehicle of the centre for territorial penetration.

Whereas the start of the process of integration tends to fragment archaic state forms which were multi-ethnic empires - although the Ottoman, Austro-Hungarian Empires and Tsarist Russia survived until the end of the World War I - it does, on the contrary, bring about cohesion in those nation-states which emerge out of pre-modern absolutist states. Even in these cases the process of integration is never complete; the centre is faced with difficulties rooted in the periphery's diverse cultures, traditions and identities, which it must moderate through various measures of territorial administration (Keating, 1996).

Nation-states may create state bureaucracies on a territorial basis, which while representing the central government in the periphery, mediate between the latter and the centre (for example, Secretaries of State in the United Kingdom and Prefectures in France). They may decide economic and fiscal policies and tariffs which protect important sectors of the economy on specific peripheries (protectionists tariffs demanded and finally obtained by the Catalan bourgeoisie). In Anglo-Saxon societies where the state is weak, even though unified, it preserves the autonomy of peripheral decentralised institutions on a functional or professional basis (such as the Kirk, the Scottish Church which for many centuries was responsible for education and hospitals).

The state may establish a clientelist relationship (that is, a utilitarian alliance between two interlocutors of unequal status in which the dominant partner offers protection to the client in exchange for submission) with peripheral elites, by placing them in a privileged position and converting them into the mediators of the peasants. The "notables" in the south and the west of France and the chiefs of clans in Corsica are the intercultural translators of rural Catholic society in non French-speaking areas, integrating them into the nation, to the great advantage of the right-wing governments of the III Republic, who viewed them as the traditional counterbalance to the working-class bastions in the north and east of France (Loughlin, 1987).

The integration of the centre depends not only on the functional co-ordination of the various political, administrative, economic and societal domains, but also on the spatial grouping of the functions. Paris is the political, economic and cultural centre of France, but in Italy the economic centre is in the north, the cultural centre is in Tuscany, and the political centre is based on the alliance between Rome and its clientelist ramifications in the south of the country.

A weak political integration could foster the role of modernising agents of the peripheries, either because of the weakness of the market (as was the case in Spain), or through fragmentation inherent in the archaic empires (Central and Eastern Europe). Nevertheless, it is often the case in the West that consciousness-raising in the various peripheries takes place in the phase characterised by universal suffrage and the appearance of mass political parties. Various factors, such as industrialisation, migration, literacy and mass media result in a move away from passivity toward mass mobilisation and the demand for greater political involvement. Competitive oligarchies may respond to these demands either by opening up the regime to the new groups by broadening suffrage (which has been almost always the case in Northern Europe) or by forcibly demobilising the masses through the imposition of a dictatorship.

States introduce a national school system as a means of political integration, (although such a measure can also be taken at an early stage of the process of national-state construction) by assuming control of the education apparatus.[IV] The language used within the schools is that of the dominant ethnic group within the state. Opposition to this measure is double: that of the Catholic Church in southern Europe, which had until this point enjoyed a monopoly of control over teaching, and that of the minority ethnic groups (which explains the pragmatic alliances between the Church and some peripheral movements, tingeing the ideology of the latter with religious traditionalism).

The fact that many nation-states do not introduce members of peripheral ethnic groups into their apparatus of repression (it is unusual to find Basques and Catalans in the Spanish army or police after the end of the Carlist Wars) makes them a powerful catalyst for a hostile response from the periphery, thereby blocking their political integration.

This situation opens up the possibility of a third response, in addition to greater suffrage and authoritarian tactics, to the demands for wider political participation: the counter mobilisation of the masses of the centre against the reaction of the periphery. The ideological apparatus of the state can encourage the formation of a "sacred national union" against particular ethnic, religious or linguistic minorities, which, after being identified as the "internal enemy", are sacrificed to the patriotic furore of the majority. This response, often found in those states which emerge out of de-colonisation and lacking in civil society, and in which the administrative apparatus is dominated by one, or several, ethnic groups, could lead to pogroms and bloody persecutions (the oppression of the Kurds in the Near East or the domination of the Tamil minority by the Singhalese majority in Sri Lanka). Although such processes have also occurred in the West,[v] they tend to adopt more subtle ways of ethnic stigmatising and stereotyping.

The Formation of the Community: Cultural Functions and Mechanisms of Legitimacy

The cultural functions of socialisation and acculturation, reinforced once the state has assumed control of the education system, seek to obtain the loyalty of citizens to the state and make them co-partners in the nation.

Through these processes the state consolidates the values, norms and symbols which constitute, in the broadest sense, the national culture. In a narrower sense, it inculcates in its citizens the basis of its political legitimacy, that is, the system of stable attitudes which create a positive outlook toward the regime. The process of social learning, which consists of political socialisation, is crucial during childhood and adolescence, even though it continues throughout the individual's life.[vi] The political culture created by the processes of socialisation expresses itself through political identities and includes evaluating, cognitive and affective factors (Schemeil, 1985; Magre and Martínez Herrera, 1996).

The formation of political identity orders the range of feelings of belonging experienced by a child or adolescent, feeling dominated hierarchically by the sense of belonging to the nation. Political socialisation structures, through the school system, the perception of this notion in the two identity levels of space and time (*a priori* forms through which, according to Kant, the human being perceives of reality). If the state is the exercise of sovereign power over the population which inhabits the territory defined by national borders (De Blas, 1988) then this territory should not only be protected by interdiction (attacks against national territory constitute crimes of treason in criminal law everywhere) but also by an emotional bond that must be built between the citizens and the territory. If the Nation is the permanence down through the generations which unite the state with the citizens, the naturalness of the continuity of the nation in time must be instilled in them. This double work of education is carried out through regular schoolwork on the national geography, history and literature. The formation of national identity emerges from the selective rejection of negative identity phenomena and the assimilation and absorption of a series of positive infantile identity phenomena (conceived of as places of belonging: family, class, town, nation) in a new configuration, thanks to which a specific society provides a young individual with an identity (Erikson, 1972).

The most successful national processes of socialisation and acculturation lead to the formation of a "civic culture", an ideal-type of political behaviour described by Almond and Verba (1970), defined as a high degree of participation in structures considered legitimate, a sense of personal political competence, or efficiency, and a broad participation in voluntary associations.

However, the process of socialisation selects its own hierarchies and turns many groups into candidates for exclusion, or at least marginalisation. The socio-economic status and ethnic origin of the family discriminate the children in the school system. Bourdieu[VII] (1977) points out that it is in the nature of teaching to ensure that those who are excluded internalise the legitimacy of their exclusion, and also to convince those who are downgraded to second class schools to accept the inferiority of their schooling. In this way the relation of the state process of acculturation to the many social or ethnic sub-cultures (whose existence may be tolerated, ignored or repressed) is often a cause of conflict, and always hierarchical in nature.

The creation of the national community also requires the same mechanisms of legitimation. The irreducible nucleus of political power rests on an alliance of coercion and legitimacy which, in turn, consists of the consent of human beings to their domination. The mechanisms of legitimation are the rationalisation, idealisation and sacralisation of power (Lagroye, 1985).[VIII]

The legitimating mechanism of the idealisation of the leader, charismatic legality, as defined by Weber, applies as much to traditional as to modern societies. It is worth referring here to the psychological principle of identification described by Freud (1921), in which the "ideal I" of the individuals forming part of a collective is personified in an external actor which embodies the functions of ego and super-ego, in other words, the leader. It is this way that the individuals are bound with a double tie: that which binds the individuals among themselves, and the identification between the individual and the leader. In contemporary pluralist societies the leader, who exercises a legalised power, is often backed by consensus rather than enthusiastic support. Nevertheless, here too the idealisation of the leader takes place, allowing the electorate to love, as well as obey him.[IX] For this reason the mechanism functions with greater difficulty the more the citizen diverges from the ethnic, linguistic or religious origin of the political leadership (a problem which Machiavelli observed when, in his calls to Italy to free itself from the Barbarians, he reclaimed the ethnic community of the Prince and his subjects).

In modern states an archaic mechanism of legitimation which also survives is that of sacralisation. It consists of a fusion between the sacred and political power (even justice and the law in the West retain the imprint, in its language, symbols, form and rituals, etc., of the period in which law was the invention of priests) (Soulier, 1985). Power is the attribute of the majority group and, as such, it also therefore constitutes, in the nation-state, the sacred centre of society. According to Balandier (1981), "in every society, political power is not completely de-consecrated [...] Whether it is visible or hidden, the sacred element is always present in the core of power. It is through this that society is conceived of as a unit. Political organisation introduces the true totalising principle: order and permanence."

When, historically, the relation with God and salvation was de-consecrated and societies became secular, the sacred did not disappear, rather it was displaced; it is in this respect that the Rousseauian doctrine of national religion applies. And the political nation, correlate of the nation-state and utilitarian organisation for the citizen, became an instrument of legitimation of present-day states, not only through rationalising mechanisms, but also through (to a greater or lesser intensity, according to each case) the older mechanisms of sacralisation.

Notes

[I]   In the Germanic sphere, for example, Saxon, Bavaria, Franconia, Swabia; in the Franco sphere, Burgundy, Aquitania, Gotia, the Tolosan region, the Hispanic March, Normandy, the County of Flanders, the Breton March, etc. (Dhondt, 1987).

[II] The figure of the enemy has been, in effect, structural in the nation state since 1792. If the 19th century is considered in its widest sense, that is ending with the first world war, to be the offspring of the French Revolution, it does not earn this title for being the century of the rights of man and citizen which the Revolution proclaimed, but rather for what it created without naming it, and perhaps without knowing it: the era of the Nation States, more focused on hostile polarity toward the enemy (exterior or interior) relying upon the huge energies liberated by the transformation of a society of subjects into a people-nation, than in defence and safeguard of liberal rights. The parable that beginning in the constituents of 1789 goes from the Girondists, Jacobins and Thermidorians to Napoleon is an evidence of it.

[III] Adam Smith preached the need for men to accept the Project of Divine Providence, which had transformed the desire for personal improvement into an element of admirable social harmony.

IV   National school systems were set up in Holland in 1806, in Greece in 1823, in Belgium in 1842, in Argentina in 1853, in Brazil in 1954, in Spain in 1857, in Italy and Romania in 1859, in Finland in 1866, in Hungary in 1868, in Japan in 1872, in Bulgaria in 1881, etc. (Hayes, 1966).

V   The persecution of blacks followed their emancipation after the north American War of Secession, not to speak of the anti-Semitic "final solution" of the Nazi regime.

VI   Secondary socialisation is ongoing in third level education, but it also occurs in the workplace, through the mass media, political parties, peer groups, and voluntary organisations.

VII   In the United States black children, educated according to norms and forms of behaviour appropriate for white children, end up with a level of schoolwork performance which is much lower than that of their white counterparts (Dowse and Hughes, 1982).

VIII   Weber (1979) defines legal-rational legitimation as that which in the relation of power is subjected to the following rules: activity of the public function defined legally; competence in the exercise and objectively delimited power of authority; definite hierarchy, which provides subordinates with the opportunity of appealing to a higher authority; legal and technical rules which govern access to public positions; finally, complete separation between the use of administrative goods and their private appropriation.

IX   This mechanism is based on identification, which links the individual separately with a higher power above him (it must be completely differentiated from the identity, a collective sense of belonging which amalgamates groups). This process has a number of perverse ways of expressing itself, by tingeing acceptance of political subordination with affection: in the psychoanalytic sphere identification consists in the state of mind of the person who begins to feel at one with another, to adopt their way of being and their attitudes, to mentally putting himself in their place, internalising and modifying himself (Reich, 1972). The I, in order to defend itself from internal and external attacks, identifies with the frustrating reality represented by the people who embody it. In this way, the aggression which had been mobilised against the frustrated person turns against the I, and this expresses itself through reactive attitudes. Lefebvre (1976) says in this respect, "Identification has within it a terrible and evil strength known to psychologists and psychoanalysts "(...) It has political consequences. It implies and presupposes a relation of domination and subordination. It annuls in an ideal and real way a social distance and a political space which at the same time it reproduces. The subject identifies with the prince, the king, the dictator [...]What happens is that the victim identifies with the tormenter".

2 The Centre (2): Types and Origin of Nation States

Centre-periphery conflicts vary according to the particular characteristics of the nation-states themselves. The states may be mono-cultural or pluri-cultural in nature, depending on whether they are the home to plural religious, linguistic or ethnic groups; they may be strong or weak states, depending on the nature of the state-civil society relationship; unitary or composite, depending on whether they have been formed from one or various entities. Equally important is the genesis of their formation, which means they may emerge from the modernisation of a traditional unitary state (essentially the model described in the previous chapter); from the break up of an empire, or from de-colonisation.

The first type has barely any relevance in terms of classification. Mono-cultural states are an exception (except for micro-states, the only ones to be found in Western Europe are Denmark, Iceland and Portugal). Their peripheries are of an exclusively socio-economic nature.

In the Far East of Asia there are two large political aggregates which are almost completely mono-cultural: Korea (divided since the war of 1950-53 into the northern and southern states) and Japan, which has only one small minority ethnic group, the Ainu. The present number of about 20,000 Ainu living in the north of the Japanese island of Hokkaido and in the south of the Russian island of Sajalin, are the descendants of an indigenous proto-Caucasian people which at one time inhabited the whole of Japan, until they were driven toward the north by the expansion of modern Japanese (Gonen, 1996).[*]

Unitary, Federal and Autonomous States

From the point of view of territorial control by political power, the two historic state models have been the unitary state and the confederation of states (Seiler, 1994). Other models have appeared in the era of the nation-states, the federal state and the autonomous state. The federal state is the result of a compromise between the unitary state and the confederation,

---

[*] Case studies have different width and size of font than the rest of the text. One line of space separates each case study.

while the autonomous state is a consequence of modifications introduced in a unitary state in crisis with the aim of preventing its further erosion. Given the incompatibility existing between the confederation of states and the inseparable bonds between nation-state and sovereignty - the Swiss Confederation has been a federal state since 1848 - the range of states can be reduced to the unitary, federal and autonomous models.[1]

According to Seiler, defining the type of state demands going beyond constitutional definitions and determining the degree of territorial control held by the peripheries; it is a question which must be examined within the framework of the corresponding historical process. History shows that the unitary state emerged with the establishment of absolutism, in the geographical space of what Rokkan (1970) describes as "maritime imperial nations": France, England, Portugal, Castile and Denmark. This was the area adjacent to the network of city-states located at the central axis which runs from north to south - and continues to do so in the present day - of the European continent. The formation of the state, which requires the absence of a dense concentration of powerful cities, was so much easier the further the state was away from the catholic counter-reformation. The strength of the French and English states derived from the existence within them of either a Protestant church - Anglican - or of a national Catholicism autonomous of Rome - French Gallicanism[II]. In the East, in the area of the "Continental Imperial Nations", new unitary states based on Lutheranism and the power of the army appeared; the Swedish model was brought to perfection and copied by the Russia of Peter the Great.

In the confederation of states - the oldest model dates back to classical Greece - the centre is created out of agreement among the different parts which delegate to it a variable number of powers related to the exterior (defence and diplomatic relations at the very least and, occasionally, currency) allowing them to withdraw freely from any aspect of the federal arrangement. Modern confederations originated in the early middle-ages, in the zone of City-States: the Lombard League, the Hanse League, the Republic of United Provinces (the Netherlands), the United States of Belgium, the Rhine Confederation, the Germanic Confederation, the Swiss Confederation, etc. It was in these leagues of urban merchants - more so than in England - where a true government of civil society took place. In the era of nation-states, the pluri-sovereign conglomerates were either absorbed by strong nation-states or suffered ongoing territorial losses (as was the case of Belgium). The Swiss Confederation, due to its setting as a defensive mountain enclave, was protected, although after the Sonderbund civil war it became federalised.

Federalism constituted, at the end of the 18th century and the first half of the 19th century, a transitional solution between a confederation of states and a unitary state. The north American federal model was a legal invention born out of compromise between Jeffersonian confederates and Hamiltonian unitarists. Federalism retains from unitarism the state nature of the centre and the formation of a single national will. From the confederation it retains the state nature of the constituent parts and the participation of these in the formation of the consensus of the central state. In this case, contrary to the confederation where only the parts are states, both the parts and the centre are referred to as states.

The autonomous state is the result of the transformation of the unitary state and represents the most recent solution to its crisis. Its genesis is, therefore, the inverse of the federal state; while the latter originated from the drive toward unity by the different parts of a confederation, the autonomous state takes note of the failure of the unitary state and sets up its decentralised solution. In any case, this makes its influence strongly felt on the autonomous state: the parts which form it enjoy autonomy, being able to exercise the prerogatives of the three state powers, but they do not participate in the formation of the will of central state, which has complete freedom of action. Likewise, the centre, which holds the exclusive sovereignty of the state, is the source of power for the parts which, for this same reason, are not sovereign. It is this absence of sovereignty in the component parts which differentiates autonomous from federal states, not the degree of authority assumed, which could be greater or lesser than that present in federal states. In the last quarter of this century the erosion of unitary states in the West has facilitated their transformation into autonomous states, which, moreover, follows the internal logic of the evolution of the welfare state.

Federalism, as a factor in the formation of a nation-state, is almost always a consequence of the presence of an external enemy. Tensions among the different constituent parts are resolved in a move toward defensive unity. Subsequent conflicts can take place if any of the original parts reinforces itself as a centre, making the others peripheral, in which case they will support a return to their initial sovereignty. If the crisis is resolved maintaining unity, federalism becomes formal and the nation-state acquires a solid centre, in which case it comes to resemble a unitary state.

The first federal states (the United States, the Swiss Confederation since the middle of the 19th century) represented a constitutional anomaly. Their structure was based, in the United States, on the specific constitutions of the member states, on a federal constitution which prevailed over them, on a two-tier parliament formed by two Houses - one for the federation and the other for the federal states - and on a federal tribunal which resolved conflicts among member states and between the central state and member states. They were not a political exception with respect to their centralising will; in this regard they were similar to other nation-states.

Present day Switzerland originates from a confederation of cantons formed in the early middle-ages against the attempts to annex it by the Hapsburgs. In 1291, the three cantons of Schwyz, Ury and Unterwalden united, integrating thirteen of them by 1513. After the defeat in 1847 of the secessionist Sonderbund League, made up of seven catholic mountain region cantons that had united to defend the system of confessional schools - among which were the original founder members of the confederation - Switzerland became the consociational democracy of today, where relations between the catholic and protestant communities and among the French, German, Italian and Romance linguistic groups are conducted by consensual elites. The only serious recent conflict in Jura was resolved in the sixties by plebiscite, on the basis of which a new canton of this same name, made up of three French-speaking catholic areas, was created. Its creation brought to twenty-three the number of cantons in Switzerland.

The United States was born out of the unification of some parts, the thirteen colonies, in which there were no ethnic, linguistic or religious differences. These colonies had very different origins: mercantile companies, fiefdoms awarded by the king to feudal lords, Calvinist communities of puritans escaping from the persecution which they suffered in Europe; they were constituted on the basis of "social contracts". The drive toward unity arose from their common struggle against the fiscal demands of the British metropolis. Until well into the revolution, the thirteen colonies continued to consider themselves independent: the *Articles of the Confederation,* in force from 1781 to 1789, proclaimed the sovereignty of each state. The federal form, defended by Hamilton and finally adopted by the 1789 American Constitution, had been accepted by the Philadelphia Convention as a synonym for centralisation. The new nation-state was consolidated in the subsequent fifty years, although tensions continued between, on the one hand, the centralism of the Boston and New York bourgeoisie, which had encouraged Hamiltonian federalism, and on the other, the aspiration toward local autonomy supported by local communities who drew inspiration from President Jefferson's thinking. The expansion toward the west marginalized the Native American Indians, who, as a result, were deprived of their means of subsistence and substituted by the colonists for black slave labour, which was imported in large numbers from the African Gold Coast. The conflict leading to the War of Secession in the years 1861 to 1865 had economic causes: the "peripheralisation" of slave plantations of cotton of the south by the Yankee industrial north of New England and the Great Lakes. The Confederate States of the south, which considered that the constitution granted them the right to secession, were politically and militarily defeated by the northern Union.

From this point onwards, the successive waves of European, Asian and Latin American immigrants turned the United States into a melting pot, relatively speaking, in the shape of an ethnic pyramid, the peak of which was dominated by White Anglo-Saxon Protestants (WASPs). The centre-periphery conflicts have been made territory-less. The Native American Indians lack the resources necessary to defend their claims. And, leaving aside the mobilisations of the black community in the sixties and seventies, which placed it above the Hispanic community in terms of social rank, together with the sudden and sporadic explosions of anger in ethnic urban ghettos, tensions are borne with apathy and despondency by marginalized sectors. This explains the high and chronic rate of non-voting in elections in the most powerful of all nation-states on earth.

The acute crisis of the unitary state can make its dissociation necessary through a federal solution, which can be brought about in a number of ways. Firstly, as a result of international pressure on the internal order (in the same way that German federalism was imposed by the winning sides of the Second World War, which worked because it was based on previous historical divisions). Secondly, through the replacement of one political regime for another, the multinational nature of which it wished to emphasise (for example, the substitution of the Tsarist empire for the Soviet Union). Or, thirdly, through the intensification of the centre-periphery conflict, making it necessary to carry out constitutional changes of territorial significance.

The recent federalisation of Belgium, a bi-polar unitary state created on the basis of the Flemish and the Walloon groupings (together with a small German minority) is an example of this latter case since it emerged out of a crisis in its previous consociational set up.

The Belgian state, with a present-day population of ten million, one million residing in the capital of Brussels, is made up of three ethnic groups: the French-speaking Walloons (3.2 million) in the south, the Flemish (5.7 million) in the north and a small German-speaking community of seventy thousand in the east. They are divided in this way, among the nine historical provinces: the Flemish are in western Flanders, eastern Flanders, Antwerp, Limburg and the north of Brabant; the Walloons are in the south of Brabant, Hainaut, west and central Liège, Namur and Luxembourg; and the Germans are in the east of Liège. Brussels is an enclave composed of eighteen urban communities, the majority of whom are French-speaking surrounded by the Flemish Brabant.

Belgium won its independence after a fifteen year struggle against Dutch rule (1814-1831), in which both Walloons and the Flemish bourgeoisie equally participated, and its Constitution guaranteed linguistic freedom. However, the Flemish language, divided into many dialects, was the language of the poor and defenceless peasants. To speak French, the language of culture of the era, was a means of social mobility for the Flemish elite. Laws were passed in French, a decree in 1830 declared it the only official language of Belgium and the administrative capital, Brussels, originally Flemish, became more and more French-speaking.

The cultural demands of the Flemish did not cease throughout the whole of the 19th century. Little by little the right was won to have Flemish literature courses taught in the University of Gante, and in 1896 Flemish and French were awarded equal rights. However, the Flemish had to continue to emigrate to the north of France and Wallonia to work in the powerful textile and steel industries and the coal mines.

Nevertheless, after the Second World War, in the fifties, the relationship of economic forces between Flanders and Wallonia was dramatically reversed. Multinational capital investments became concentrated in the Flemish north in the electronic and automobile industries, at the same time as the coal mines and steel industries in Wallonia suffered a decline and came to a standstill. Wallonia then began to demand regional decentralisation in the economic decision-making process. In this way, while the Flemish national demands were focused on the requirements of a politico-linguistic *community*, the demands of the Walloons were for an administrative-economic *region*.

In the same way, the primitive system of parties underwent a double transformation. Belgium politics had been dominated in the first half of the 20th century by three traditional parties: the Catholics, the socialists and the liberals. Until the fifties the ethnic factor did not play a role as the overriding criteria of the parties and the structures of the state remained as centralist as they were in the times of Napoleon. But from this decade onwards, and in a first phase, the growing sense of group identity gave rise to ethnic parties, new ones in the case of the Walloons, reborn in the case of the Flemish.

In the inter-war years there were several Flemish parties. In 1919, the *Front Partij*, later the *Vlaams Nationaal Verbond,* emerged, but it was discredited after 1945 due to its relations with the Nazis. It was then that a modern party, the *Volksunie*, entered parliament in 1954 with 3.9 % of the Flemish vote, rising to 18.8 % in 1971. In Wallonia the extreme left Mouvement Populaire Walloon appeared in 1961; in 1968 *Rassemblement Wallon* emerged. The law of August 1963, which generalised regional mono-linguism with the creation of four linguistic regions - Flemish, French-speaking, German and the capital, a bi-lingual Brussels - gave rise in its turn to *Front Democratique des Francophones* in 1965, which aimed to defend the rights of the French speakers of Brussels, and which soon won 10 % of the votes.

Even more decisive than the appearance of parties which were overtly nationalist, was the polarisation of traditional Belgium parties along the lines of communities which took place in a second phase: the catholic party, split in two parts since 1965, adopted in 1968 the names of *Cristelijke Volspartij* in Flanders and *Parti Socialiste Chrétien* in Walloonia. The socialist party held out longer, but it finally split in 1978 into the Flemish *Socialistische Partij* and the *Walloon Parti Socialiste* in Wallonia.

As a result the unitary policy of the Belgium consociational type could not continue to work; it is for this reason that at the end of the seventies a review of the Constitution was set in motion. This was carried out in three phases. In the first, from 1970 to 1971, a constitutional reform aimed to institutionalise the Flemish gains, protect the French-speakers and put direct pressure toward more radical change of a federal nature. It also aimed to introduce (with a Solomon like approach to satisfying the demands of both parties) two concepts which did not agree territorially: the three linguistic *Communities* - Flemish, French-speaking and German - and the three administrative-economic *Regions* -

Wallonia, Flanders and Brussels - with their respective regional councils. With the exception of the Prime Minister, the ministerial offices were divided in two: Flemish and Walloon, and two groups were also created in the House of Congress and the Senate.

In the second phase of the review of the Constitution, from 1980 to 1983, the three communities acquired powers in areas such as health, social services, family rights and emigration, while executives were created both for Communities and for Regions; each was supplied with their respective source of expenditure. However, this solution which favoured the Walloons, who, in spite of being fewer in number, had two French-speaking regions, Wallonia and Brussels, did not satisfy the Flemish. This latter group viewed the Belgian state in terms of two large communities and not three regions. (In this second phase the German community was awarded the same autonomy as the two other communities).

The third phase of constitutional reform, which began in 1988, was approved by a slim majority of the Belgian parliament in February 1993 and ratified by the Senate in 1994, transformed Belgium into a federal state. This phase established the direct election of the members of the Regional and Community Councils, who up until then had been delegates of the National Assembly, which were then given internal autonomy.

But this complex process of decentralisation at two levels - distinct in this sense from all the other European decentralising processes - still did not satisfy the Flemish community, mainly due to the configuration of Brussels as a region (Seiler, 1982; Wakabayashi, 1994).

Strong and Weak States

It is not possible to identify the strong state with the unitary state and the weak with the federal, or autonomous, state. France and the United Kingdom, strong and weak models respectively, are both unitary states. While the criteria which distinguishes these states from the federal and autonomous model is territorial control, that which differentiates strong from weak states is the relation of the state with regard to civil society. Strong states, differentiated, are distinguished by the presence of a military civil bureaucracy that is meritocratic and universalistic, the importance of public legislation, a system of control over schools and universities, unitary administrative territorial administration and by the importance of the high public function in state power. Weak states, hardly distinguishable from civil society, are defined by a higher degree of democracy, loyalty shown by the territorial and social groupings toward the centre, the simplicity of its bureaucracy, the late arrival of public law, a private school and university system, and by its self-regulated territorial administration (Birnbaum, 1985).

The continental tradition of the strong state - France, Spain, Germany - and the Anglo-Saxon tradition of the weak state - Great Britain, the United

States - derives from diverse historical situations and theoretical bases. In Europe the process of state construction did not geographically coincide with the development of capitalism. The strength of civil society in Holland and the United Kingdom (the exchange of goods by private owners, public opinion and a network of voluntary associations not controlled by the state) prevented the full differentiation of the state. The thinking of Locke and Rousseau, liberal and democratic respectively, was initially the consequence of and then later the cause of the diverse paths followed in both spheres. The Lockian minimum state, the commonwealth, emerged to defend society (understood as an ensemble of individual rights, especially economic ones, inherent to the state of nature) from its dangers. The active state of Rousseau, the "general will", emerged to regenerate a society corrupted by social differences created by the inequality of wealth. Locke's thinking did not defend actually existing states: any collective of individuals could have founded a commonwealth by consent, in the way that the Puritan pilgrims of the *Mayflower* signed a social contract before disembarking in Massachusetts. Instead, Rousseau's thinking, introduced by the Jacobins, was soon rigidly statised (although initially it did not indicate such a tendency: the ideal framework of the compact "general consensus" was, for him, not an extensive monarchy such as the French, rather a homogenous *politie* in the style of Geneva, the canton of his birth).

In this way the different forms which the state assumes in the two traditions can be understood. The preservation by the United Kingdom of the civil institutions of the British periphery (Wales and especially Scotland) acted as a substitute for decentralising political measures which were never implemented. Paradoxically, the tradition of the weak state has maintained the specificity of a culturally French-speaking society, that of Quebec, in the heart of the British colony of Canada.

Scotland, which occupies the northern third of Great Britain and islands of the area, including the Hebrides, today has a population of five and a half million (about 10% of the British population). The linguistic assimilation carried out by the British was successful in the Lowlands, where the main cities of Edinburgh and Glasgow are located, but partially failed in the Highlands and the Islands (although today there are no more than 80,000 speakers of this language).

From 1603 onwards James I united under him the Crowns of the Kingdoms of Scotland and England. With the Act of Union in 1707 London put an end to the Scottish state, transferring its powers to the Westminster parliament. Since then the doctrine of pact has formed part of Scottish political culture and, moreover, the survival in Scotland of civil society - the church, legal and school systems - was made possible thanks to the singularity of the United Kingdom. This, in effect, had historically left large areas of public life in the hands of self-regulated institutions; the absence of centralisation and uniformity prevented the introduction of absolutism in the 16th century and of Jacobinism in the 19th century. Later, education, professional life and labour relations remained outside of state control.

For this reason, the disappearance of their parliament in 1707 did not worry the Scottish that much; they were much more interested in the survival of their laws, based on Roman law, contrary to that of the English, which was based on common law. Their reaction is understood all the more if the over-representation awarded to the Scottish in the London parliament as compensation for the loss of their own parliament is taken into consideration, accentuated as a result of the relative decrease in population (Scottish affairs in Westminster are still dealt with by special commissions in the present day). Likewise, the separate Church of Scotland (*Kirk*) held an important role, based on the "doctrine of the two kingdoms", in the spheres of regulation of teaching and social life. Until the end of the 19th century there was, then, an informal Home Rule in Scotland, with most of the administration being in the hands of the *Kirk* or local government. In addition, in the era of industrial revolution, the British empire offered Scottish and Welsh industrialists and mine-owners access to the British and imperial markets, which encouraged the loyalty of Scotland to the Empire (Moreno, 1995; Keating, 1996).

Quebec currently has a seven million inhabitants, a quarter of the Canadian population, of whom 82 % are French-speaking, 8 % are Anglo Saxons and the remaining 10 % are immigrants of various origins. In addition, there are sixty thousand "registered" natives. Quebec is one of the ten provinces of Canada (the rest are Anglophile). Ontario is the most populated with ten million inhabitants. The four coastal provinces of the Atlantic - Newfoundland, Nova Scotia, New Brunswick and Prince Edward Island - have a population of two and a half million people. The four provinces of the west - Manitoba, Saskatchewan, Alberta and British Columbia - have nine and a half million inhabitants. There are, in addition the two vast and scarcely populated territories of the Yukon and the North-West.

The French occupation of the provinces of Acadia (the future New Scotland and New Brunswick), of Louisiana in the southern United States and "New France", or Quebec, on the banks of the Saint Lawrence River, had introduced into these areas a fief system based on feudal agriculture and the key role which the land-owning Church held as organiser of social relations. After the British conquest in 1713 of Acadia and New France in 1760, and as a consequence of the reaction of the nobility and of the Church hierarchy, in 1774 the United Kingdom recognised French civil law and the feudal system, tolerating Catholicism. Members of the French-speaking elite, loyal subjects of the British crown, rejected the French Revolution, and the monarchist *fleur-de-lis* continues to feature in the Quebec flag.

The American Revolution brought about an influx of loyalists from New England to the north of the Great Lakes. With a view to protecting them the United Kingdom divided the province into two colonies in 1791: Upper Canada (a word of indigenous origin), Protestant and English-speaking, which is present day Ontario, and Lower Canada, Quebec, Catholic and French-speaking. In order to prevent the spread of revolution London withdrew from its representative institutions - occupied by the bourgeoisie and business class -

all control over the executives (in the hands of the nobility and the clergy in Quebec). This situation gave rise to a Movement of Patriots of Quebec that fought in the name of the "Canadian Nation", which was defeated in the years 1837-1838.

In 1840, England united the two provinces, making the French-speakers into a minority, as a first step toward the formation of a state, an objective in which economic motives also converged. Since the English conquest, Canada had been planning to act as a commercial intermediary between New England (Boston and New York) and London and to create a canal out of the Saint Lawrence River, which turned out to be impossible after the United States built the Eire canal.

At that time the idea emerged of forming a Canadian state with a protected market that would be united by a railway system, the Canadian Pacific, a railway line running from coast to coast. For this reason (and also out of fear of Yankee nationalism, on the increase after the War of Secession) a British parliament law in 1867 created the state of Canada, which it called a confederation in order to placate the fears of the French-speakers, although in legal jargon it is referred to as a federation. However, the founding fathers not only had to confront the resistance in Quebec, but also the suspicions of the maritime and eastern provinces, who feared losing their autonomy in a centralised Canada. The regime offered the provinces (a colonial term) the option of having their wealth entirely at their disposal and the power of veto over federal decisions; it was, in effect, a compromise solution that would not prevent future conflicts. An arrangement which satisfied Quebec would give rise to similar resentment in other provinces against central Canada - in which the main cities of the state (Toronto and Ottawa in Ontario and Montreal and the City of Quebec in Quebec) were grouped together in a small radius. At the same time the demands of the Quebec people for self-government would easily come into conflict with those of the aboriginal native peoples.

The failure to create a unified market encouraged, in this way, centrifugal processes. Canada was not structured from east to west, as had been anticipated, but on north-south axis, in terms of American economic trends. It was a colony of luxury, exporter of raw materials, particularly wood, and an importer of consumer goods, with one of the highest standards of living in the world. Its three or four economic spheres of activity were focused on the United States: British Columbia depended on the prosperity of California; the east, initially poor, experienced growth through the growing oil industry; while the centre was going through relative decline, especially Quebec; Ontario, whose borders run from New York to Minnesota, benefited from the Yankee expansion.

The liberalism of the Canadian state meant that management of social and cultural affairs was left in the hands of private bodies. In Quebec this led to a consensual distribution of power, which lasted almost a century. Under the English-speaking high bourgeoisie, the Church reaffirmed its leadership in the administration of institutions such as schools, hospitals, asylums, etc., while the urban petit-bourgeoisie took charge of representative bodies. At the same time "the Canadian nation", called for by the rebels of 1800-1840, became the

Catholic French-Canadian nation whose distinguishing characteristics were the French language and the Catholic religion (Bourque, 1995).

The Break up of the Empire: The Russian, Austro-Hungarian and Ottoman Empires

The characteristics of historical empires are the agglutination of a plurality of peoples, the organisation of a vast space and the tendency toward expansion. They are multi-cultural in a sense that is different from the unitarian pluri-ethnic states: they form centres that accept the diversity of the peoples of which they are composed, meaning that they are self-limiting in their ability to penetrate peoples. Their legitimating principles operate more on a level of identification (the legitimating mechanisms based on the political loyalty of the subjects to imperial authority are often of a religious kind) than on a level of identity (as understood by the unification of norms, values and symbols of the different peoples) (Badie, 1985).

All this contributes to the break-up of empires when modernity befalls them: the role of religion is displaced from that of a legitimating mechanism of a distant power to that of a community identity marker and, therefore, of potential confrontation among groups. The processes of creating the market, of integration and socialisation are not viable in the imperialist heterogeneous framework. This process must also be considered in the international heterogeneous context. The European empires were defeated not only for internal reasons, but also because of external pressure from rival nation-states, anxious to complete their political and, often, their military victory, thereby bringing about the break-up of the defeated.

With regard to the Russian, Austrian and Ottoman empires, it is important to make brief reference to the clash of civilisations which took place in eastern and central Europe during the course of more than a thousand years. The diverse political and cultural nature of each empire arising out of this explains the specific process of transformation of the peripheries in nation-states. It is usual to divide Europe into two areas, east and west, which have their roots in the Constantine' division between the Roman West and the Byzantine East; Seiler (1993) adds to these a third zone of transition, of an intermediate character. The oldest post-Constantinian border in the West was that of the Carolingian empire, defined by the Elba, Saale and Leith line; from within it, through the mixture of Latin and Germanic traditions, the feudal world developed. The religious schism which in 1054 separated the Latin from the orthodox civilisation, pushed the old border toward the East, displacing it from the River Elba to the Carpathians and from the Baltic to the Adriatic, creating a zone of transition between the old and new boundaries within which Poland-

Lithuania, Bohemia, Hungary and the greater part of Croatia were situated. To the east of this boundary Eastern Europe was located, which continued to draw inspiration from the Papal-Caesarism of Byzantium, ignoring the Augustinian division between temporal and spiritual power and turning its back on Renaissance individualism, which found expression, above all, in the Protestant Reformation, but also in the Catholic doctrine of natural law. The Russian empire and the European part of the Ottoman empire drew nourishment from this common culture of the East. The empire of the Austrians, born in the West, expanded toward the zone of transition, which was occupied on different occasions by one or other of the eastern empires; these left their influence on it, even though it remained essentially western.

> The primitive *Russian* centre grouped around Kiev and made up of the Rus (Scandinavians who had become Slavs and had set up economic relations with Byzantium) moved in the early Middle-Ages, as a consequence of Genghis Khan's invasion and the hegemony of the Gold Horde toward the Duchy of Moscow. Its logic of state construction was military (as it also was to the east of the Elba, in Prussia). Both states supported their nobility who were willing to maintain the tradition of servitude, which was at the same time disappearing in the West. However, while in Prussia the monarchy allied itself with the feudal *junkers*, the Tsars became autocrats who set up and dissolved at their whim the aristocracy of the Boyards, making it into an element of their personal entourage.
>
> After the conquest of Constantinople by the Ottomans, Ivan III in the 15th century, and, a century later, Ivan V assumed the politico-religious inheritance of Byzantium (the word Tsar, imported from the near East, was related to the Byzantine *basileus*). Ivan III symbolically wedded Zoe, inheritor of the dethroned Paleologists. At the same time the orthodox metropolitans declared Moscow the "new Constantinople" and the "third Rome". After the conquest in 1552 by Ivan the Terrible of the Tartars towns of Astrakhan and Kazan - the centre of the latter being the Volga Valley - inverted a millennial tendency: the Russian empire began the conquest of Siberia, advancing along the enormous steppe corridor linking the coasts of the Japanese sea with the plains of the Don passing through the north of China, India, Persia and Caucasus in the opposite direction to that taken in the invasions by the Manchurians, Altaic-Turks and the Indo-Europeans which had come successively since the times of the Roman empire.
>
> The peoples conquered by the Tsarist empire did not, in the main, practise the orthodox religion. In the 18th century the Baltic countries, Catholic or Lutheran in religion, were occupied. In the Caucasus, conquered at the beginning of the 19th century, many mountainous peoples lived who were Islamic or belonged to Christian churches some of which, like the Georgian or Armenian, dated back to the early centuries of Christianity. The peoples of central Asia, colonised in the middle of the 19th century, were homogeneously Muslim.

The October Revolution saved Tsarist empire from the fate suffered by the Austro-Hungarian and Ottoman empires. Apart from Finland and, provisionally, the Baltic Countries, the Soviet Union preserved the territory of the old empire. After recognising its multi-national diversity, it was structured in the form of a Federation of fifteen Republics which were guaranteed their cultural-linguistic autonomy. However, among them there was Russia, an enormous federal republic that was dominant, which transformed the non-Russian territorial collectives of its hinterland into autonomous republics - the Tartars, the inhabitants of the Volga Valley, the Ugro-Fins, the Siberians - and many of those living in border areas with the Federal Republics, as was the case in Caucasus. In addition, the federalism of the Soviet Union was more formal than real; the ethnic republics lacked political power, which was concentrated exclusively in the hands of the Communist Party of the Soviet Union (CPSU). The authoritarian political culture typical of the Tsarist autocracy reappeared through Stalinist totalitarianism as a natural extension of itself.

Can a "Soviet empire" be spoken of? Yes, if the definition of the term is broadened. Contained within it is an open recognition of the diversity of peoples; an imperial centre also existed, the Moscow Kremlin, the consolidation of which contributed, once Lenin had died, to the victory of the Stalinist thesis of the building of socialism in a single country, as opposed to Trotsky's thesis of permanent revolution. Up until the fifties a legitimating mechanism worked which allied the archaic with the most ferociously modern, consisting of consecrating a theory, that of the inexorable triumph of socialism in the world arising out of the scientific laws of historic materialism, the exclusive interpreters of which were the CPSU and, more precisely, the Stalinist leadership. The thesis referred to would likewise legitimate an expansionist "imperial" tendency in the territorial sphere: the states formally independent in the east and centre-east of Europe governed after World War II, as a result of the military triumph of the USSR, by communist parties, were subordinate to the Moscow imperial centre (which contributed to the prevalence of a false idea of European bi-polarity.) The crisis suffered at the end of the eighties by Soviet Marxism-Leninism (planned economy and the infallibility of the party) substituted, certainly, by a banal acceptance of the logic of the most savage form of capitalism, has had the same effect in dissolving the empire that the laicisation of the religious legitimating mechanism had on the historic empires (Goehrke *et al.*, 1884; Carrère D'Encausse, 1991; Seiler, 1993).

The Austrian empire developed from the East March (*Ostmark*), created by Charlemagne in order to prevent new invasions. Converted into a hereditary Duchy, the March became the property of the House of Hapsburgs in the 13th century. Since the time in the 15th century, when Frederick III concentrated in his hands the titles of ruler of the House of Hapsburg and of emperor, Austria became the inheritor of the sacred Roman-Germanic empire. Maximilian I equipped the empire with solid structures and extended its territory. His grandson, Carlos V, the last of the emperors who dreamed of a reunified Christianity, bequeathed the Austrian part to his brother, Ferdinand. While his

Spanish successors constructed a modern imperial monarchy, Ferdinand built, after the annexation of Bohemia and Bulgaria, a multi-cultural empire which brought together a vast combination of peoples. His heirs, who successfully defended Vienna against the Ottoman attack in 1683, later took Transylvania from the Turks, together with the Low Countries, Naples, and Milanese from France. Religious tolerance was unknown; the Hapsburgs brutally imposed the Catholic counter-reformation at the beginning of the 17th century on Protestants in Bohemia and Hungary, and later in Transylvania too. Moreover, the *Staatsrechte* - the rights of the state - were only respected in the case of large formations such as the kingdoms of Bohemia, Hungary, Croatia and Polish Galicia, and not among minorities such as Slovaks, Slovenes, Rumanian, Ruthenians and Jews. Above all, the government favoured the Germans in every sphere of the imperial administration.

At the root of their first confrontation with the world of the nation-states, on the occasion of the war unleashed against revolutionary and later Napoleonic France, the Hapsburgs symbolically renounced the crown of the sacred Empire, transforming their possessions into imperial Austria. Austria, nevertheless, continued to dominate Europe from 1815 to 1848 and, faithful to the idea of a Christian empire, it opposed German reunification.

After the attempt to transform the empire into a Germanic nation-state during the early decades of the long reign of Franz Josef (1848 - 1916), Austria reached an agreement in 1867 with the Kingdom of Hungary on the basis of which a dual Danubian monarchy was constituted with two capitals, Vienna and Budapest, and a single ruler, the Austrian emperor. Hungary assumed responsibility for the administration of the eastern territories of the Transleithania (Croatia, Slovakia and Rumania) while Vienna became responsible for Cisleithania. The process was, however, not symmetrical: Hungary was intent on the formation of a unitarian nation-state, while the Austro-Germanic territories, which represented only a part of the German linguistic sphere, remained fragmented. Czechs and Slovenes soon moved from legitimism to nationalism; Rumanians and Slovaks reacted against the policy of Magyarisation, without however converting to separatism; Croatia, which felt betrayed by the compromise of 1867, hoisted the banner of Yugoslavism, that is, of national union with the Slavs of the Ottoman empire.

The defeat of the Austro-Hungarian empire in World War I brought about the formation of a small Austrian nation-state and the reduction of the territories of the Kingdom of Hungary, which left a number of pockets of the Magyar population in Rumania, in Vojvodina, Yugoslavia, and in the Slovak part of Czechoslovakia. (The allies favoured the reunification of the Czechs of Bohemia and Moravia and of the Slovaks in a single state; but the two parts, linguistically close, diverged politically and culturally, having formed part, respectively, of Cisleithania and Transleithania) (Seiler, 1993).

The Ottoman empire has given its name to an extreme case of patrimonial regime, "sultanism", in which the resources of the state, the bureaucratic apparatus and even the bureaucrats themselves are the private property of the Sultan and the instruments of his will. Even though its legitimacy was religious, this Islamic empire engaged its European possessions in a subtle political game of client alliances with the heads of non-Muslim religious communities, which facilitated the emergence in the 19th century of a number of Christian nationalisms inspired by the Byzantine-orthodox tradition.

The Uguzes, western Turks originally from the Altaic region and founders of the Ottoman dynasty, settled at the end of the 13th century in Byzantine Anatolia. After spreading into the Asiatic part of this empire, they entered the European Balkans in 1365, taking Constantinople in 1453. Once concluded the early Christian resistance, out of which emerged legends such as that of Prince Vlad Tepes (Dracul), populations of Greek, Albanian-Illyrians, Rumanian, Bulgarian and Serbs from the Balkans exchanged, without too much regret, the harsh repression unleashed by the Latin Crusades for the more bearable rule of the Sultan. In Europe the Ottomans reached the height of their splendour after the conquest of Hungary in 1526; their defeat on the threshold of Vienna in 1683 and the Hungarian liberation of 1691 signalled their decline.

The Ottoman regime combined two political cultures, Islamic and Byzantine, adorning them with their own sultanist features and moderating them with a notable pragmatism in their relations with other beliefs. The administration of the Sultan in Istanbul was based on the system of *devchirme*, consisting in taking Christian children from the Balkans who, after being converted to Islam, became part of the bureaucratic elite of the "Slaves of the Door", out of which came almost all the Viziers; personal property of the Sultan which allowed him to disregard the Turkish nobility in order to govern. The empire was a religious state whose authority rested on the *saria*, that is, the law of the Koran. Only the believers were, then, political subjects of a whole body. However, the *Zimmi*, or "Peoples of the Book", that is Christians and Jews who had not resisted Islam, were respected with regard to their property and freedom of worship. For this reason, the conquest was not accompanied by massive conversions, except in Bosnia where the "Bogomile" heretic slaves, related to the Catharians, who were cruelly persecuted before the arrival of the Ottomans, converted *en masse* to Islam when they appeared.

The organisation of the Millet under the leadership of *Rum Millet*, patriarch of the Greek Church of the East who resided in Istanbul, made the Christian (and also Jewish) religious leaders the administrators responsible for their community under the rule of the Ottomans. The self-government which Millet preserved in the spheres of religion, the family, teaching and welfare facilitated the survival of the orthodox religious cultures; Serbs and Bulgarians retained their Otthodox churches, Cyrillic alphabet and Slavonic liturgical language. This made it possible for these people to become conscious of themselves in the future as national groupings.

Christian rural villages constituted units responsible for the payment of taxes, being ruled by leaders called *akón* in Greek, *knez* in Serb and *corbadji* in Bulgarian (although the anti-Ottoman movements of the 18th and 19th centuries rejected these leaders as collaborators). However, ethnic mobility brought about by migrations, forced upheaval and military events turned some territories (Vojvodina to the north of Serbia, Macedonia to the south) into complex ethnic mosaics. The ethno-religious mix was, above all, the rule in the main Balkan cities (especially in Bosnia) where an urban Ottoman elite ruled neighbourhood complexes, regrouped according to religious belief.

The Ottoman despotism was unable to learn; it provoked the greed of Western powers, who became involved in the "Eastern question" during the 18th century, the century of nationalisms, causing the empire to explode. The processes of the emancipation of the Bulgarians, Serbs, Greeks and Rumanians coincided at the end of the 19th century and at the beginning of the 20th century with inter-Christian conflicts which brought Serbia into confrontation with Austria-Croatia for control of Bosnia, and Bulgaria, Serbia and Greece for control of Macedonia. The absence of civil society inherent to the orthodox political culture converted the Balkans into a powder keg waiting to explode. In parallel, the defeat of the Ottomans fostered the hard-line nationalism of the "young Turks", who were intent on creating out of the empire a mono-ethnic nation-state, a project for which first the Armenians, and later the Kurds, would pay the consequences.

In the European map of states which emerged from World War I the interests of the winning powers prevailed over the will of the peoples affected. The concentration of "Yugoslav" peoples (that is, Slavs of the south) from the defeated Austro-Hungarian and Ottoman empires in the kingdom of the Serbs, Croats and Slovenes under Serbian hegemony froze a conflict that would explode at the end of the 20th century (Castellan, 1991).

The nation-states which emerged from the break-up of the empires often had in common the artificial nature of their borders, especially in those cases where the international context had decided its location. Likewise, the weakness of civil society limits, with respect to the West, the role that socio-economic elites can play in the construction of the state; it sidelines them in favour of political elites and fosters authoritarian solutions.

De-colonisation: Pluri-cultural States of the Third World

These characteristics define even more convincingly states which emerge from the processes of de-colonisation. Colonisation ("imperialisms" are a sub-category of the general phenomenon) consists of the permanent occupation of a territory by a foreign power whose aims are always economic, frequently political, and also, on occasion, cultural. This converts the metropolis into the political centre of a distant colony, which is condemned as such to always being a periphery. Colonisations can be carried out by pre-capitalist powers (Spain and Portugal), or capitalists (Holland, France and England). The colonies may be occupied dominions, where a colonial elite has military control of the territory which continues to be inhabited by local people (the British in India and the French in Vietnam), or they may be populated, in which case the colonisers settle permanently in the area, evicting their primitive inhabitants in the process (the Anglo-Saxons in America and Australia).

De-colonisation consists of the transformation of the old colony into its own centre, which frees it from the metropolis, and changes the liberated territory into a nation-state. The process of emancipation could be the work of the colonisers and achieved through struggle against the metropolis out of which they originate (which constitutes the norm on the American continent). It may also be achieved through the work of the colonised peoples, struggling against the colonial occupiers or through concessions gained from the metropolis. (This is the double origin of almost all states of the Third World; for which reason this sub-category constitutes something of a muddle, with innumerable variants.)

Pre-capitalist states (such as the Spanish monarchy in America) created archaic colonial structures which, like the historic empires, fragmented with the struggle for independence; a process intensified in these cases by pressure from external powers (such as England, especially in Latin America). However, contrary to the European empires, the Ibero-American colonial empires did not see themselves as a combination of peoples, rather they perceived of themselves as spaces composing a common universe of Christian salvation. In this way, the ethnic specificity of the indigenous groups was not taken into account during the processes of emancipation, neither was there any recognition of autochthonous characteristics in the construction of national identities. For this reason the lines of state fissure did not correspond to ethno-cultural divisions, rather they coincided with administrative boundaries. This is why the Creoles who led the struggles for independence within the different Latin American states initially shared very similar cultural and political tendencies.

The colonies created along the lines of capitalist models adopted instead a position of opposition to any kind of fragmentation of the market and, therefore, to the creation of internal borders. The case of the United States was a prime example in that its colonies rapidly expanded to the Pacific. It is worth pointing out the different treatment meted out to the indigenous peoples in each case: survival in conditions of servitude imposed through neo-feudalism at the behest of a white elite in Latin America (although survival was possible in powerful pre-Columbian civilisations, notably decreasing in the rest of the region); the indigenous peoples faced relentless eviction from their lands in North America, suffering "philanthropic genocide", in the words of Tocqueville.

Spanish neo-feudalism in America gave rise to a mixed colonisation of dominion-occupation and settlement. The large administrative units were the viceroyalties. In 1538 Carlos V created the first viceroyalty of New Spain, with its headquarters in Mexico. It embraced the area conquered by the Spanish in

North America, with its centre on the Mexican High Plain, Central America, the Antilles and Venezuela; here the Viceroy represented the charismatic authority of the monarch. After Pizarro conquered Peru in 1543, a second viceroyalty of the same name was created with its headquarters in Lima; it stretched throughout the rest of South America. Its territories were so extensive that in the 18th century it became necessary to create two new viceroyalties: New Granada in 1739, with its headquarters in Bogota, which included Columbia, Ecuador and Panama; and Río de la Plata in 1776, with its headquarters in Buenos Aires, which included Argentina, Uruguay, Paraguay and Bolivia.

Within the viceroyalties *Audiencias* were established, these were collegiate authorities with judicial and political functions which became a counterbalance to the viceroyalty's authority. In the viceroyalty of New Spain four *audiencias* were established in the 16th century: Santo Domingo, Mexico, Guatemala and Guadalajara. In the three viceroyalties of South America nine *Audiencias* were created; six in the 16th century (Panama, Lima, Santa Fe de Bogotá, La Plata de las Charcas, Quito and Chile) and three in the 18th century (Buenos Aires, Caracas and Cuzco). Their jurisdictional boundaries often became the state borders of the future republics (Konetzke, 1984).

The Wars of Independence had at the outset a unitary spirit. The Creoles, opposed to the mercantilism of the metropolis, spurred on by the example of the North and contaminated by French revolutionary ideas (which they had become well acquainted with in Madrid), took advantage of the weakness of the colonial bureaucracy in 1811 during the Napoleonic invasion of Spain in order to initiate hostilities. The movement of Bolivar's troops in the north, spreading out from the Caribbean, together with those of San Martin in the south, expanding with British support from Río de la Plata, culminated in 1824 with the defeat of the Spanish in the Battle of Ayacucho. The liberation of Mexico took place in 1821 as a result of a Creole revolt of an anti-liberal nature against the republican Spanish government of Riego.

Nevertheless, Bolivar's dream of Hispanic American unity, which he wanted to bring to fruition in the Congress of Panama in 1826, came up against the localism of the *Audiencias* and the interests of the United Kingdom, which were entirely hostile to the idea. The provinces of Río de la Plata, Chile and Brazil abstained from participation in the Congress. Great Columbia divided into the three states of Columbia, Venezuela and Ecuador. The Central-American Confederation separated from Mexico, breaking up later into Costa Rica, Nicaragua, Honduras, Guatemala and El Salvador. In the south, three small states formed between Argentina and Brazil: Bolivia, Paraguay and Uruguay.

On the other hand, the emancipation of the Portuguese colonies of Brazil (the new independent state maintained the unity of its vast territories) constituted proof of the autonomy of the political. Napoleon's threats against Lisbon for not respecting the continental blockade forced the transfer of the entire Portuguese Court to Brazil. It brought with it an administrative and military outfit of about ten thousand people, a modernising graft which impelled the centralised construction of the new state from the court in Río.

When Pedro I declared independence in 1822, Brazil, contrary to the old Hispanic possessions, operated as a unitary government (G. and H. Beihaut, 1986).

Settled capitalist colonies, once liberated, carry out the construction of the centre without any problems, since they themselves are a product of modernity. In these cases, modernisation is a factor in the exclusion of autochthonous peoples, marginalized from the process of national construction and often reduced to the condition of a-national groups.

Multiple variables are apparent in their situation. Deculturisation can be so profound that it deprives these groups of any ability whatsoever to resist (such is the case of the indigenous peoples of America). Autochthonous liberation movements may emerge in independent colonies (for example, black Africans against white domination in South Africa). The foundation of a settled colony may constitute in itself the act of establishing a centre by a group which lacked one, thereby creating a "national homeland" which then becomes the object of dispute with native inhabitants (as happened in Liberia, where native black inhabitants eventually opposed the domination of the free black slaves in a state created specifically for them).

The Jewish group is a religious collective spread throughout the world. It is as pluri-ethnic as the societies of which it forms part and only became "ethnified" through the historic persecution suffered by its members at the hands of Christians. Nowadays its extreme diversity is reflected in the presence in the state of Israel of Ashkenazi Jews, originally from central and eastern Europe (who represent 85 % of the Jewish population, which, in its turn constitutes 80 % of the total population); Sephardim Jewish, Castilian-speaking descendants of those expelled under the Catholic Kings from the Spanish "Sepharad" (5 %); and "eastern" Jews, Palestinian and Ethiopian, among others (10 %). The resulting linguistic plurality made it necessary to revive a liturgical language as the official one.

The Jewish creation of the state of Israel through the physical act of occupying the Biblical lands inhabited up until then by Arabs (the Zionist justification for this occupation is not very different from that which encouraged the English puritans to establish the "Promised Land" in North American territory) represents the model for the foundation of a modern "national homeland". The need to establish the new territory as Israeli explains the policy of purchase and colonisation of Palestinian land carried out by the Zionist organisation Keren Kayemeth Leisval (KKL). This organisation categorically denied Jews the option of transferring the land acquired, which became national property, in order to ensure that the land did not revert back into Arab hands. Until 1947 voluntary sale and purchase did not produce the results desired as up until this point since the KKL had not managed to acquire more than 6.6 % of Palestinian land. Nevertheless, after the foundation of the state of Israel and up until the war of 1967, the ban on the transfer of "state lands" was applied to 92 % of the territory of the new state. Likewise, agricultural work was promoted in order to establish bonds between Jewish colonies and the land. The KKL forbade salaried work, predictably Arab, on these lands with the aim of preventing the re-establishment of links between the Palestinians and the land. This was the political significance - more than socialism - of the collective agricultural work of the *kibutzim*, where private ownership of the land was rejected.

Israel was not formed as the state of its Jewish or Arabic citizens, it was rather the state of an unspecified Jewish people, the majority of whom did not live in it. The result was that it favoured Jewish immigrants and discriminated against Arabic citizens. Left-wing Zionists applied this logic to the sphere of labour relations. The main Zionist trades union, *Histadrut,* only accepted the membership of Jews and pressurised industry to import machinery, making Arab manual work unnecessary. This was accompanied by a social division of work in which skilled labour was reserved for Jews and unskilled for Palestinians.

Until 1977, the Labour party promoted a policy of Jewish settlement in Galilee, the northern part of Israel where the majority of the Arabic population lived, with the objective of making it Jew. With this same objective a chain of settlements was established along a north-south axis of the West bank (Nablus, Jerusalem, Belen, Hebron) as a means of preventing the homogeneity of a future Palestine state in Gaza and Transjordan (a possibility which they did not support but neither did they dismiss). Right-wing Zionism, as espoused by Begin and Aaron, would not nevertheless accept this option. From their point of view the lands occupied in 1967 also belonged to the Jewish people in accordance with a principle of divine legitimation; the historic-biblical right to dwell in *Eretz Israel*. Jerusalem, the sacred heart of Israel, was also inalienable; for which reason the Arabic part was annexed in 1967.

The government of Begin therefore interpreted the "Camp David Agreement" of 1978 to the effect that Palestinian "persons" could become autonomous from the military administration, but not the "lands" where they lived, which would remain open to Jewish settlement. This interpretation was compatible with the establishment of a Palestinian "Bantustan" endowed with cultural autonomy but which the Jewish state would ignore with regards to social welfare loans and support. Nevertheless, the army would remain in place defending Jewish settlements and guaranteeing them free access to water supplies on the West Bank of the Jordan.

However, the huge financial demands of maintaining the Zionist structure of the state, promoting Jewish immigration, avoiding Jewish emigration, creating jobs and colonising the land has made the Israeli state very dependent on Western capital. All of this represents a double-edged sword in that it has placed the West, particularly the United States, in a position of being able to put pressure on Israel.

The Oslo Pact is a consequence of such pressure. Yet there is still uncertainty over whether the autonomy of the West Bank and the Gaza Strip will lead only into an Arabic "Bantustan" controlled by Israel or whether it will become an embryonic Palestinian state juxtaposed to the Jewish one. Besides, the dialectic of reciprocal acknowledgement, the basis of which was established by the Pact, could undergo setbacks. What gives immediate cause for concern in this respect is the victory of the right-wing Likud over the Labour Party (Diner, 1987).

The construction of the state in African and Asian societies, which essentially took place between the years 1945 and 1975, was strongly influenced by the colonial period. The territorial boundaries of the new states rarely coincided with the pre-colonial divisions. In some cases entire civilisations were lumped together, as was the case of India. In others, especially in Africa, divided societies were juxtaposed inside arbitrary borders drawn in response to the changing rivalries within Western powers in a period which is now remote, the years between the 1885 Conference of Berlin and World War I.

The colonial period has, in the same way, been responsible for the economic dependence of the new states with regard to the West. This is not a transitory phenomenon, brought about by the absence of a national bourgeoisie who with rapid industrialisation could resolve the situation. It is, instead, a permanent structural fact: the Western powers control like "enclaves" the sectors of production which they consider profitable. In the processes of state construction, more intensely so than in the cases of imperial break-up, this has privileged political elites emerging from the colonial administration or from the intellectual strata, since economic elites either do no exist or they are foreigners. This voluntarist predominance of the state over a civil society which it must create has made authoritarianism the norm in the new states of the Third World. In this case, the formation of the centre does not adhere to the scheme, put forward by the theories of development, of the teleological transformation from the traditional to modern. Traditional societies themselves become vectors of modernisation, determining the specific mode of state construction.

This portrays a particular configuration of centre-periphery relations in the Third World, characterised by "cultural pluralism". According to Badie (1985), this consists in a mode of organisation arising from the juxtaposition of distinct and irreducible ethnic groups characterised by strong internal integration, which results in a vertical structure in society. The centre of the system is monopolised by a group which dominates society; this group could be a section of a tribal or clan-based society, although this is quite unusual. The dominant group usually comes from a non-tribal elite because the political mobilisation brought about by de-colonisation has tended to dissolve tribal identities and create new ideological identities, paradoxically increasing the degree of hostility among the older groups, once they come into contact with each other and the reasons for confrontation multiply.

The neo-patrimonial system is based on clientele practices, which are structural, and not exceptional, like the "notables" network in France. The centre penetrates the periphery by associating the traditional tribal or clan chiefs with the system, thereby making them into a local rung of power. Neo-patrimonialism and clientelism tend, therefore, to unify the functioning of political systems of the Third World, defining authoritarian practices of "limited pluralism" in the sense described by Linz (1965). The "proto-corruption" which prevails in these systems derives from the lack of distinction between the public and private spheres. The omnipresent bureaucracies integrate the elites into the network of clients with the aim of preventing the emergence of competing groups, especially of economic elites who could become rivals to them. The important political capital which military bureaucracies accumulate as supporters of technological progress and defenders of a unitary conception of government, represents a constant temptation to stage military coups.

In these pluri-cultural societies where one or several groups takes political power for its own ends, the causes of centre-periphery conflict are multiple. They may emerge when the clientele system fails because the traditional leaders have not been sufficiently integrated, or when a territorial economic elite to the margin of the centre calls for political self-government. They may also emerge when a violent seizure of power by a minority provokes a counter mobilisation by the groups excluded, or when an existing "national homeland" opposes the *hinterland* against the centre which monopolises it. Or, finally, when cultural, religious or civilisational differences are revealed as insurmountable.

The examples which are referred to from this point onwards are located in two of the extremes of the Third World, in India and sub-Saharan Africa. On the Indian sub-continent the strong pre-colonial civilisational identity, linked to the extreme ethnic and linguistic diversity and to the existence of an old and solid political party formed by native elites, who had been trained since the 19th century by the British administration, facilitated the formation of a centre which, through its tradition of compromise, brings to mind the functioning of the consociational Western democracies. In this case the religious factor has been the causal element of the centre-periphery conflicts. It has been at the root of the religious self-determination of Islamic Pakistan, the discrimination of the Muslims in Kashmir in India, and the persecution of the Sikh religious minority in the Punjab state of India.

Since it won its independence from the British in August 1947, India, a colossus of 800 million inhabitants, has been closest to a state-civilisation. Contrary to the Chinese model, it lacks a dominant ethnic majority. The Indo-Arian languages, originating from Sanskrit, are spread throughout the sub-continent into six groups which include more than a thousand dialects.

Attempts to make Hindu the official language of the state came up against the fact that the 200 million people who speak it represent only 28 % of the population. In the south, 100 million people speak Dravidic non Indo-European languages. In the north-east there are many peoples who speak Tibetan-Burmese.

Hindu is the majority religion, 82 % of the present population of India subscribe to it compared to 12 % Muslim (90 million), 2.4 % Christian (20 million) and 2 % Sikh (15 million). Nevertheless, it has not become the key element of Indian nationalism because of the absence of dogma and, above all, because of the social structure of the caste system that it has given rise to, which has created the "untouchables".

The Indian Congress Party, which emerged in 1885 from the core of Indian Anglophile elites, was profoundly modified by the powerful and charismatic Mahatma Ghandi between 1920 and 1948, the year of his death. The party took into consideration the multi-ethnic and pluri-religious nature of India and from 1920 onwards incorporated into its programme non confessionalism and the future division of India on the basis of the main linguistic differences, making abstractions of British colonial structures. This, however, did not prevent the influence, against the wishes of Ghandi, of aspects of nationalistic Hinduism, such as that of Hindu Mahasabha.

The Muslim League was founded in 1906 to demand separate electoral registers; its influence was felt in areas such as Sind and Bengal, where the majority of the population were Islamic. The growth of the League was due not only to the favours of the British, who were loyal to the principle of *divide et impera*, but also to the Muslim fear of the growing influence of Hindu on political life following the victory of the Party of Congress in the regional elections prior to World War II.

After the end of the war, Indian opposition to the British presence ran in tandem with the electoral successes of the Muslim League, which prompted Lord Mountbatten to authorise at the end of 1946 the establishment of two dominions, India on the one hand, and Muslim Pakistan (previously the Sind) on the other. Bengal, like the Punjab, chose to be divided into two states. However, even though the division of Bengal, where Gandhi resided, was peaceful, the division of the Punjab was particularly violent.

In June 1947, one month prior to independence, a massacre took place in which hundreds of thousands of people were killed. Sikhs and Hindus escaped into India while the Muslims took refuge in Pakistan. For some months afterwards macabre processions of trains full of corpses crossed the borders in both directions with signs saying "A present for India" or "A present for Pakistan". However, the Sikh-Hindu alliance which had been built up over the years was weak, as events in the future would reveal.

The tensions in Jammu and Kashmir, situated in the territory to the north of Punjab and bordering the Himalayas, have arisen from two unresolved problems: the Indo-Pakistani division and the integration of princely states into an independent India. British India was divided into two, the part that was under direct rule and the part that was the principalities under British control, of which there were a total of five hundred, with 45 % of the territory, and where 25 % of the Indian population were governed. In 1947, all the principalities accepted integration, except for the largest three, among which was Kashmir, where a Hindu dynasty governed a majority Muslim population. After the independence of India, the Patans of Pakistan crossed the border to support their co-religionists; the Hindu Maharajah fled, but not before announcing the annexation of Kashmir by India. Mountbatten consented, although he promised a plebiscite in the future. A war between India and Pakistan broke out and the truce of January 1949 led to a *de facto* division of Kashmir, the terms of which were not recognised by either of the two states. Subsequently, India opposed the plebiscite, citing as an official explanation that a secular state could not recognise the right of secession on religious grounds, although the real reason behind its opposition was a fear of losing the plebiscite. The denial of the right to self-determination to the people of Kashmir provoked a violent conflict between India and Pakistan in 1965 and continuous uprisings in favour of independence until the nineties, which were suppressed by the army.

The configuration of India since 1950 as a macro-state composed of twenty-one federal states did not mean the end of ethnic tensions. These were firstly linguistic, then ethno-centric, and finally, nationalist. The mobilisations of the sixties calling for the formation of mono-lingual states could not be rejected by the Congress Party given that they were demanding the enactment of one of the oldest points of its programme. Local coalitions were formed in Bombay with the objective of dividing it into two states in accordance with the two languages: Gujarati and Maharashtra. The local sections of parties of the Indian side, including the Congress Party, participated in these coalitions. Once their objectives were accomplished they dissolved. This phase culminated in 1969, when the Punjab was divided into three states: Himachal Pradesh, Haryana (a Hindu state) and Punjab (a Sikh state), which satisfied this latter group. Haryana and Punjab shared the same capital, Chandigarh, which would become a subsequent cause of conflict.

The movements of the second phase, during the seventies and the beginning of the eighties, claimed, from an ethno-centric standpoint, that the main economic advantages and 80 % of jobs in mono-lingual states were reserved for the population which spoke the language of the state. At the outset, the mobilisations in Assam were against the Indian Bengalese while those in Bombay, promoted by the *Shiv Sena* party, were against Indians from the south, but they soon polarised against the "external enemy". In Assam this referred to the Bengalese refugees from Bangladesh; in Bombay, it referred to the "pro-Chinese" communists who were therefore "foreign". This meant that while the movements were becoming more and more fascist they were, at the same time, gaining legitimation throughout the state.

The eighties and nineties saw the emergence of nationalist parties in the Dravidian south: the Telugu Dessam of Andra-Pradesh and the Dravida Munnetra Kazhagam (DMK) of Tamil Nadu. The Tibetan-Burmese peoples of the north-east, such as the Nagas and Mizos, took part in insurrections The conflict with greatest repercussions was that in which the Sikhs of the Punjab opposed the Indian state (Gupta, 1995).

In sub-Saharan Africa the centre-periphery conflicts which are described here are of a very diverse nature. Confrontations in South Africa between the settled Boer colony and the colony of British rule, were followed by the *apartheid* of the Africa population. (The apartheid concealed the conflict between Zulus and Xhosas, which has survived African emancipation in the tensions between the African National Congress and Inkatha). In the Sudan, there is the incompatible bi-polarity between the Muslim north and the black Christian south. In Sierra Leon and Liberia there are confrontations between "Creole" elites of the freed slaves and the native inhabitants of the respective hinterlands. In Nigeria there are multi-ethnic conflicts among the Christian Ibos of the oil-rich south, the Muslim Fulanis and Hausas of the north and the Yoruba of the east, which have resolved by an unstable federal compromise. In Rwanda and Burundi there was a reaction of the Hutu peasant majority against the long-standing military domination by the Tutsi; the attempt at power-sharing failed tragically in Rwanda in 1994 (Geiss, 1987).

The expansion of the Bantus in South Africa preceded the foundation of Cape Town by the East India Company (1652). This set the scene for the settlement of a Dutch colony, the Boers. Subsequent conflict between the Boers and Bantus, both of whom were semi-nomad farmers, gave rise to the war of the "Kaffirs" at the end of the 18th century. Later, the United Kingdom established on Bantu lands a dominion colony; at the same time the Boers founded two republics, Orange and Transvaal, which wrested their independence from the British in 1881. After the dramatic expansion of the Zulus under the command of Chaka at the beginning of the 19th century (the *mfekane*), the British protected three territories inhabited by Africans who had fled from the Zulus: Bechuanaland (currently Botswana), Basutoland (currently Lesotho) and Swaziland.

The discovery of gold and diamonds in the last third of the 19th century drew numerous whites to the Johannesburg area claiming equality of rights with the Boers. This provoked a British-Boer War from 1899 to 1902. During these years a system of *apartheid* began to be imposed on the black African masses living in industrial areas. Once peace had been agreed in 1902, the British abandoned the Africans to their fate at the hands of the Boers. From the foundation of South Africa in 1910 as a British dominion, which did not prevent it acting in practice as a sovereign state, until the electoral victory of the radical Boers in 1948, the *apartheid* system was being organised on the basis of the theory of the separate evolution of the four "races": blacks, whites, mixed race and Asians. The latter group, of whom Ghandi was one, had been imported from India to work in the sugar cane plantations of Natal. Eight "Bantustans" were created together with the three existing Protectorates, which the theoreticians of *apartheid* justified by presenting it as a "positive" process of African re-tribalisation.[III]

Black workers were brutally segregated into ghettos of the large cities (Soweto in Johannesburg being an example of one such ghetto). Initial resistance was shown by Ethiopian Christian churches. In 1912 the African National Congress (ANC) was founded, inspired by the Indian National Congress, and like its role model, the ANC was multi-racial in nature. In 1950 all African national organisations were banned. Two years after the declaration of independence of the Republic of South Africa, in 1963, the government arrested the leaders of the "Spear of the Nation", a youth group which carried out violence against objects but not people. Nelson Mandela was among those detained. The revival of the black movement in the ghettos, which expressed itself through the *Black Consciousness* movement, was the target of brutal white repression which led on to the Sharpeville and Soweto massacres in 1970 and 1976 respectively. The year 1975 saw the beginnings of an armed struggle with its base in Mozambique, but it did not lead into guerrilla warfare owing to the fear which the base state had of South Africa.

That same year an alternative African movement emerged, with its base in the Zulu "Bantustan" of Kwazulu, which supported the liberation of the whole South African population through Inkatha (Association for National Liberation). This organisation, which had 150,000 followers (as against 100,000 ANC supporters) did not win support in the ghettos and its leader, Buthelezi, was rejected by *Black Consciousness*. The ANC, having set up links in exile with the South African Communist Party, kept up a prudent relationship with Inkatha based on collaboration.

The movement for black freedom in South Africa was successful in the period from 1989 to 1994 thanks to the wisdom of the ANC and its charismatic leader, Nelson Mandela (who was, in many ways, like Ghandi). The success of the ANC is apparent in the abolition of *apartheid* in 1991, in the De Klerk-Mandela government of national consolidation in 1993, and in the electoral victory of the ANC. Even so, the black majority still continues to live in inhuman conditions and the buried conflict between the Zulu and Xhosa peoples remains in the form of rivalry between Inkatha and the ANC (whom Inkatha accuse of being controlled by the Xhosa).

The consociational formation of the new South African state is, in any case, exceptional in sub-Saharan Africa. Present-day Sudan is made up of two different areas. There is the Arabised Muslim north, linked to the Mediterranean civilisation by the Nile and Egypt, and there is a black south spurned by the north. This is the land of the historic Nubians, inhabitants of the marshlands whose monophysitic Christianity withstood Islam until the 16th century.[IV] The Mahadist Islamic revolt was crushed by the British, who created in the Sudan a theoretical Anglo-Egyptian con-dominion in 1899. The United Kingdom closed the south to Islam and re-Christianised it with missionaries. For this reason the south modernised much more quickly than the Muslim north, which stagnated. After the independence of the Sudan in 1955, the north engaged in an attempt to forcefully assimilate the south by withholding its political rights. This provoked long term armed resistance which brought about the provisional autonomy of the south in 1972. The imposition in 1989 of an

Islamic government with plans to apply the *saria*, or the Koran Law, throughout the state has caused a new revolt in the black south.

The origin of the states of Sierra Leone and Liberia is located in the transatlantic trafficking of black slaves and the prohibition of the trade in 1807 by the United Kingdom. The colony of Sierra Leone was founded in 1787 by well-intentioned British abolitionists who wished to offer a home to ex-slaves resident in England or returning from America. From 1807 onwards a British squadron, based in Freetown, intercepted all slave ships in Western Africa and unloaded the cargoes of slaves at their base where they were converted to Christianity. In this way, the liberated Africans became a stratum of Europeanised "Creoles" who ruled the "tribes" of the hinterland (which was made a Protectorate of the United Kingdom in 1896). The tensions between the colony and the hinterland erupted after independence in 1961 and the *coup d'ètat* of 1967 and 1974 put an end to Creole rule in favour of the peoples of the interior of the country. There have, however, been more military *coup d'ètat* since then.

Liberia was founded as a state in 1847 to become the "national homeland" of ex-slaves and mulattos repatriated from the United States. This parasitic minority of American-Liberians exercised a quasi-colonial control over the people of the interior. They subjected them to hard labour in a regime of barely disguised slavery, which provoked rebellions in the hinterland in 1919 and 1930. The start of the 1980s saw a military revolt of African peoples, led by Sergeant Doe of the Krahn ethnic group. The revolt ended with his death in 1989 but it opened the way to a bloody multi-ethnic civil war between, on the one hand, the Krahn, and on the other, the Dan, Mano and Kissi peoples. Intervention in the war by the States of Western Africa in the attempt to reach a peaceful outcome has had limited success.

Since its foundation in 1960, Nigeria has been the most populated multi-ethnic state of Africa. It has ninety million inhabitants divided among the three main groups: the Ibos of the south-east, the Fulani and Hausa of the north and the Yoruba of the west. These groups dominate numerous smaller tribes and peoples. Historically, the Ibos have been the defenceless victims of the slave-hunting Islamic peoples of the north. The Fulani sultanate of Sokoto had begun to wage a successful "holy war" at the start of the 19th century against the Hausa city-states of the savannah and another, unsuccessful in this case, against the Yoruba kingdom of Oyo, which was surrounded by an impassable jungle and supported by the British. The United Kingdom subjected the sultanate to a system of indirect rule, although it did leave its original structures intact. During the century in which "colonial peace" lasted, the Ibos converted to Christianity, broke out of their powerless situation and became the most economically developed part of Nigeria.

In 1960, the Fulani and the Hausas of the north, which were the politically dominant groups, announced their plan to spread the Koran into the coastal regions. The announcement provoked a reaction by the Christian south. After a period of opportunistic coalitions among various parties united against the Yorubas of the west, in 1966 the central conflict between the Muslims of the north and the Christian Ibos of the south broke out. The island Ibos of the north were massacred, which in turn, encouraged their desire for secession. The Ibos attempted to convert their habitat in the delta of the Niger River, which contained rich oil reserves, into the new independent state of Biafra. One and a half million people died in the course of a bloody civil war. General Gowon called an end to the war in 1970, preserving the unity of Nigeria through the division of its three large regions into nineteen federal states; a solution which has not, in any event, prevented subsequent political instability.

The kingdoms of Rwanda and Burundi have, since the 16th century, formed part of the chain of Hima states of East Africa, which Buganda (currently Uganda) also belonged to. In that era there was a wave of immigration of pale-skinned nomad warriors, the Watusi, or Tutsis, who had come from the Ethiopian Horn of Gold. They established kingdoms governed by a military elite which exercised their authority over the Hutus, Bantusan peasants who were banned from cattle-breeding, which provided a profitable income. Subsequent colonial powers, first the Germans, then the Belgians, left these warring aristocracies intact. After independence in 1960, the Hutus, who made up 85 % of the population in both states, wished to exercise their political rights as the majority population. In Rwanda, under the protection of the United Nations, a plebiscite was held in 1961 which resulted in the abolition of the monarchy. However, the Tutsis imposed their rule in Burundi, where they carried out massacres of Hutu peasants in 1965 and 1972. It was not until 1988 that a majority Hutu government was established in Burundi and in 1992 a multi-ethnic constitution was passed.

The murder of the two Hutu presidents of Rwanda and Burundi in 1994, unleashed in Rwanda a wave of tribal killings of an unprecedented ferocity. The Hutu militias, called *interahamwes* (those that kill together), were responsible for the death of almost a million people, Tutsis in the main, but also Hutu who wanted to share power with them. The reconquest of power by the Tutsis in July 1994 led to an enormous exodus of six hundred thousand Hutus toward Tanzania, in the south, and more than half a million fled west to Zaire. The refugee camps there were devastated by epidemics and death.

This situation has created a serious political crisis in Zaire. The Banyamulengues, that is, 400,000 Zairians of Tutsi origin who have been inhabitants of the south of the Great Lakes for more than a century, found themselves under threat by the radical Hutus militia which controlled the refugee camps. They united in November 1996 with the Zairian opposition in the Alliance of Democratic Forces for the Liberation of Congo-Zaire with the aim of fighting against President Mobutu and his armed forces, made up of 25,000 disorganised and corrupt soldiers, in an insurrection which ended with the victory of Kabila. The defeat of the Hutu *interahamwes* in the refugee camp of Mugunga, in Goma, led to the massive return of more than half a million Hutus to Rwanda (where 90,000 people are in jail awaiting trial for massacres committed during 1994). Tanzania also put pressure on the Hutu refugees to leave its territory; the decision by the UN to go ahead with international intervention to protect the refugees was secondary.

Notes

I The regional state is in fact an autonomous state - as was the case of the Spanish Republic of 1931 - although it could also be represented as a type of administrative configuration of the welfare state in which the region is the basic unit of its territorial economic activity. The confusion arises from the fact that in the European Union the same term Region is used for federal *Landers*, for the territorial parts of autonomous states and for those units of administrative intervention.

II The religious factor divided central and western Europe into four zones: state Protestant Churches (Iceland, Norway, Scotland, Wales, England, Denmark, Hanse, Prussia, Sweden and Finland); mixed territories (Holland, Switzerland, the Rhine countries, the Baltic countries, and Bohemia); Catholic national zones (Ireland, Brittany, Lotaringia, Burgundy, Bavaria and Poland); and zones of the counter-reformation (Castile, Portugal, Catalonia, Belgium, Italy, Austria and Hungary).

Rokkan (1970) combines the religious factor with those of the strength or weakness of the formation of the centre and the density, high or low, of the network of cities in order to draw up a "conceptual map" of Europe between the years 1500 and 1800. This facilitates an understanding of the differing pace of the transformation of traditional European societies into modern societies, the chronological imbalance in the appearance of nation-states - of almost one and a half centuries - as well as some of the historical centre-periphery national conflicts: *maritime peripheries*: Iceland, Norway, Scotland, Wales (Protestant), Ireland and Brittany (Catholic); *maritime empires*: England and Denmark (Protestant), France, Spain, Portugal (Catholic); *Europe of the City-States*, integrated into broader systems: Lotharingia, Burgundy, Catalonia (Catholic); consociate formations: Holland, Switzerland (mixed), Belgium (Catholic); fragmented until the 19th century: Hanse (Protestant), Rhine countries (mixed), Italy (Catholic): *terrestrial empires*: Prussia, Sweden (Protestant), Bavaria, Austria (Catholic); *terrestrial confines* - with respect to eastern European orthodox societies - : Finland (Protestant), Baltic countries, Bohemia (mixed), Poland, Hungary (Catholic).

III South Africa currently has thirty-six million inhabitants, of which twenty eight million are Bantus. The largest sub-group of Bantus is the Nguni, of which there are seventeen million; this figure includes the Zulu, Xhosa, Swazi and Ndebele. There are five million members of the Shoto sub-group. The Tsonga and Venda are the smallest sub-groups, with one and a half million and half a million members respectively. The early black inhabitants, the Khoi-San, Bosquimans and Hotentots, are today a small group. Whites total five million inhabitants, of which three million are Afrikaners and two million are of English descent. There are three and a half million members of the coloured group, or mixed race, and one million Asians.

IV The Arabs of the north, who represent more than half the population, include the Jaaliyin, Guhana and Kawahla groups; non-Arabic groups, such as the Beja and Kakwa, have been integrated into Arabic customs and Islamic beliefs. In the south there are the Dinka, the Neur Etiopes, the Shilluk and the Azande.

3 The Periphery: Ethnic Groups and their Mobilisations

The construction of a national culture by the centre is likely to provoke the resistance of ethnic groups of the periphery that differ from it in religious and linguistic terms. However, these factors operate at different levels. Even though religious and linguistic discrimination constitutes a denial of the central legitimating elements of the nation-state, which would foster the belief in equal access for all citizens to the national culture, both have an inverted relationship with the change from traditional to modern society. Linguistic discrimination is a result of modernity and leads to the selection of a language of the state as a vehicle for cultural and economic communication and to the marginalisation of other languages. Religious discrimination, on the other hand, is a consequence of the failure of modernity, if modernity is assumed to be the substitution of traditional-metaphysical legitimating principles for instrumental-rational ones. The fact that the religious element becomes an element of marginalisation of specific groups reveals a return to the archaic, to legitimating mechanisms based on the consecration which exalts a state religion.

Nevertheless, language and religion taken separately would rarely give rise to specific mobilisations. These take place when they are part of an ensemble of identity markers (or *Gestalt* identity phenomenon) which singles out a reactive ethnic group in relation of the centre. For this reason the linguistic and religious facts must be examined in the context of specific centre-periphery conflicts.

Linguistic and Religious Groups

The fact that legitimating mechanisms often use historic religious beliefs explains the numerous tensions which exist throughout the world and increase the potential sources of centre-periphery conflict. While there are not many more than a thousand ethnic categories, the actual number of ethnic, religious and linguistic groups reaches almost ten thousand.[1] Moreover, the conflict becomes de-territorialized; if there is not always an exact relation between ethnic group and territory (consider the Gypsies), this relation becomes even less common in the case of religious groups (the most well-known example of which is the Jews). Liberalism, which aimed to convert metaphysical principles into private beliefs, loses the battle; and religious difference not only generates ethno-religious hierarchies in liberal states, it also gives rise to enormous counter-mobilisations of a religious nature in the Third World against the West (Islamic fundamentalism reappeared powerfully in the eighties after the decline of modernising Arabism).

Language is the most important of the symbolic complexes transmitted from generation to generation through the social reproduction of human groups. The facts of language are the determining elements of "us", the ethnic personality in the historical long-term. Language may even be consciously chosen as a badge of collective identity by members of a group who do not speak it, who share a feeling of belonging without being co-partners in the sphere of mutual comprehension. Conflicts related to languages, normally perceived as a hierarchy of languages, represent an important element in the fracture lines between groups. In France, for example, the term *patois* is a euphemism which indicates socio-cultural systems under domination.

In the long period of the official mono-linguism of unitary states of the West, which have lasted, in many cases, well into the 20th century, the language of the state is the fundamental factor in the processes of national acculturation and socialisation. The language of the education system is that of one group most associated with the pre-national state; it is usually the group that speaks the "most prized language", the language which has a written form or shares it with a family of "powerful languages," and which has a well developed body of literature. Nation-states have not, therefore, created the more powerful languages, what they have done is make them into the "national language" (Poulantzas, 1978).

In these situations of mono-lingualism the nation-state magnifies the distance between the dominant culture and the peripheral cultures, restricting educational work within the latter to the family while stigmatising their linguistic cultural systems. Monolingual schooling, generates, with respect to peripheral linguistic groups, an insurmountable contradiction between the agents of primary socialisation, the family and the school. The result is either the cultural submission of the dominated group (the diglot, an expression of linguistic discrimination, is a result of the use, according to the circumstances, of a familiar language of less prestige, or of the official language, considered learned and valued) (Martinet, 1970) or the assertion of ethno-cultural rights with rejection of the national-state political culture.

The social use of the mother-tongue or the state language could become optional in those complex states with various linguistic and education systems. On the other hand, even though in contemporary non-religious societies of the West belief is a matter of choice, the decision as to whether to practice one or other religion, or none at all is in the hands of the individual adult. The reality is that in many societies membership of a religious organisation is transmitted through socialisation within the family with an intensity and automatism practically identical to language. If this is taken into account together with the fact that religions often prohibit mixed marriages, it can be seen that the fusion of an ethnic conflict with religious differences deepens the trenches which separate the groups in conflict. From the point of view of minorities, religious discrimination represents an intolerable attack on their identity. From the point of view of the majority, putting in question religious hegemony constitutes an attack against the sacred centre of society, and therefore, an attack against the basis of their power.

Linguistic differences are often associated with peripheral conflict in many states of the world, although analysis of their geo-political distribution refutes any direct identification between language and national culture. It is not unknown for groups speaking the same language to engage in diversifying processes of state construction; the American Spanish-speaking states are the most prominent examples. There are, on the other hand, peripheral-national conflicts against centres endowed with a powerful language, where the periphery has suffered the acute undermining of its ethnic language. A case in point is that of deep-rooted Irish nationalism in the British Northern Ireland, where the use of Gaelic has practically disappeared (although it is very much on the increase in republican areas).

A language can be dominant or dominated, central or peripheral, depending on the position its speakers occupy in the social-ethnic pyramid. Spanish, the state language in Spain and a large part of the American continent, is the language of the marginalized Hispanic community in the United States, submitted therefore to the same limitations afflicted on the US Hispanic community. In the same way French, once the universal language of culture, must protect itself from the hegemony of the English language in Quebec. Languages of the colonial metropolis, symbols of oppression during de-colonisation, have become the obligatory official languages of many third world multi-cultural states, with the objective of preventing the linguistic (and hence political) dominance of certain ethnic groups over others: French and English have therefore become the official languages of a large number of sub-Saharan African States. In independent India the substitution of English by Hindi as the official language was widely rejected by the non-Hindi peoples of the subcontinent.

Although Western societies have laicised intensely over the past half-century, religious factors continue to mark out differences in Protestant and Catholic civilisations. This historical religious division marks out ethnic groups in conflict even when religious practice is considerably diminished.

This happens in Northern Ireland where Republicans and Unionists continue to be identified, and to identify each other, as Catholics and Protestants, despite many of their activists being personally agnostic.

In Western European terms, however, Northern Ireland is an exception to the rule, as linguistic differences tend to outweigh religious factors in national conflicts. Examples include the conflict between the Walloon and Flemish groups in Belgium, the Catalan, Basque and Galician nationalisms against the Spanish centre, and the Welsh and Scottish nationalisms against the British centre (if Scottish Presbyterianism and Welsh Methodism are not taken into account). The use of ethnic languages is widespread in Catalonia and Galicia, moderate in both parts of the Basque country, weak in Brittany and Corsica, considerably reduced in Occitania, Scotland and Wales, and non-existent in the Canary Islands. In Italy the ethnic consciousness of groups in South Tyrol and the Aosta Valley is based on German and French respectively, this is also the case in Sardinia with Sardo, although this language is less widely spoken. Northern Italy is experiencing efforts to revitalise the native languages in the ancient area of city states known as "Padania." In the conflict between Quebec and the Canadian state, linguistic and religious differences run together in the confrontation between the French speaking Catholic and English speaking Protestant societies.

Religious differences are more important in national conflicts in Eastern Europe. The official atheism of the former Soviet Union exacerbated their use as identity-markers, which reinforced the anti-centralist tendencies inherent in its extreme multi-lingualism. Whilst Lithuanian Catholicism, Estonian and Latvian Lutheranism and Georgian and Armenian paleo-Christianity have become factors in differentiation, Islam is the most important element of identity in several now independent former Federal Republics ("independent" in many cases in name only): this is the case too in some non-Russian autonomous Republics of the current Russian Federation. Islam is a major force in the conflicts within parts of the Caucuses - above all in Chechnya and Igra. It is a moderate force in the States of Asian Turkestan (with the exception of Tajikistan, where religion is an essential factor in the conflict which has unravelled into a bloody civil war) and a weak factor in. There it is religion rather than language that divides up the "Southern Slavs": the Orthodox Serbs and Catholic Croats settled their differences in Muslim Bosnia. A political-cultural convergence movement led Serbs and Croats to merge their languages into 'Yugoslavian' in the 19th century (the Slovenians and Macedonians have their own languages). A movement against Serbian hegemony in the former Yugoslavia has now led the Croats to assert their claim to a separate language from Serb-Croat. In the separation of Czechs and Slovaks, historic and economic factors have taken precedence over any minor linguistic differences.

National conflicts within the Third World present complex religious, linguistic and political diversities, which render detailed analysis of such conflicts beyond the scope of this discussion.

Ethnic Groups

This discussion exposes the difficulties that emerge when attempting to define the nature of ethnic groups. Would it be correct to include in the same category the isolated Indian groups of the Amazon jungle, the 2,300 Lhoba people who live in the Southeast of Tibet, or the huge "Han" Chinese majority formed over three millenniums by the fusion of around 3,000 different peoples? Before precisely defining what ethnic groups are, we should define what they are not.

Firstly, ethnic groups are not biological groups, demonstrated by their historical formation in Europe. The Western European groups come from three distant legacies: ancient tribal structures, the remnants of the Roman political superstructure and the invasions of barbarian "hordes" (Vilar, 1980). Subsequent to this, political factors such as conquests and matrimonial links shaped the territorial set-up of Kingdoms and placed ethnic groups into dominant and dominated hierarchies, modifying their limits and setting the base for future integration of some groups by others. (Geographical factors in the mixing and absorption of groups cannot be overlooked, these being strongest in the plains and weakest in mountainous regions). In the regions of the Central-Eastern European empires, forced or voluntary migrations and movement of peoples due to war have been commonplace. Nielsson (1988) rightly includes in his enumeration of 1,305 ethnic categories, macro-groups such as the Chinese, the Russians, French, British and Spanish. However the historical construction of such mega-groups is the result of the politics of monarchies and empires.

Existing definitions of ethnic groups reveal their cultural character. According to Moreno (1995), the ethnic group is a self-conscious autonomous community characterised by: a set of shared beliefs and values; a common language; a sense of belonging; and an association, real or imaginary, with a specific territory and history, which gives them particular characteristics in relation to other groups which recognise them as such. A. Smith (1979) bases his definition on the following six conditions: 1. a common denomination and a collective sense of solidarity; 2. a subjective belief in a common, contrasted or mythical, history; 3. an ascription, real or symbolic to a territory; 4. shared cultural assets that identify the group, such as language, religion or folklore; 5. a self-ascriptive consciousness of a sense of belonging; and 6. such differences identifying the group to others.

The diverse scope of cultures explains the different dimensions of the ethnic groups. Seiler (1982) argues for the use of Levi-Strauss's simple definition of culture: "All ethnographic sets that present, in relation to others, as an effect of investigation, significant differences" (1961). In this sense "multi-cultural societies" and "multi-ethnic societies" are the same thing; whilst language is the privileged vehicle of ethnic culture, it is by no means the only method of its communication. (Language and culture are therefore not necessarily the same.)

Many of the conceptual difficulties disappear if we distinguish, for the purposes of analysis, two moments in the formation of ethnic groups: their conditioning over history in the long term, and their current identity over the short term.

In the long-term ethnic groups are socio-cultural groups, a product of a succession of generations that were constituted in social collective formations. Through such formations the patterns of social practice were transmitted. In pre-modern societies this organisation of experiences was determined both by geographic and historical perspectives, and by economic means and ends. These experiences, transmitted as the corporal experiences of the child, become part of his or her "I". In modern multi-ethnic societies this is the result of interaction, conciliatory or conflictive, of the different identities of the groups, between each one and with regard to the state-national identity (which is often confused with that of the dominant ethnic group.) These groups allow the self-location in time and space of individuals through their constitutive essence that is their personal identity.[II] Collective identity, as a base of personal identity, complements the past and future simultaneously in the individual and in the group and mixes the realities of past life with what the future promises.

The space-time dimension of ethnic identities distinguishes them from almost all other political identities, for example those of a party political nature. Whilst political identities are fruits of the short and medium-term, ethnic identities on the contrary are long-term products. They can also be distinguished by their degree of voluntarism; the individual conscientiously possesses a political identity, yet the individual appears to be possessed by his ethnic identity.

What this means is that identities are not static; we do not even find fossilised identities within ethnic groups previous to modern political mobilisations. In this state there is also interaction of the ethnic group with neighbouring groups, which is dictated by the double movement of differentiation versus convergence and of ethnocentric opposition versus desire for recognition (ethnocentrism is made up of vulgar non-systematic judgements over outside groups, which tend to be degrading; in conflict situations become instruments of combat (Rodinson, 1975). The collective identity of ethnic groups is far from being an immobilising factor, it is characterised not only by continuity, but also by the transformation of the "us"; the dynamic and inter-convertible elements that make its totality can decompose. This allows the ethnic group to recognise itself in the distinctive phases through which it travels (Berque, 1978).

The same can be said of the space and time experiences of the group, which are both social constructs. The spatial linking of a social group changes as its control of an ethnic territory transforms over time. As far as time is concerned, this is lived by pre-modern groups in an imaginary sense, recomposing in a mythical way the genealogy of the group - whether this is clannish or not (Balandier, 1981).

In pre-modern ethnic groups and in modern groups that have hardly started processes of self-identification, ethnic personality is predominant over identity. According to Devereux (1970), ethnic personality is constituted by two sets of data: firstly, the observable behaviour of the group, and secondly, the self-generalisations of the members of the group themselves. Ethnic identity, on the other hand, is a means to single out and label, and does not develop except by a confrontation with "others" who, for whatever reason, are attributed with a different identity.

In modern societies the inevitable bringing together of ethnic groups in situations of reciprocal hostility that industrialisation (and to a greater extent colonialism) bring about, provoke ethnic categorisation. This is both selective and voluntary, and can activate whichever elements or groups of elements of the pre-modern ethnic personality, giving them a new form. Barth (1976) believes ethnic groups are categories of ascription and identification used by individuals to organise interaction between themselves. The focus of investigation therefore moves away from the internal constitution and history of these groups to centre on the ethnic boundaries and their persistence. The features that are taken into account by members of ethnic groups are not the sum of "objective" differences, rather those that are considered significant, whose order of precedence cannot be predicted beforehand. Therefore, the cultural ethnic aspects of a group such as dress, speech and housing are signs which individuals exhibit to indicate their identity. In this sense the cultural exposure of ethnic groups through cultural contacts in the modern world does not necessarily mean a reduction in organised ethnic identities. [III]

In modern societies however, ethnic personalities do not create processes that arrive at a unique identity. This is because in one person there is not one single collective identity, but rather multiple identities structured in the form of concentric lines of community consciousness. Each grouping constitutes a place in identification; the family, the local community, belonging to a tribe, the definition of a homeland, and recognition of a nationality. When the individual possesses a large number of sufficiently diverse identities, each one becomes an *instrument*, and its totality an *instrument case* that forms its personality pattern or unique model (Devereux, 1970).

But in a world ruled by nation-states primacy is given to one of the concentric identities, the national state identity. Today, Gallisot (1976) states, "because of the dominance of the nation-state the national identity prevails (national passport); the other sentiments of belonging do not disappear entirely, but are reduced and marginalised".

This situation can slow down the dynamics of identity elements. In this case the domination of the ethnic personality over identity pushes towards the *"picturesqueness"* of groups. This means the features of their historic personality, fixed once and for all in their pre-modern appearance, appear "painted-on" to the national society. Such features can be deliberately stigmatised as inferior, but this is not always the case. What frequently does take place is a process of overvalued compensation of such features (e.g. customs, dress, landscapes) being described in highly favourable terms and incorporated like museum pieces into the imagined centre of the nation-state.

Ethno-Political Identities

Ethnic identity can be limited to the exhibition of group symbols, therefore remaining politically inactive. Its conversion into an ethno-political identity (an indispensable requisite for the transformation of an ethnic movement into a national one) is the result of a dual reaction within the ethnic groups: firstly, their reaction against the creation from the core of a national society and community, and, secondly, the willingness to create their own ethno-national society and community, in a process of imitation, or mimesis, antagonistic to the centre.[IV]

The emergence of this dual process is by no means inexorable. Especially in the West many ethnic groups end up losing their dynamism, or by being absorbed by the state national identity through *picturesqueness*. In regard to the ethnic identities that do endure (most of which will have a territorial base although this is not obligatory) their process of differentiation and their selection of identity-markers almost always has the nation-state as an opponent and as a point of reference.

Dynamic ethnic groups mirror-image the national-state process which involves the creation of the community fundamentally through discourse. From this surges the need for community shaping nationalist ideological elites. Without such elites the construction of national identity would be impossible. This provokes however the process of stereotyping and devaluation to which they are inevitably subjected to from the nation-states own ideologists. This mimetic process, as for its state anti-model, gives rise, firstly, to the task of singling out the marking factors of national identity and, secondly to the recreation of a sense of identity and, over time, continuity for the ethno-national group. This involves the "invention" of tradition and the production of a national history, in the same way that, since the French Revolution, ideologists, particularly historians, have recreated or "invented", national "traditions" in each nation-state: a recreation in which the new national symbolic order pulls itself up on the ruins of past symbolic constructions, shaping a new "us" - a symbol marker formed by the emblems of existence, above all by the name of the nation.

The process of producing a national identity is not arbitrary, neither is it determined. With greater intensity than in the pre-national phase, the discourse of the nationalist elites activates ethnic elements which include language, historical memory, customs, symbols, religion, objective and subjective ascription to a territory. This activation in highly varied, giving them a new form which recreates them in a new identity construct or *gestalt*. This autonomy of discourse explains the well-known fact that from one single pre-national ethnic personality a number of complex ethno-national identities can emerge, and hence separate nationalist movements are derived. (It is impossible to explain the contemporary nationalism, or nationalisms, moderate and radical, of the Basque Country without taking this into account.)

The ethno-national processes frequently develop into a rational strategy aimed at the acquisition of their objectives. Territorial demands are presented and the construction of their own institutions is demanded, to ensure that the survival of their ethnic identities is set in motion. It is effectively the jump between ethno-cultural demands to political-territorial petitions which converts the process of ethnic identity formation into a national movement.

In the first phase of their mobilisation ethnic groups lack the political power required for the creation of their own national society (that is to undertake the political-administrative integration and the unification of their territories' market). Their cultural character confers on them a greater possibility of recreating the community (understanding as such the processes of socialisation and the mechanisms of ethno-national legitimisation), but the procedures followed are also different to those of the nation-state. Socialisation through schooling, the mechanisms of rational legitimacy and identification with the power of the state are substituted in ethno-national movements by the consecration of the "imagined nation" and by a sort of acculturation derived from ideological intensity and effervescence of identity.

It is rare that the citizen who lives in conformity with his state national culture needs more than a system of beliefs. By contrast the members of an ethno-national movement who act against the flow of a state need to remanufacture their culture. Because of this, ethno-national cultures are born on the ground that official political cultures are weakened or rejected, constructing patterns for the creation of the conflictive collective consciousness, and being maps of problematic social realities (Geertz, 1973).

In pluri-ethnic societies numerous cultural interactions between diverse groups are produced: acculturation, de-culturation (generally driven on by the nation-state) and antagonistic acculturation. The modalities of acculturation reflect the cultural composition of states. A number of situations can be distinguished. Firstly, contact with global societies through invasion or colonisation (those who invade tend to adopt the culture of those who they invade, whereas those who colonise usually impose their own culture). Secondly, through contact with private groups of individuals of different nationalities, like missionaries or soldiers (acculturation is unilateral). Thirdly, through contact with national groups with equal status (e.g. the Walloons and the Flemish) where a-culturation is reciprocal, yet remaining largely blocked. Fourthly, through contact between numerous ethnic groups compromised into a state (e.g. Indians) - here the acculturation is multi-lateral but is blocked by geographical and social compartmentalisation (Selim Abou, 1981). Finally, if the ethno-national groups are demographically unequal or politically hierarchic, the majority dominates in the economic, political and cultural spheres. In this case, acculturation, whilst reciprocal, is unequal and susceptible to generating counter-acculturation attitudes, or attitudes of antagonistic acculturation, from the dominated groups.

Devereux (1970) distinguished three types of antagonistic acculturation: defensive isolation, the creation of new means destined to fulfil existing goals, and the negative disassociate acculturation that consists of the creations of cultural constructs deliberately in opposition to the culture of the dominant group. This is what most frequently occurs in the peripheral movements reacting to state national culture.

Ethno-National Mobilisations

Western Europe has historically produced three successive identity reactions of ethnic groups (Seiler, 1996). The first is the legitimist-reactionary mobilisation that began at the end of the 17th century in the United Kingdom with the Jacobites (from 1688 to 1750), continued with the *chouans* of Brittany in the era of the French Revolution, and allowed, in the first half of the 19th century (from 1833 to 1876), the Carlist movement that took up arms against the weak liberalism of the Spanish national state. The big chronological gaps are due to the differing dates in which the respective national and industrial Revolutions were initiated in the separate states. The second reaction, which takes on a populist nationalist form, consists of the use of the arsenal of nationalism by the periphery in its struggle against the centre. This second phase produces the emergence, towards the end of the 19th century, of autonomist-nationalist parties, defined by Seiler as those who "originating themselves from centre-periphery divergence aim to obtain an ethno-territorial community that is self-governing". The third form of identity mobilisations, the progressive-nationalist, relevant in the sixties and seventies of the 20th century in diverse areas of the West, is born out of the combined effects of the economic expansions of those years and the repercussions for Europe of the de-colonisation process. This is therefore the consequence of contradictory factors: the discrediting of racist nationalism provoked by fascism, the peak of the Third World mimesis, European Federalism etc. This mobilisation fostered the post-industrial values of the generation of May 68 and the alternative movements - ecologist, feminist, and anti-militarist- which were born in its mist. It is necessary to add a fourth type that consists in the neo-centralist reactions of economically developed peripheries against a backward centre, directed by national bourgeoisie and involving local capitalism.

Centre-periphery conflicts are generated by cultural, economic and political factors. Cultural factors can surpass ethnic limits: entire civilisations that have been marginalized by past or present colonial processes develop reactive mobilisations whose manifestations are similar to those of ethno-national movements (this is the case with Islam at present).

Economic factors can give rise to situations of "internal colonialism" (Nairn, 1979) related to the economic subordination of the periphery to the centre, but also in the aforementioned inverse phenomenon of the economic colonisation of the centre by the periphery, giving rise to neo-centralist movements. A multi-centred state can have a number of local bourgeoisies: either because the state belongs to a network of city-states, totally (e.g. Belgium) or partially (e.g. Po Valley in Italy) or because it has an extensive territory which is economically globalised at a continental level as opposed to state level (e.g. territories of Canada which integrate on a North-South axis with outlying areas of the United States.) The periphery can have a strong local bourgeoisie that does however align itself, for market reasons, with the centre: Norwegian nationalism until independence was the fruit of an anti-bourgeois alliance of middle class urban radicals with rural populists; in the Basque Country the financial and iron/steel oligarchy, hostile to Basque nationalism, has always opted for Spain.

Because of this the local bourgeoisie can be the protagonist of reactivated neo-centralism: on facing a less developed centre (e.g. Catalonia up to the Spanish Civil War), or shaping a techno-structure that wishes to break away and become autonomous of a developed centre, or both at the same time. The second phenomenon (observable in Quebec) emerges in dependent societies whose local bourgeoisie is aligned to the centre or in the service of foreign interests; the state however is complex enough to allow, either that local powers manage the economic development of their country, or that it sets up nucleuses of peripheral leaders, integrated to the central institutions of the state. Once the lack of interest or efficiency of the state has been demonstrated, the techno-structure (consisting of civil servants, specialists and experts) makes the nationalist cause its own with the object of creating its own modern state, centralised and capable of guaranteeing economic development (Seiler, 1982).

The origin and make-up of nation-states explains the nature of peripheral reactions. In the West, when the political integration of the state is still not complete, it is the traditional society that on occasions stands up against the centre, with the social hierarchies and the most high ranking within the society at the forefront. In multi-cultural states of the Third World it is the vertical tribal or clannish societies (or at times ideological groups) that react against the persecution or discrimination suffered at the hands of the centre. In the case of fragmented federal states brought about by the decomposition of an imperial ideology (Soviet Union, Yugoslavia) it is the political federal elites that reclaim for themselves the prerogative of statesmanship and head processes which frequently have more to do with political secession from above than with a national movement from the grassroots. However, in those nation-states that have finished the processes of political integration and mobilisation, peripheral nationalist movements are born out of the political activity and identity construct of a self-created ideological elite.

The Intersection of Church-State and Class Cleavages

The expression of centre-periphery conflict also depends on its intersection with other cleavages defined by Rokkan and Seiler, in addition to tradition-modernity conflict; the church-state cleavage, specially in catholic countries, and the class cleavage. Periphery movements are frequently situated on one side or another of these dividing lines, which gives these conflicts an ethnic nature.

In the early stages of the penetration of the state in the periphery, ethnic groups often invoke a respect of tradition, of the "old order". However if the state is not a factor of modernity (or is not a sufficient factor) it is possible that ethnic movements will take upon themselves the construction of modernity against the backward centre. In these cases such movements emerge earlier and develop with more strength than those within strong nation states. At the same time, the modernity-tradition conflict moves inside the movements themselves, antagonistically diversifying its progressive and conservative tendencies. This is relevant to the differences in the nature of the peripheral reactions to the Spanish and French centres during the 19th century and at the beginning of the 20th. In Spain these reactions presented a wide range of tendencies: liberal, federal, traditionalist; whilst in France they were unanimously conservative.

The centre-periphery conflict will also interact with the state-church cleavage. This emerges when the lay state attempts to take away the church's monopoly of education in a phase after the start of modernity but before mass political participation: which is why the liberal and federal tendencies of the Catalan and Galician movements appear before traditional tendencies. Where the church maintains its control of the population, its attitude to centre-periphery conflicts is ambiguous: it can become the bastion of their resistance to the state (as in Spain, when they were useful allies in the struggle against the state), or, alternatively, it can unite with the local elites to integrate the peripheral masses in the state order, as in France. (The successful construction of the French nation effectively permitted, in the conservative and de-industrialised peripheries, the survival of a dual identity whose ideological space was filled and integrated at a later stage to the institutions of the 3rd republic by the social doctrine of the Church (Loughlin, 1987.)[v] The differences that exist between the Catalan, Galician and Canary Islands movements, on one hand, and those of Occitania and Brittany on the other, are revealing in this sense.

The Hispanic unification undertaken by the Catholic Monarchs and the Hapsburgs was based on religious monolithism. The peculiarities of distinctive Kingdoms and peninsular political entities had survived, a diversity that the Bourbons had lessened without eliminating it. In the panorama of 19th century nationalisms, Spain is a case study of a weak and dependent market, of the failure to create a modern bureaucracy, and of liberalism both rhetorical and forcibly imposed by the military. The weak construction of the Spanish national centre explains the vigorous peripheral nationalisms, above all in the Basque Country and Catalonia.

Catalan and Galician nationalisms both underwent an intense cultural-linguistic renaissance in the mid-19th century, which was followed by a process of ideological crystallisation in which the liberal, federalist-leftist and traditionalist tendencies entered into conflict.

In Galicia, the Lugo Declaration of 1846 led by Faraldo (although consequence of the battle between Spanish liberals and moderates) presented a clear national perspective; the Galician Upper Assembly he created categorised Galicia as a *Colony of the Court*. The Galician *Rexurdimento* of the 1860s brought out a trio of excellent poets: Rosalía de Castro, Pondal and Curros, and a generation of romantic historians led by Martínez de Murguía that brings together the "Celtism" felt to be an important factor in the Galician nation, with Italian inspired liberalism.

The Galician republican federal nationalism born in the heat of the 1868 revolution was short lived. From 1880 to 1916, alongside the Galician nationalism of Rosalía and Murguía - pushed along by sectors of the bourgeoisie-, a Galician anti-liberal traditionalism emerged, strong in the Carlist and agrarian mediums. Brañas gives this shape in his 1889 work *El Tradicionalismo* (The Traditionalism*)*, in which he defends a decentralised organisation for the Spanish state and repudiates modernity.

In Catalonia the pre-*Renaixença* exalts from 1835 Catalan art and history. Aribau published his "Oda a la Patria" (Ode to the Homeland) and Milá and Fontanals, inspired by the Provençal Felibres, organised the Barcelona Floral Games in 1859, which attracted Balearic and Valencian poets. In 1869 Amirall, leader of Catalan federalism, created the Aragon, Catalan, Valencian and Balearic Republican Committees. His federalism was rejected by the conservative regionalism and defence of tradition of Mañé y Flaquer and of Torras y Bages. However, the economic drive of the country made the convergence of all the Catalan tendencies possible. Successive Catalan Congresses defended Catalan civil law against the Castilian laws in 1880, they also produced in 1883 the protectionist Memorial de Greuges that denounced the perceived threat to Catalonian interests of trade treaties with England and France.

From 1882 Prat de la Riba undertook a synthesis of these trends working on a programme of self-government ("the Bases of Manresa") and an ideology which whilst was nationalist, did not call for independence, set down in *Compendi de la Doctrina*. In 1902, he created the Catalan Regionalist League. According to Prat this should have re-addressed the great imbalance between the living and dynamic Catalonia and the dead, bureaucratic, agrarian Spanish state; the Catalan market should extend itself, which would allow Spain to become a modern imperial power.

Catalan nationalism triumphed at the beginning of the 20th century by creating a "national bourgeoisie" that succeeded in unifying conservative Catholics from a rural origin with the liberal urban bourgeoisie, under the objective of "*catalanising*" society. This led to the standardisation of the Catalan language and the aim of modernisation of the Spanish State in which Catalonia could incorporate itself on a federal basis (Ferrán Soldevila, 1978; Rubiralta, 1988; Mansvelt, 1994).

By contrast, Galician nationalism did not manage to consolidate itself as a stable political force. Both its weak industrialisation and the delays in urbanisation, added to the existence of an isolated, very poor peasant community (who was subject to dependence on local rulers) impeded the emergence of a synthesised "National Class" that could overcome the irreconcilable nature of liberal urban and traditional agrarian goals.

The conquest of the Canary Islands in the 14th and 15th centuries wiped out the ethno-cultural characteristics of the native *Guanche* population to an even greater extent than in the Americas. The islands were integrated in the 19th century, at the hands of the British, in a circuit that united the Indian and Atlantic commercial channels with Europe. The segregationist stance of the local national bourgeoisie pressured the Spanish state to recognise the characteristics of the archipelago's commercial enclave; this ended with the 1852 Bravo Murillo decree which declared some exempt ports on the archipelago. From then on the nature of Canary Islands nationalism was linked to the workers movement, being clearly leftist (Alemán et al, 1978; Gari Hyek, 1994).

In Galicia political Galicianism was expressed from 1916 in the *"Irmandades de Amigos de Fala"*, both republican and supportive of an Iberian federalism that would include Portugal. Galician culture found a niche between 1920 and 1936 in the *Nos* ("Us") Magazine, run by cultured rural notables such as Risco and Otero Pedrayo, dreaming of a world that had disappeared. On the setting up of the Second Republic, Castelao founded in 1931 the Galicianist Party, a progressive synthesis of the Irmandades and the Nos group. Despite being a minority party they succeeded in their goal of Autonomy for Galicia. The Statute of Autonomy was approved by referendum in June of 1936, but was never brought in due to Franco's victory. (Millares, 1986; Maiz, 1988).

Occitania, the territory of the Oc language, is a broad linguistic-cultural space that spreads across from the Atlantic to the Mediterranean, from the Massif Central to the Pyrenees, and over the historic regions of Aquitania (Gascogne and Guyenne), Limousin, Dauphine, Languedoc, Provence. Lacking a centre, or more precisely being based on two centres, the Aquitaine-Atlantic and the Provençal-Mediterranean, Occitania is thus a typical example of an imagined country. This does not mean that it is insignificant, the South regarding itself as having a glorious literary and historical past.

Provence is the Roman "province" of Marseille, of the the Waters (Aix-en-Provence), open to all the Mediterranean cultures it adjoins, with Italy to the east (whose valleys it penetrates) and with Catalonia to the south (in fact it takes in the Rosellón, which had been part of Catalonia). Aquitania is the meeting point of Basque-Pyrenean ethnic terrain with the Romans and Gauls. Today the Occitanian groups are Gascognes or Bearns who share the Atlantic Pyrenees Department with the Iparralde Basques.

Occitan is a literary language, older than French; the excellence of the poetry of its singer-songwriters made it Europe's principal lyrical language in the 11th and 12th centuries. In the 13th century the Kingdom of the Isle of France occupied Occitania under the pretext of Manichean Albigensian heresy. The church created the Holy Inquisition in support of France over Occitanian cultures. Guyenne administrated by the English (which stretched from Bordeaux to Bayonne), was recovered by France in the mid-15th century, on this occasion by negotiation. Memories of resistance against the invading army of Simon de Montfort gave birth to epic Occitanian literature whose expressions have lasted into the modern age. The ethos of defence of an Occitanian identity continued to echo during centuries of French domination and were still heard in both Revolutionary and Napoleonic France.

Explicit affirmation of Occitanism emerges, in parallel to Catalanism, in literature. In 1854, a group of poets, led by Frédéric Mistral, founded the "Felibrige", a literary movement that defended federalism and the Occitanian community defined by a language, sharing with the clerical bourgeoisie and the traditional peasants a nostalgia for the lost order of the rural world.

The Felibres kept themselves apart from social struggles that took place mostly among the small vine growers and the wine wholesalers. In 1907, when the vine growers united alongside a Unionised agricultural proletariat, the ageing Mistral refused to join the arranged demonstrations. This leads to the "red" Felibres republicans to turn back to France.

After the Great War the old tensions between "whites" and "reds" continued, and a new tension emerged. Mistral's provençalists opposed Catalan-influences which reproached the "whites" for their "provençalist imperialism". The Provence Party was formed in these years. Federalist and of Mistralist orientation, it rejected class struggle; in parallel with Brittany, the Pétain regime, supposedly in favour of local languages, attracted a sector of the Felibres (Touraine et al. 1981; Pourcher, 1992).

The Bretons are descendants of primitive inhabitants of the island of Great Britain. When the waves of Germanic tribes (Anglo, Saxons, Jute) invaded the island in the 5th and 6th centuries, the Bretons arrived in mass to the Armorica, which had been populated until then by peoples who were also of Celtic origin. The Bretons installed themselves in Armorican land and it became Brittany. A kingdom was created which lasted until 1532, the year in which the French King François I united both under one Crown with the Nantes Edict, although there was respect for the customs and practices of the Brittany kingdom. In the modern age (as in the northern Basque territories) the counter-Reformations had a considerable influence over the country; at the same time movements emerged in defence of old customs and against Bourbonic centralisation (e.g. the *Bonnets Rouges* of 1675; the Marquis of Pontcalleck in 1720). Such movements were cruelly repressed.

In 1789, Catholic Brittany had no confidence in the French Revolution and opposed the civil status of the clergy. A new peasant and popular movement emerged, the *Chouans*, pre-empting the Carlists by forty years, which was won over by the exiled nobility. As a result of this Brittany lost its unity and was divided in 1799 into five Departments. During the 19th century Breton society was conservative in nature, with cohesion provided by the church, and in which the clergy and nobility managed to integrate the peasant masses into French national structures by the end of the century. At the same time the centre power undertook a systematic "de-Bretonisation". The Bretons were treated with the most disdain in comparison to any citizen in France. The Breton language was outlawed and its use in schools punished. If some customs did survive they did so as folklore. Brittany did in fact have two native languages: Breton, a Celtic language, and Gallo, a Latin (but not French) based language, spoken in Upper Brittany. Between 1870 and 1914 the Bretons received the same treatment as is today given to the Maghrebians, although did not stop them from being sacrificed in the battle fields of the First World War in which 240,000 young Bretons died. One in four soldiers in the French army who died were Breton, although the Breton population is one eighth that of France.

In the last third of the 19th century an incipient nationalist movement had emerged, largely of literary character in the grain of Romanticism. The Viscount of Hersant de la Villemarqué had compiled the Barzaz Briez, a group of popular songs that became the catalyst of a renaissance of the Breton language and culture. The Breton mobilisations reached the political podium in the inter-war period. The Breton Nationalist Party was created in 1925 and expressed itself through the Breiz Atao (Brittany Forever) newspaper. Shortly afterwards a secret society *Gwenn ha Du* (White and Black) was formed and launched symbolic attacks on French monuments. The French leftist parties (with notable exceptions such as some Breton Communists) were the fiercest adversaries of regionalism in this period, which was seen as an attack against a France that should continue to be Jacobin. As a consequence all autonomists and federalists were to be found in right-wing groups (with the exception of Maurras's *Action Française*, which was as reactionary as it was centralist).

In addition, the rediscovery of the Breton past brought with it the rediscovery of Celtic traditions and Pagan mythology. This would convert some of its cultivators into easy prey for the propagandists of fascist National Socialism, which presented itself as the force behind the renaissance of pre-Christian myths. As a consequence, the far-nationalist right collaborated with the Nazis, feeling that they would liberate Brittany from their marriage with the French and grant it independence. Although minorities, such sectors left the claim for Breton identity in an uncomfortable position when France was liberated. Because of this the subsequent purge was more violent and unjust than anywhere else (Markale, 1985).

Class cleavage is always present in ethno-national movements: given that the national state is the fruit of interaction between the industrial and national revolutions, modern societies, and of course their ethnic communities, are by definition class societies. This conflict can take effect within ethnic movements in two ways: defining the movement's political orientation according to the social class hegemonic within it; diversifying the movement into various tendencies whose *Gestalt* responds to separate class ideologies (which are at the same time translated into a discourse of identity). This can give rise to situations in which an ethnic "social division" of work is produced fundamentally re-defining the centre-periphery conflict as a class conflict. According to Hroch, such a relationship existed between the ethnic Czech and German groups in Bohemia until the First World War. The clear-cut nature of the class struggle between the Czechs (a dominated majority) and the Germans (a dominant minority) turned the Czech national movement into a model for others in Central-Eastern Europe. (The same cannot be said for the Slovaks, a periphery of the periphery, whose veiled opposition to the Czechs has expressed itself in their recent secession).

The Czech Republic, made up of Bohemia and Moravia, sited on the western part of the former-Czechoslovakia, has a current population of 9.7 million. Of these, 84 % are Czech, and 14 % Moravs. The Slovak Republic, to the east, has 5.7 million inhabitants of which 84 % are Slovaks, the minorities being 10 % Hungarian and 5 % Gypsy.

The Czech people have historically organised themselves into two political entities, both Slavic: the Kingdom of Bohemia, and the Margraviate of Moravia. This state, formally integrated since the 10th century in the Roman-Germanic Holy Empire, maintained its formal independence until the 16th century, when it became part of the House of Hapsburg. Its relative liberty ended in 1620 with the Protestant revolt of the Bohemian states; the Czech Lutheran aristocracy was forced to emigrate, and was replaced by alien German speaking Catholic landowners. The counter–reformation persecuted the Czech language for heresy and imposed German as the language of the government.

Bohemia is a geographically closed entity. It has depended on stable borders from the 10th century, and has a single political centre, Prague. By the end of the modern period Czechs made up 67% of the population and Germans 33%. Yet the Germans, the richest and most powerful group, failed to learn Czech and obliged Czechs to learn German. Moravia, however, was geographically open to the south with lower Austria. German influence was felt here more intensely, and when the Czech national movement emerged it did so alongside strong Moravian nationalism. Czech, however, was not only spoken by peasants but also by the urban middle classes. Both languages were used in education in the Austrian empire and because bi-lingualism was never permitted in higher education (it was 1882 before bi-lingualism was accepted in the University of Prague) it became a strong motive in national agitation. According to Hroch the social conflicts that had a strong national significance became integrating factors in Czech nationalism. The conflicts he refers to were those between Czech peasants and non-Czech landowners, between Czech workers and German or German-Speaking Jewish capitalists -in fact, the Social Democratic Party split in 1890 between the small Czech petty bourgeoisie and the great German industrialists and traders; tensions were also felt in the Czech intelligentsia who had no access to top jobs or to the best paid professions, that were all in German hands. The role of the Catholic Church was characterised by its adoption of a neutral position, not opposing the involvement of grass roots clergy in the Czech national movement. Towards the end of the 19th century the movement became a model for other Slav and Balkan nationalist movements (Slovaks, Slovenians, Ukrainians, Latvians and Lithuanians).

Land inhabited by Slovaks, who were ethnically and linguistically similar to Czechs, became, from the late Middle Ages, part of the Hungarian Kingdom, integrating the Slovak with the Hungarian nobility. Following the Ottoman invasion of the 16th century, Bratislava became the administrative capital of the Hungarian Kingdom, which intensely accelerated the assimilation process. Subsequently, the Slovak nobility participated in the Hungarian nobility's resistance movement against the Hapsburg centralism. But the ethnic Slovaks maintained broad contact with Czech culture, and the Catholic majority spoke local Slovak dialects alongside Latin, the kingdom's official language. Towards the end of the 18th century the Clergy began to develop the written Slovak language using the dialects as a base. However, Slovak territory had never been a single political entity, since geography played a disintegrating role, with the mountain ranges of Slovakia divided into four major valleys and numerous minor ones. Bratislava did not develop a national self-consciousness, and there was no major urban centre with a Slovak speaking majority in the 19th century. The policy of the imposition of the Magyar language by the Hungarians from 1870 led to the assimilation of the Slovaks, thereby making the emergence of a Slovak intelligentsia more difficult. Despite this, a Slovak nationalism did emerge, pushed by intellectuals and by Lutheran clergy (more so than their Catholic counterparts) although the participation of artisans and peasants in the movement was not, however, significant. Other dividing factors were the fact that the majority of Czechs, along with many Lutheran Slovaks, did not accept Slovak "secession" in the Czech "common nation". In addition, the Catholic church supported the project in Hungary to create a Magyar-speaking nation and (with the exception of a sector of the grass-roots clergy) opposed the Slovak national movement.

After the defeat of the Austro-Hungarian Empire in the First World War in 1918, the Allies pushed for Czech-Slovak unification into a republic presided over by the Czech philosopher Masaryk. Slovak was accepted as the language of education at all levels – including in the University of Bratislava. However, the majority of teachers and bureaucrats that went to Slovakia were Czechs who, convinced that Czechs and Slovaks formed one nation, used the Czech language in government, schools and in the army. This led to Slovakian anti-Czech nationalism in the inter-war period which emerged out of the opposition of Slovaks to the Czech elites and to the losses suffered by Slovak artisans at the hands of Czech industrial imports, and the widespread hostility to the police (then Czech). Nationalism was fuelled by the Catholic clergy which opposed the Czech-Slovak secular state and mobilised the peasant masses against it.

In this way Nazi Germany, having annexed the Sudetes in 1939 created a Bohemia-Moravia protectorate, yet in Slovakia set up a puppet government presided over by the Catholic Monsignor Tiso who did not think twice about persecuting Jews and other minorities. In 1945 he faced trial for war crimes (Hroch, 1994).

Conditions for Ethnic Mobilisations

The conditions that give rise to ethno-national conflict and which promote the development of ethno-national consciousness are diverse, involving the following: the existence of a territorial contiguity of the ethnic group; the presence of elites that produce an ethno-national discourse; the image of a nation (imagined or based on a historic proto-nation); a sense of urgency that can be caused by: state monopoly of education and the imposition in education of an official language which is not that of the ethnic group, or the mass immigration of alien groups; a sense of relative frustration when the political expectations generated by economic status are not satisfied - or vice-versa; the diffusion of similar national movements, etc.

Territory gives an ethnic group recognition (given the universal acceptance of the philosophy of the nation-state, which is based on territory), group identity (one of whose principal markers is the association of a people with a place, although the homeland can be constructed, or a source of conflict between two or more groups: the Biblical Zion of the Jews is the Arab home of the Palestinians), and human and material resources that can be mobilised and used in defence of the characteristics of the ethnic group.

The political value of ethnic territory comes into conflict with the new meaning that territory acquires in the era of the nation–state, that is, the space that defines the area of sovereignty of a state over its citizens, the claim over institutionalised ethnic territory calling into question the sovereignty of a state.[VI] The territory chosen during the process of identity-construction is commonly that which has been historically inhabited. Sometimes this relation is non-problematic (often with mountain peoples). But the selected territory can also be that of an imagined community because it hosted historically diverse groups and political formations, such as Occitania with an area that covers countries and cultures as diverse as Aquitania and Provence; or because it aspires to unifying areas made-up of dense networks of cities which are polycentric by definition- as in the project of "Padania". The territory can be claimed by two or more groups, who set their own historic and/or religious motives against each other. This takes place in the current Israeli state, or in the former Yugoslavia where the Bosnian Muslims, Bosnian Serbs and Bosnian Croats dispute Bosnia. A territory historically inhabited by one group can become occupied by another that makes it the territorial nucleus of its state: this happens with settler colonies. When the balance of force between the settlers and the indigenous population is excessively favourable to the colonialists (the European colonialists and Native Americans, the Anglo-Saxon settlers and the aborigines in Australia), colonialists present the relationship of these peoples with their territories as a phenomenon more related to nature than to politics.

Territorial clams are facilitated when ethnic groups live in neighbouring areas, which is why immigrant groups rarely claim territory. This rule does have exceptions - there are some identity-based movements that do not have a territorial base (e.g. the black community in the United Sates). Exiled communities, or diasporas, sometimes become the principal focal point of a national movement, either because the ethnic group has suffered forced movements or expulsions (the Armenians in Turkey for example) or because they have had a historic territorial claim to a national homeland (the Liberians, the Jews). Archipelagos frequently develop multi-centred ethnic movements due to the existence of many peripheries of the periphery (the main Canary Islands autonomy party is called the Canary

Islands Coalition, due to the strong political personalities of each of the separate islands).

Ethnic groups tend to make huge efforts to obtain the former's territorial contiguity, while the states try to prevent them. Despite the ethnic territories being along side each other, if the group is divided by a border and hence has been split over different states (either over the long term as with the Basque Country or in the short term as in Kurdistan) its national movement will be divided into the same number of parts as states. This is even more so if the territory of an ethnic group is geographically separated, as with the Tatars who are situated in very different zones, from the Volga Basin to the Crimea in Southern Ukraine. On occasion it is states themselves that cause ethnic territories to be geographically separate (the autonomous Palestine regions of the Gaza strip and the West Bank lack a linking corridor across the Jewish state); in other instances they align with sub-peripheries in their struggle against the peripheries (the Russian and Southern Osetian against Georgia), or cause separate periphery groups to clash amongst themselves (Armenians and Azerbaijanis became enemies over the Armenian enclave of Nagorno-Karabaj, situated in Azerbaijani territory).

A vital pre-condition is the existence of an elite that shapes the discourse of identity, and has the will and power to transform the ethnic group into a national movement. It is not problematic whether the peripheral elites have had an historical relationship with the centre or not (the Basque coastal maritime periphery in the 16th century was the advance party in the Oceanic expansion of the Spanish empire, but that did not weaken the force of its peripheral reaction in the era of Spanish National construction).[VII]

Given that peripheral elites question in their discourse the processes of national socialisation and acculturation set in motion by the nation-state, they are systematically devalued and subjected to stereotyping by the centre's elites. However, the greatest hostility toward them comes in the cultural phase from the traditional hierarchies, intermediaries between the state and the masses at a local level. These networks of the local "notables" react against the new nationalist elites with more intensity when the new elites are closer to them (they frequently come from the same social background) perceiving them as the most direct threat to the source of their power. Their virulence is at a maximum when they are denounced by them as anti-patriotic and collaborationist.

The labour needed to create a periphery nation is facilitated by the existence of an historic proto-nation that is easily recognisable and linked to the present. Those proto-nations, by definition, lost the battle of constructing their own state: this explains how the historical events which they commemorate are not only those that brought good fortune but also the defeats, such as national commemorations which have a sadness about them. Sometimes, however, the proto-nation projects itself in a predominantly linguistic and literary manner. The clash of civilisations can, on the other hand, defuse the image of a historic proto-nation as an active element in the taking up of consciousness-raising.

Catalonia was born out of its resistance, as the "Hispanic March", against a Muslim invasion. The earldom of Barcelona, which brought together a number of Catalan Earldoms, became independent from the Franks in the 10th century. Two centuries later a single sovereign held the titles of King of Aragon and Earl of Barcelona, which politically consolidated Catalonia. During the Middle Ages, the Catalans conserved their language, customs posts and specific institutions with their representative bodies. The territorial expansion into Occitania, Valencia and Majorca was followed by the conquest of the Islands of Sardinia and Sicily (and subsequently Naples) and by the foundation on Byzantine lands of the Duchy of Athens and Neopatria. But following the Union of the Kingdom of Castile and the Crown of Aragon in the persons of the Catholic Monarchs, Catalonia became part of the Imperial monarchy. In the 17th century the decadence of the Empire and its absolutist nature provoked the war of Catalan separation, also known as the war of *Els Segadors (the Harvesters)*, which ended in 1659 with Spanish victory on Catalan soil and the French annexation of the Roussillon and Cerdaña. The war of Succession that broke out in Spain at the beginning of the 17th Century between the Hapsburg and Bourbon dynasties came to a head on the 11th September 1714 with Barcelona being taken by Franco-Spanish Bourbon troops. This was followed in 1716 by the Nueva Planta decree which abolished the self-governing bodies in Catalonia for having been aligned to the losing side. The powerful ethno-national movement of the 19th century (driven by the economic boom initiated in the previous century) selected patriotic symbols of tragic ethos: the national anthem would be *Els Segadors*, and the Diada, the National Day, the 11th of September, the date national liberation was lost.

The Scottish proto-nation experienced real independence as a Kingdom between the years of 1329 - when the wars of independence culminated- and 1707, the year of the Treaty of Union with England. The events of those now distant wars have come to occupy an important place in the Scottish national identity, in a dual pathetic and glorious dimension. From this is derived the national significance of the Stone of Destiny (the site in the High Middle Ages where historically Scottish Kings were crowned) which was taken away by King Edward of England. Scottish nationalism also has roots in Robert Bruce, an astute King of Norman origin, who after having shown respect for Edward, defeated his successor at the Battle of Bannockburn in 1314. William Wallace is also an important figure, the minor Scottish noble who galvanised anti-English resistance a few years before Bruce. He was captured by King Edward in 1305 who ordered him to be hanged, drawn and quartered (Moreno, 1995).

An important role was played in the emergence of a Welsh national consciousness by reminiscence of the literature of Celt bards (some of their epic poems, for example the Taliesin, date back to the 6th century. In Welsh sagas the mythical figure of King Arthur and his Court appear, who enjoyed a long descent in the European chivalresque literature of the renaissance). In the 19th century pre-national Welsh romanticism resuscitated the *Eisteddfod*, bards meetings that fanned the patriotic flame until World War One. Cultural activists such as Michael Jones dreamt of a second Wales on North American soil, an idea which was given up for fear of English-speaking hegemony. In 1865, a Welsh colony was finally established in Patagonia, Argentina. Saunder Lewis, founder of Welsh nationalism and President of the Welsh nationalist party Plaid Cymru from a year after its founding in 1925, was himself an excellent literary author and Welsh Professor at University until his expulsion in 1936 for his political activities (Graham Jones, 1990; Davis 1993).

Few territories exist in Western Europe with an ethnic personality as strong and as recognisable from outside its borders as Andalusia. Its land has successively played host to three rich cultures, the most shining cultures of their respective eras on the Iberian peninsular: Tartessos in the first millennium b.c. alongside the Guadalquivir river, Boetica in the times of the Roman Empire, and the Islamic Al-Andalus from the 8th century until 1492. The cultural substratum of each of these groups enriched the culture of the groups that followed them. The structural breakdown occurred as a result of the clash between the Islamic and Christian civilisations that peaked between 1212 and 1492, the year in which the Catholic Monarchs conquered the Kingdom of Granada. Castilian feudalism meant Al Andalus became New Castile, in the same way American territories would subsequently become New Spain and New Grenada. The Cultural Court and the American irradiation in the Modern era of the House of Contracts (Casa de la Contratación) of Seville erased the memory of the Mudejar rebellion of the Kingdom of Seville in the 14th century, and the rebellions of the Albaicin and Tierra Bermeja that took place following the end of the Kingdom of Granada. The insurrection of the Moors of the Alpujarras in 1570 was a prelude to the mass expulsion of the Moors from across the Imperial Monarchy in 1609. Given such events one can understand the Andalusian anti-centralist revolts of 1651.

In the 19th century Andalusia became an exporter of Spanish statesmen and military, whilst Spain portrayed itself through Andalusian *picturesqueness*. Andalusian culture became alien to its people. At the same time the rural large landowning structures blocked growing Andalusian industrial development, creating misery among the peasants. The uninterrupted peasant outbursts that took place from 1880 right up until the Spanish Civil War were fuelled by millenarian anarchism, which lurked on the edges of ethnic consciousness.

Such consciousness has had three main expressions: the federalist Republican of Tubino, that lasted from 1860 until the Cantonalist insurrection of 1873; the revivalist of Blas Infante (father of the Andalusian homeland), who worked on the Cordoba Statute in 1933, and was executed by Franco in 1936; and the Andalusian Socialist Party (PSA), founded in 1976 before the establishment of the Spanish Autonomous State. In all of these movements, historical memory has been replaced by claims such as those for parity of esteem for Andalusia with the other Spanish regions, its cultural reconstruction, and the end of its underdevelopment (although it must be added that, in the case of the PSA, its opposition to Catalan and Basque nationalisms is, on occasion, stronger than its opposition to the centre) (Acosta, 1978).

Ethno-national consciousness can be brought about by a sense of urgency, caused by a sudden deterioration in circumstances. A large-scale immigration of alien groups in the ethnic territory frequently gives rise to such consciousness. The two decisive moments of Basque self-conscience, at the end of the 19th century and in the 1950's and 1960's of the 20th century, coincided with years of strong industrial development accompanied by an intense immigration of workers. The two forces that today continue to be the main protagonists of Basque nationalism, the Basque Nationalist Party (PNV) and ETA, emerged from these two moments separated by half a century in distance. Taking this into account, governments can try to manipulate these situations, although this is not simple in a free-market. Neither is it easy to predict how much change to the ethnic balance will provoke an ethno-centric reaction, nor whether the ethno-national group will win over the immigrant groups to its objectives.

The sense of a deteriorating situation also increases when the state takes control of the education system or imposes on it a language which is not that of the ethnic group. This is particularly so if the ethnic group, in relation to its language, culture or religion, is subject to discrimination or persecution, at the hands of other ethnic groups or the state. This situation has arisen most frequently in Third World societies and in the former empires of Eastern Europe, than it has in the West. Often the persecutions are the result of a re-composition of forces within the ruling group that alters the multi-ethnic equilibrium that had been maintained up until that point. Repression can, however, be so brutal that it eliminates the possibility of a backlash. This was the case in the deportations to Central Asia and Siberia of Chechens and other Soviet minorities undertaken by Stalin, based on a supposed collaboration with the Nazis, and mutely suffered by the groups in question.

Ethno-national consciousness is awoken, even more so than by sudden deteriorations in the group's situation, by the abrupt interruption of the development of their political, economic and cultural conditions, thereby frustrating previously held expectations (these processes have been analysed by Gurr in his theory of "relative frustration").

Imitation by diffusion of geographically close or politically similar ethno-national movements plays an important role: Seiler (1990) argues that peripheral nationalisms are produced in waves. The occurrence of such waves is also related to varied phases in the national construction of nation-states; although diffusion can function mimetically even when the model and its imitators are completely different. The Cuban, Vietnamese and Algerian Revolutions (through the writings of Fanon and Sartre (1968), the latter were the spark for numerous ethno-nationalisms that considered themselves to be anti-imperialist; for example those in Brittany, Occitania, the Basque Country, Quebec, Corsica etc.)

Notes

[I] The impossibility of making each group into a Nation-State, as the corollary of the German thinker Herder would be, is due not only to the ethnic heterogeneity of many present-day states, a result of the migrations brought about by industrialisation, but also to the large number and diverse nature of these groups. Nielsson (1988) has counted 1,305 different ethnic categories present in the almost 190 states currently existing in the world (575 categories if those living in different states were grouped together). Out of the total number, 381 ethnic groups (67%) have fewer than one million people belonging to them and less than 157 have fewer than one hundred thousand members. However, this position can be turned *sensu contrario* against those who defend the "naturalness" of present-day states, be they liberal or orthodox Marxists, who disregard the fact that these have arisen out of a long-term historic struggle of various groups.

[II] Ericson (1972) defines identity as *sameness*, a subjective and tonic sentiment of personal unity; and *continuity* as continuance through time, accompanied by the perception of the fact that others recognise this personal unity and continuance.

[III] This construction brings together primordialist, situationalist and modernist ethnic theories, which in this sense they are not incompatible. In the primordialist thesis, Smith (1979) and Gertz (1973) principally take into consideration the ethnic personality. In the situationalist thesis, Barth (1976) analyses the processes of ethnic identization. In the modernist thesis, Geller (1988), Anderson (1983), and Hobsbawn (1983) study the condition of the emergence of identization in the general conditions of the industrial revolution and the era of the nation states.

[IV] It is this second process which confers peripheral nationalisms with their "mirror nationalism" character. This expression was first coined by Yves Person and used by Seiler. The term has been employed in the title of this book as it is clearer than its synonym expression "mimetic nationalism", used sometime back in my doctoral thesis.

[V] This intersection of *cleavages* produces an intermediate phase in the historic reactions of the periphery, which is situated between the legitimate and the populist, the Catholic "regionalist" (which proceeds from organist theories that articulate intermediate bodies, whose antecedent is the medieval theory of the "mystic body of Christ" and the Thomist tradition). This can be just as much used by a particular conservative state nationalism as by those seeking peripheral autonomy, as it conceives the centre as an organic and hierarchic mix whose different regions should have their own life and be doctrinated by the church, seeing it as incompatible with state control of education. This organised Catholic regionalism is the "humus" that in Spain unifies the theoretical production of Spaniard Menéndez Pelayo with that of the Basque regionalist Campión, being also the predecessor of the nationalism of Sabino Arana. Where peripheral identities are powerful, this phase runs into its immediate successor which is populist nationalism (as occurred in the case of the Basques.) In France, by contrast, the peripheral movements continued to be almost unanimously conservative-regionalist until the Second World War which gave Vichy's regime the opportunity to play the card of organic state nationalism before them.

VI   Modern approaches inherit the religious meaning that was present in the attachment of human groups with territory in pre-modern times. Eliade (1965) explains that for human being religious space is not homogenous. There is a strong consecrated space and other non-consecrated profane spaces. The consecrated space constitutes the centre of the world, giving the pre-modern human being the aspiration to live as close as possible to the centre. The secularisation of religious values present in modernity, he adds, does not constitute a break in continuity with the sacred, the profane being nothing more than a new manifestation of the very same constitutive structure of humankind. Effectively all states consecrate their territory, so do ethno-national movements, but lacking co-active force they do so symbolically.

VII The 16th century was in effect the Basque century. A group of rich merchants arose –exporters of iron and wool, tradesmen with business in the Indies, shipbuilders favoured by the war campaigns of the Austrian dynasty, who needed boats of large tonnage, most of which were built in the Bilbao shipyards etc. The Basque second sons worked for the domestic and foreign political objectives of the Spanish imperial monarchy. Many of them found in the American colonies an adequate terrain for their desire for glory: the conquistadors and explorers Juan de la Cosa, Elcano, Legazpi, Lope de Aguirre, Urdaneta, Oquendo, Blas de Lezo, Garay... Some of them became wealthy in America with the Spanish feudal system and came back to the country as rich "indianos"; some others entered the church – St. Ignatius de Loyola founded an Order, the Jesuits, defenders of the imperial dream of Catholic Counter Reformation against the emerging Protestantism; others followed a career in literature. Some of these key people – St. Ignatius for example – have been elevated to the status of national glories by the first Basque nationalism.

4 Peripheral Nationalisms (1): Identity and Ideology

Ethno-national mobilisations in the West are social movements. Each social movement presupposes the existence of a conflict which emerges with consciousness-raising (which has a double ideological and identity dimension) and brings about the collective mobilisation of a specific group.[I] Some of these traits may be weak or absent in certain ethno-national processes: those processes which are instigated by traditional peripheries that react against a modern centre; those which are led by local political elites in federal states; or those led by tribal or clan hierarchies in vertical states of the Third World. In these cases the construction of identity is not particularly intense, and the conflict is settled exclusively by elites of the centre and periphery with very little, or no, involvement on the part of the masses. On the contrary, the traits which define social movements are purest in those cases where peripheral reaction is based on support for a national initiative against an integrated unitary state.

In recent decades, alternative social movements have given rise to different interpretative approaches: pragmatic, culturalist, unitary, etc., theories which focus on different aspects of ethno-national movements.[II] The theory of frames (Gamsom, 1992; Maiz 1995) emphasises the role played by discourse and, in particular, the selection or alignment of frames in the construction of collective identities. This alignment gives rise to a principal frame which becomes the central axis in the us-them confrontation. Discourse plays, in effect, a crucial role in national movements, both in their pre-national cultural phase and in their phase of national political maturity. The development of national identity culminates when a principal frame emerges around which the us-them opposition is constructed, differentiating and opposing, on a scale which runs from maximum reconciliation to maximum opposition, the feelings of belonging to the ethno-national nation or to the official state nation.

The construction of identity shapes the internal flow of social movements, while the mobilisation of resources directed toward the goals of social movements shapes their external flow.[III] (However, these two flows do not operate in watertight compartments. With respect to national movements, the constant failure to achieve their instrumental goals may have a double consequence: it may weaken the collective identity or, on the contrary, it may lock it into an identity domain which operates on a closed circuit in which the us-them opposition appears dominated by the hostile form of the enemy).

Both the construction of identity and the instrumental-rational mobilisation of internal and external resources are more complex and diverse in ethno-national movements than in other social movements. This is because they mirror-image all the subjective mechanisms and attempt to use many of the objective resources activated by the nation-state in the process of constructing a national community and society.[IV]

The identity dimension of national movements embraces, for the purposes of study, three domains: the identity proper, the ideological and that originating from the scale of political legitimation (although the three domains are amalgamated in a *Gestalt* which operates on a unitary mode). The identity domain consists of the selection of identity markers: spatial (which relate the group to an ethnic territory); time-shaped (which outlines the continuity in time of the group, of a utopian or nativist type); and cultural (which roots the ethnic language and the sense of belonging to a historical or religious group, among other aspects, in the national social capital). The ideological domain, amalgamated with the identity for practical purposes, involves the use of philosophies or theories, whether they be profound or superficial, for advancing the aims of the group; this domain represents the most dynamic aspect of the *Gestalt*. The third domain mirror-images the diverse legitimating mechanisms activated by the nation-state; these include those of a rational type, those based on identification, or on sacralisation, and they determine the degree of conciliation or confrontation in the relation between the national movement and the nation-state.

The Identity Dimension

In each social movement, including ethno-national movements, the importance of identity is, at the outset, crucial. The new theories of identity[V] state that the individual's actions only make sense when the person is located in a stable manner in a community with whom he/she shares common values, leading to an inter-subjective recognition within the core of the community. Personal identity arises out of the intersection between this horizontal community dimension and the time dimension made up of the series of successive individual belongings. The bonding of a community of inter-recognition amounts, therefore, to joining a new world of values. If the new social movements, which experience new cultural codes and types of relationships and hold alternative concepts of the world, are considered, the initial symbolic challenge, often appearing as a "conversion" to the individual who adheres to it, is especially relevant to ethno-national identities, as immersion in them has the effect of a conversion for those in question.

Nevertheless, in the case of ethnic groups, collective identity assumes the form of a space-time projection which is much more intense than that present in other political identities. Moreover, what is noteworthy is the fact that the building of frame alignments, culminating in the us-them national opposition, takes its values from the ethnic personality constructed in the historic long-term and from symbolic capital which is placed at its disposal by the ethnic personality.

The symbolic capital of pre-modern groups includes all property, material or otherwise, which is rare and sought-after: kind words, respect, challenges, insults, honour, power, pleasure, distinction (Bourdieu, 1972); folklore, historic memory, customs, tradition, etc., must be added to these. Symbolic capital, which is transmitted from one generation to the next, is mixed with regard to the historic memory, with outlooks, myths, collective psyche, etc. (Vovelle, 1985; Marcuse 1982).[VI]

Contrary to symbolic capital, which is inherent in pre-modern societies and operates on a long term basis, identity development acts on the boundaries separating groups, is the result of modernity and operates on a medium or short-term basis. However, in the case of ethno-national groups, the choice of identity markers does not spontaneously arise from within the ethnic group in its entirety, it is instead a process initiated by peripheral elites. Their discourse is what selects the elements of historic-symbolic capital, thereby modifying it.

In the process of its construction, the nation-state uses the cultural arsenal of the dominant ethnic group (for example, the Castilians in Spain) in such a way that, with the exception of its language and high culture, its particular historic features, especially the ethnic community features, are often more side-lined than those of peripheral groups. For this reason the ethnic periphery does not reject these features in the early stages of the cultural phase; they only react against the official language (when it differs from the ethnic language) once the state begins to assume control of the education system and imposes a monolingual educational policy. From this point onwards, peripheral elites develop their own cultural counter-values which are deliberately opposed to the national state processes of socialisation and acculturation. This is when the ethno-cultural identity activated by the peripheral elites starts to become ethno-national in nature.

The selection of identity markers is, in this way, a reaction to the nation-state and therefore political in nature; although it feeds off the social processes of counter-acculturation which are characteristic of the conflicting interaction of ethnic groups, of which the discourse of the elites is the most developed expression.

This antagonistic acculturation assumes the form, in the time identity dimension, of nativisms or utopias. Both are universal forms of reaction by collective identities when they believe themselves to be under threat, either external or internal. Groups consciously react in order to assimilate or reject the cause of the threat; their endeavour to maintain their perception of time may reach into the past, thereby bringing about a "return to the origins" (which, in any event, modifies the ethnic historic memory), and out of which nativisms emerge. On the other hand, antagonistic acculturation may reach into the future, bringing about an imaginary projection of a "beyond" which focuses the hopes of the group; and utopian ethnic movements emerge out of this attitude. Nativism and utopia are, therefore, the two polar reactions of groups engaged in the processes of antagonistic acculturation.[VII]

In the cultural phase, ethnic self-awareness leads to the transformation of symbolic pre-modern capital into socio-political capital of modern ethno-national movements; on occasion, the process is transparent. This occurs with the re-evaluation in nationalist terms of Basque symbolic capital, which had been the aim of the first redefinition of identity during the Carlist Wars, carried out again by Sabino Arana at the end of the 19th century. This redefinition was determined by a double ethnic reaction against the construction of the Spanish state and against the immigration resulting from intense Basque industrialisation, and fuelled by the ideological arsenal of the church in its conflict with the state.

> The Basque collective psyche has its roots in the early Middle and Modern Age. From the 13th century onwards the Archbishop of Toledo Ximénez de Rada attributed a noble origin to the Basques, by claiming their descent from Tubal, son of Japeth and grandson of Noah. In the 16th century it was claimed that the Basques descended from the Cantabrians, an unconquered people who had never been subjugated by the Romans or the Arabs. The mythical explanation of the origin of political power, according to the most elaborate Biscayan version, and that which Sabino used in order to refute it, went back to the "Pacto con el Señor" (Covenant with the Lord). On the basis of this theory of the Covenant the New Biscayan *Fuero* (special law) of 1526 established a range of obligations of the Monarch of Spain and proclaimed the theoretically voluntary nature of taxes (as donations). From the end of the 15th century onward, historians of the Castilian monarchs claimed that the Basque language was "the oldest in Spain." It was elevated to the rank of language of the Paradise by Basque *fuerist* theoreticians of the 18th century such as Larramendi and Astarloa. From the 16th century onwards, writers such as Zaldibia asserted the timeless monotheism of the Basques. Legal bodies such as the New Biscayan *Fuero* proclaimed the "purity of blood" of the Biscayans and related it to collective nobility in order to discourage the immigration of outsiders, specially Jews or moors. Various elements of a non-legal nature

complement this symbolic capital: the cult of the homestead, genealogical foundation of the nobility and point of entry of the Basque into the family sphere, and honour as a criterion of social relations and rank.

Sabino Arana recovered this symbolic capital in nationalist terms. The myth of egalitarianism was used to extol the immemorial lost Basque democracy and to condemn the contemporary Biscayan class society, and later, Basque society in general. In this way, in 1893, he spoke of "anteiglesias", rural Biscayan communities, as "completely free and independent and, at the same time, harmoniously and fraternally united; these small political entities, governed by laws which arose out of their own society and based on religion and morality, were perfectly happy with their existence, since it never crossed their minds to expand their domain into new lands." The Basque past became mythologized and every trace of class struggle disappeared, being referred to as an outside invention of the Spanish in complicity with powerful Biscayan elements. As a consequence, Sabino Arana condemned, from a moral perspective, the most important Biscayan capitalists for their ambition and greed and for their corrupt electoral practices. He argued that, on bringing foreign labour into the country they were responsible for the introduction of immorality, impiety, socialism and anarchism. Likewise, his opposition to socialism was absolute; he viewed it as a "maketa" (foreign) doctrine and as a threat to the Basque personality.

The genealogical basis of the universal nobility of the Basques underwent a total reformation. Basque-Iberianism was completely denied and it was argued that the Basque race, which was original and unclassifiable in the context of other races, did not arise out of a mix in the same way as the Spanish. The invincibility of the Cantabrian-Basques became an implicit yearning for the past and a reason for action in the present in the context of the Carlist wars, and as an explanatory reductive factor of the complex relationship between Spain and the Basque Country as one of conquering and conquered countries. This colonised view of the Basque Country opened the door to the more modern and progressive aspects of Sabino Arana's thinking: his strong anti-colonialist feelings, contrary to the war which the Spanish were waging during that period in Morocco.

Peasant society, based on the now abolished institution of the "mayorazgo", the rights of exclusive inheritance of the homestead in favour of the eldest son, was behind the idealisation of the rural. The opposition which existed during the 19th century between the Carlist countryside and the liberal cities reinforced Sabino' support of the rural communities and against that which was urban, and led to the crystallisation of "rural" as being the essence of the Basque nation.

The theory of the Covenant with the Lord was reformed in some respects and refuted in others by Sabino. The complete break with regard to the *fuerist* tradition and the consequent conversion from allegiance to the *fueros* to nationalism took place in the context of the *fueros* themselves. These were the "national codes" of the Basque states which were in force when they were independent. Contrary to what the Carlists claimed, to be loyal to the *fueros* was to be nationalist.

Primitive monotheism is related to the influence of religious traditionalism over Sabino. The Sabinian religious project incorporated three aspects: independence of the Church and state, priests must preserve their political neutrality and could not become actively engaged in politics; harmony between the Church and the state; and the subordination of the state to religious obligations. This project is summed up in the slogan "We for the Basque Country and the Basque Country for God".

Race is, in Sabino's thinking, the concept which encompasses all these characteristics and crystallises their "difference". It is an affirmative and non-somatic racism, affirming the moral high-ground of the Basques. Race is the "essential" factor in the nation, more important than language. For Sabino, the Basque language, *Euskera*, is a defensive instrument which is used as protection against the "maketa" invasion. Without race, there would be no Basque homeland. Affirmative racism is combined with a racist policy of exclusion with respect to immigrants. The main enemy of the Basque race is the "maketa" invasion. In Sabinian discourse Spain is "Maketania", and the invasion, with the help of Basque "maketophiles," enters into political, administrative, industrial and commercial life; it takes over mines and industry and corrupts the noble traditions and customs of the country. This racist policy of exclusion is more rooted in the historic memory of the "purity of blood" as an excluding myth (a Castilian myth directed against Moors and Jews) than in the social theory of Gobineau or Vacher de Lapouge.

The Ideological Level and the Scale of Legitimation

After moving into the political phase, ethno-national movements engage in a struggle to impose on the centre, and eventually on other ethnic groups, their institutional-territorial political objectives, and to initiate the work of political agitation and mass mobilisation. Given that ideologies are the dynamic element which fulfil such functions in every political culture,[VIII] the ethnic movement needs them even more than in the previous cultural phase. However, since ideologies are also a factor in social divisions, it is not unusual for them to appear in tandem with, and often precipitate, the break-up of the ethno-cultural reaction, which was unitary at the outset, into a plurality of ethno-national movements eventually organised into political parties. In these cases the feed-back phenomenon takes place within the identity domain. Each of the movements modifies its collective space-time perception, thereby giving rise to the well-known phenomenon in which different nationalist complexes emerge out of one single ethnic group. These complexes are opposed in their respective ideological-identity *Gestalt*, in their lived-in experience of the us-them polar opposition, and in the scale of intensity of the conflict which ranges from a critical co-existence with the state centre to the creation of an alternative centre.

Nationalisms are not only ideological; their nature as social movements means that they cannot simply be reduced to an ideology. Their symbolic-cultural dimension changes them into concentrating moulds of many different ideologies in which the ideological appears fused in the *Gestalt* with identity and the collective psyche. If each political ideology is constructed out of a range of different philosophies and theories, then nationalisms (of the centre and the periphery) amalgamate different political ideologies.

The three most important nationalist ideologies in the West are based respectively on: 1. the political-juridical nation; 2. the cultural nation; and 3. the political-elective nation (all of which have emerged from French-German rivalry during the period from the French Revolution to World War II). They are the axis which supports the processes of amalgamation.

1. When sovereignty is symbolically transferred from the monarch to the citizens (as was the case in with the French Revolution at the end of the 18th century) the "political-juridical nation" becomes the symbol of the tie between the citizens, the state and its territory, and the expression of collective sovereignty embodied in the representatives of the "people". However, this idea of the nation incorporates different theories (among them Spinoza's liberal-individualism, as well as those of Locke and the utilitarian Anglo-Saxons, and the democratic collectivism of Rousseau, which the Girondists and Jacobins transformed into ideological instruments of collective action).

2. The concept of the "cultural nation," originated in the German sphere, is devoid of political and territorial unity and was nourished by Herder's concept of *Volkgeist* or "spirit of the people." It is a term which refers to a community of sentiments and customs born out of ethnic kinship, developed organically, and consolidated in the language spoken by a specific people and through respect paid to certain institutions, beliefs and traditions. Through its expansion into other cultural spheres the idea incorporates other theories: Montesquieu's, which attributes "a spirit of the law" to each specific society; and the concept of tradition developed by Burke. The works of Fichte among others would transform these theories into nationalist ideologies as a reaction to the Napoleonic invasion of German territories.

These two concepts were enriched by other thinking which either brought them closer together or separated them further throughout the 19th century. Savigny's culturalist school of historic rights theorised the law as a specific product of each people: to want to introduce into Germany an alien institution such as the Napoleonic code represented, he said, a crime against the fatherland. (This ideological tendency which promoted the view of the nation as a "person", the result of the continuity of a conjunction of factors over a long historical period, especially language, would be welcomed by peripheral nationalist elites). The "principle of nationalities" as well as the Italian Mancini's national concept were fostered by both national macro-ideologies.

3. The concept of "political-elective nation" was developed by Renan out of the defeat of France by Germany: the territories of Alsace and Lorraine, annexed by the Germans, were no longer part of France and, in the case of Alsace spoke a Germanic language, but continued being French through the choice of the inhabitants. The organic notion of the continuity of "land and ancestors" was fused with the Renanian idea of a plebiscite of present and past generations. The plebiscite aspect of the political-elective nation separates it from the concept of cultural nation, while its mass participation aspect separates it from the political-juridical nation. This type of nation could, in effect, be non-participatory and, in fact, in the 19th century it adapted to elitist and exclusivist political systems in which the nation was identified with small groups of representatives constituted in parliament (Seiler, 1992). The "political-elective" nation which emerged in nation-states in the phase of mass political participation had to be, on the other hand, participatory. However, the peripheral "cultural nation" is also participatory in nature when it becomes a project against the centre driven by mass mobilisation, thereby making it into a political-elective nation.

The centrality of these three national ideologies is in conformity with their supra-ideological aspects. Juridical and elective nations are an expression of legitimating mechanisms: legitimation of the state in the former case, and of the state or its peripheral alternative in the latter. For its part the cultural nation expresses the identity phenomenon, in relating itself to the perception of time continuity of a particular human group. For this reason the three concepts are ideal-types rarely appearing in isolation; real nationalisms combine these three macro-ideologies into one ensemble in which one usually predominates. (The peripheral elective nation may become a political-juridical nation if it conquers a state).[IX]

Neither is there uniformity within each national macro-ideology; the theories and ideologies present in them exhaust the repertory of the classics of political thought and contemporary ideological tendencies. Echoes of Kant, Adam Smith, Bentham, J.S. Mill, Tocqueville, etc., may be found in the ideological mechanics of liberal nationalisms. In subsequent authoritarian nationalisms, which were supporters of state force, may be found, albeit somewhat distorted, the thinking of Hegel, Donoso, Comte, Spencer; noble ideologies such as Burke's conservatism and disreputable ones, such as those of List, Gobineau's racism,

Chamberlain's imperialism and social Darwinism. This thinking also combined alternatively with aspects of Pangermanism, nostalgia for imperial Rome, Nietzsche's thinking (which was grossly misrepresented) and philosophy of life appeared in the explosive ideological cocktail of central European inter-war fascism.

Peripheral nationalisms also   employ ideological liberal or conservative arguments; organicism, Thomism and the social doctrine of the church foster Catholic nationalism/regionalism. It is of little relevance that the ideology of centralist and peripheral nationalism in struggle coincides on occasion. What separates them is the us-them opposition in the identity domain and their antagonistic location on the scale of legitimation of political power.[X]

Can we speak of dominant ideology with regard to centre-periphery conflicts? Certainly not in the way in which Marx theorised it in *The German Ideology,* as a mechanical relationship between the dominant social class and ideology; rather we can do that as means of production of the legitimation of political power. In these conflicts, the dominant ideology is that of the centre, which fosters the state process of socialisation and acculturation, subjecting the social movements of the periphery to stigmatisation and stereotyping. This ideological perspective has inspired those intellectuals who have written, since the twenties and until quite recently, the history of peripheral nationalisms, building up the "black legend" (*la leyenda negra*) around them. Even though this tendency has become more modified in recent years, the "black legend" continues to dominate studies in states facing strong peripheral reactions.[XI]

Nationalist ideologies, which separate human beings into nations, clash with large-scale religions that preach messages of salvation that do not take borders into account. Uniting the former human beings within a nation, ideologies of class appear to be incompatible, since they perceive of societies as divided by social class. Yet the all-encompassing power of the centre-periphery conflict over the remaining cleavages allows nationalist ideologies to amalgamate religious beliefs and class theories. Religion, an element of the pre-modern personality of ethnic groups, becomes a crucial factor in the identity of the groups in the modern era. Likewise, nationalisms employ subversive class ideologies; particularly the enormous macro-ideology which is Marxism, that has been diversified into many micro-ideologies.

The *Communist Manifesto* of 1848 was emphatically anti-national, arguing that "workers have no country." However, such statements must be seen in the context of the time, when working class was excluded from any participatory political life. (Sieyes argued in the times of the French Revolution that only active citizens who paid taxes, described as "true shareholders of the great social enterprise of the nation", had to be given the vote. Such conceptions became the basis for the bourgeoisie regimes of the 19th century, which effectively excluded workers from suffrage categorising them as "passive citizens"). The concept of nation evolved in later Marxist thought seeing the nation as a historical category that corresponds to the needs of capitalist development and as a civilising process that liquidates the *Ancien Régime* and puts an end to the "barbarism of primitive nations." However, when Marx formulated this he only had in mind the large European nations that were well defined, viable, and within which historical progress was present (Gallissot, 1980).

Engels systematised the Hegelian distinction between "historic" and "non-historic" nations. Engels felt that the latter group, whom he defined as "peoples without a bourgeoisie incapable of developing their own political and cultural life," could "go to hell".[XII] Engels additionally argued that the nation was a category in a transitional phase, within which the expansion of free trade inherent to capitalist development would bring down national borders, and hence capitalism would itself precipitate the end of nations. He argued that national idiosyncrasies would weaken with the development of the bourgeoisie, who themselves would disappear when the proletariat took power.

A new theoretical phase was opened up from 1867 onwards with Marx's reflections on the Irish question, in which he argued that the impotence of the English proletariat was due to their hatred of Irish peoples and Irish immigrants, making the English working classes an accessory to the oppression that the British aristocracy and capitalists exercised over Ireland. This analysis led Marxism to accept that national movements could be liberating in nature, and to the introduction of a fundamental distinction between dominant nations and oppressed nations (Haupt, 1974). However, this conceptual progression within Marxism was not widely known.

The II International covered the central and eastern European empires, which obliged it to address the serious national problems with which they were faced at the time. Otto Bauer was the first to analyse class struggle in cultural terms, therefore bestowing national culture with a positive sense. Bauer believed that oral civilisations arose out of from rural collectives and were expressed and transmitted through language to produce popular culture, whereas in contrast to this, written culture generated an elitist culture, being therefore their nations a "unit of culture of the dominant classes." Bauer believed only socialism would guarantee the full development of national cultures, defined by him as "the integration of all people in the national community of culture, the total conquest of self-determination for the nation, and the growing spiritual differentiation of nations." This led him to argue, in contradiction to Engels, that non-historic nations were allied to the proletariat.

Kautsky supported the orthodox vision of the nation as an economic phenomenon, and of culture as a linguistic one. He argued that cultures converged towards universality, national idiosyncrasies tending to disappear. Accordingly, he believed that whilst languages were long-term cultural realities, they would also converge into a single universal language.

For Kautsky there was therefore no such thing as a cultural community. Rosa Luxemburg, who was heavily influenced by the *Communist Manifesto,* regarded all nationalism as a bourgeoisie contamination of the proletariat. She believed that the self-determination of the proletariat was what created the self-determination of nations.

Whilst Stalin and Lenin's Bolshevik thinking followed on from Kautsky's concept of the nation, some of their viewpoints would be used by subversive nationalisms. Stalin's writings on the national question fused the common base of orthodox Marxism with Bauer's notion of national culture. From this he derived his definition of nation as "a historically constituted mentally formed stable community of language, territory, economic life that becomes a community of culture." The Stalinist conception of the nation as a historical past with a territorial base turned his work into the bible of the Communist movement of the 1930s during which the USSR recovered its "national values" (Gallissot, 1980). It also subsequently became a source for national liberation movements (that turned a blind eye to the brutal national policy carried out by Stalin whilst in power).

Lenin, who also owed much of his positions to the Communist Manifesto, felt the nation had an instrumental value and therefore made no distinction between it and the state. His 1914 works were guided by tactical considerations that were geared to obtaining the complicity of national movements against the Tsarist Empire, which he described as the "prison of the peoples". Although, in his opinion, large states could deal much better with economic problems, the working class should maintain an attitude of neutrality with respect to nationalism, but at the same time support the right of each nation to self-determination (right that Lenin defines in plain language as "the national separation of alien national collectives" and "the formation of a national independent state") and lead the struggle against national oppression.

Leninist thought did however develop during the course of the First World War. Lenin concluded that imperialism created a division between oppressive and oppressed nations. He applied Marx's distinction on Ireland globally, and put his hopes into the national movements of the east. He consequently argued that the full sense of the right to self-determination was only fully realised when it operated as an ally to the cause of the proletariat in the struggle against imperialism. This interpretation will be shared by anti-imperialist nationalisms. (The elitist revolutionary theories of Lenin in his work "What is to be Done?" would also be used by armed national movements, who would use them to justify their vanguardist policies).

The theories of the Irish Marxist James Connolly, who was executed by the British after the 1916 Easter Rising, examined the socialist character of national liberation, which he analysed in relation to social classes. He felt that the bourgeoisie always betrayed the nation, that the petty bourgeoisie was sometimes patriotic, and that only the working class was revolutionary. His work that speaks of the humiliation of the barefooted and oppressed pre-dates the thinking of Mao (1966) Ho-Chi-Ming and Fanon (1968).

During the inter-war period the theories of the Italian Gramsci (1978, 1983) saw civil society as a set of private organisms within which the ideology of social classes was elaborated and disseminated. Such social classes, he argued, produced a stratum of organic intellectuals, who gave the classes homogeneity and consciousness of their role. The Communist Party, he said, was the organic intellectual of the working class. In the West, the strategy of the proletariat to obtain hegemony had to be a war-of-position, which would allow him to strip the bourgeoisie of its power in the ideological apparatus of civil society and in the representative bodies of political society.   The revolution therefore had to adopt a "national character" as only the working class of each respective country knew from which trenches to attack and in which order this should be undertaken in the economic, political and military spheres. Gramsci's thinking, influenced in this area by Machiavellian ideas and the Italian *Risorgimento*, mixed with state socialist patriotism and sweetened, influenced the Euro-communist strategy of conquering cultural and parliamentary power quotas, founding a particular type of left state nationalism.

In addition to Connolly's Irish Labour Party, class ideologies in the inter-war period influenced some peripheral nationalisms, albeit more through empathy than through theoretical adhesion. An example of this is the thinking of Eli Gallastegui, the head of the break-away Basque Nationalist Party in the 1920s, which expressed both anti-imperialism and solidarity with the communist workers.[XIII]

The presence of class ideologies increased in importance in the New Leftist nationalisms of the 1960s and 1970s, which appeared alongside and in opposition to historic nationalisms. However, given that by definition all peripheral movements are an alliance of groups and social classes against the centre, when the alliance diverges into a system of nationalisms, the generation of a left-right divide does not come from class conflict (as historically happened in the systems of state-wide parties in the West) but from identity-ideological differences.[XIV] Even if radical peripheral nationalisms can see themselves as class movements, there are further divisions: the social inclusion or exclusion of the movement, its "us" and "them" configuration, its orientation toward utopia or nativism, and its stance in relation to the legitimation of state power.

There is a tendency in right wing (or moderate) nationalisms for pragmatic self-inclusion in the state socio-political system. In terms of "us and "them" their concept of "us", set down in the movements discourse, possesses an inter-class content (although the highest strata of the class present in the movement is those who dominate it in practice). The group's conceptualisation of "them" does not merely refer to the centre, but also (and sometimes especially) the forces of peripheral subversion. The movements' projection of identity across time is nativist in form, and consists of a "return to roots" (which is not incompatible with a grand operative pragmatism). In terms of the scale of legitimation, the movement does not present itself as an alternative that excludes the centre.

Leftist (or radical) nationalisms relate to the socio-political exclusion from within which they operate, either because the centre pushes them into this position, or because the movement deliberately places itself in such a situation through demands that are unacceptable to the system. Leftist movements conceptualise in discourse "us" as a national class alliance led by the proletariat (in extreme cases it is presented as a class front) although in reality the movements largely shelter a mixture of the marginalized and socially excluded. The conceptualisation of "them" is polarised by the centre which is perceived by the movement as the enemy, but also tends to extend itself to moderate sectors within the same ethnic group that are felt to be accomplices and collaborators with the centre. The group's time identity is orientated towards utopia, towards a future that will fulfil the presently unobtainable desires of the group, in which the myth of the "inverted world" will become reality and the pariahs will become the salt of the earth. In terms of the degree of legitimation granted to the state, the movement will completely de-legitimise the centre and accordingly construct an alternative centre, in political and symbolic spheres, within which the members of the movement will be united through mechanisms of identification and often of sacralisation.

Radical nationalisms make good use of class ideologies, as they are a powerful tool in the creation of groups. The contrast between conservative and revolutionary[XV] ideologies becomes clear in the clash between the nation-state and radical peripheral nationalisms, (which have replaced the class movements of the 19th century in their anti-system stances). This polarised opposition generates an ideological distortion of the radical group that prevents it from acknowledging any other confrontation that is not the clash between the movement and the state, meaning it ignores other nationalist groups that often have greater social support from within the ethnic group than the radical movement.

New Nationalisms and Social Movements

The new nationalisms that emerged in the West in the 1960s and 1970s were varied in nature. Some nationalist movements were born out of exactly the same identity phenomena reaction as the new social movements, ecologist, feminist, anti-militarist, expressing a desire to return to authentic and natural communities, which was manifested in social explosions like that of May 1968 in France. In these cases peripheral movements emphasised values such as the return to the countryside, self-management, and personal self-realisation, rather than the need for a new nation state.

The new ideological nationalisms - "ideological" to distinguish them from identity phenomena - are simultaneously fostered from two sources: the "hard-line" Marxist-Leninist class ideologies and the anti-imperialist theories of national liberation. Both nationalisms place an emphasis on state independence and the construction of socialism. (Moreover, the identity and ideological dimensions can converge, compete amongst themselves or successively alternate within a radical nationalism.)

The new identity based nationalisms are, in common with social movements, a product of a social crisis of values. Both are opposed to double cultural and institutional-political phenomena: in a cultural sense, to the ideological thesis that proclaims "the end of ideologies"[XVI]; in an institutional-political sense, to the weakening of the old class movements in favour of a society with ideology removed, (expressed through the substitution of mass parties by a new model of electoral party). In the West, above all in Europe, the former mass workers' parties have effectively been transformed in to *catch-all* parties, some of which have obtained state power. Additionally, trade unions now often share the responsibility for important macro-economic decisions with governments and pressure groups, in a neo-corporatist fashion. The resulting ideological edulcoration regulates and empties the ideologies of both the trade unions and the catch-all political parties. This leaves parties unable to discuss new global problems, paving the way for the emergence of the so-called "new social movements" (Martínez Sospedra, 1996).

These movements inherit the same level of ideological-identity fusion that was present at the birth of the workers movement, although they no longer believe that they are able "to change the world", which leads them to concentrate attention on solving the problems of specific sectors of society. Moreover, the new movements are at best distanced from the new ideology-free and institutionalised contemporary workers' organisations, regarding them as "old social movements." This leads them to advocate a political alternative to the system of parties and unions, an alternative that would involve the full participation of the masses in decision-making, the transcending of conventional means of action, and the taking on of demands that are unacceptable to the system. The new movements are critical of what they see as an overwhelming state presence in everyday life and try to present themselves as a third political alternative that falls between the institutional and private worlds in the sense of being "non-institutional politics" (Ibarra, 1995). A number of authors argue that such movements grow on the post-materialist values of the young generation that was born within the welfare state (Inglehart, 1991.) Other critics (Touraine, 1978; Offe, 1988) believe that the movements attempt to reconstruct the unity that has been fragmented by modernity, lifting the subject that has been claimed back by it, but buried by instrumental-rationalism activated also by modernity.

An understanding of the aspects that new social movements and new nationalisms have in common allows us to better comprehend their differences. A decision common to both is that of placing their best hopes for success on the establishment of themselves as movements and not political parties; social movements, because of a deficit in their capacity for globalisation, and nationalisms, because of an excess of this capacity. In effect, the discourse of social movements, in contrast to that of political parties, rarely aspires to be universal; whilst parties accept themselves as a part of the whole, national movements tend to see themselves as a whole in their own right.

Both will also have to confront, for the reasons given, the hostility of workers' organisations. This is more noticeable in the case of new leftist nationalisms, many of which define themselves on class grounds. When leftist nationalisms encroach on trade union territory, their penetration tends to be minimal for a long period of time, and falls well below their political presence. When the new nationalism does begin to gather strength in this area, it is an indication of the cultural-identity phase giving way to a rational-instrumental phase.

The crisis experienced by both the new movements and new nationalism is broad-ranging. Social movements, on incorporating themselves into public space which they share with political parties and pressure groups, common rules and the normalised activity of the media, tend to become institutionalised, a by-product of which is the smoothing over of the edges of their identity dimension. Born as alternatives to parties, but also as an instrument in their regeneration, they can themselves eventually become parties, albeit small and inadequate ones (Ibarra, 1995).

However, it is unlikely that this will occur with new nationalisms, whose identity-based reaction in the short term is based on the long-term identity over space and time of the entire ethno-national movement. This allows them to develop a second identity discourse within which alternative values double with ethno-national values. For example, ecologists are equated with defending the common national territory; pacifists and anti-militarists with resistance against the violent apparatus of the state; and feminists to the condemnation of patriarchy identified with the authoritarian centre. In this sense neo-identity values are enveloped by these nationalisms in the centre-periphery conflict.

Their crisis stems from a specific cause that consists of the shutting off and closure of their identity domain, which happens when the 'us-them' confrontation ends up being dominated through the dialectic of the enemy. Ideological nationalisms are more inclined to this shutting off than neo-identity nationalisms; the latter emerge as a response to social conflict, whereas ideological nationalisms react when faced with political conflict or on many occasions state repression.

This element, alongside historical factors, explains the contrast between Spain and France. The violent Franco dictatorship caused the various peripheral nationalist movements to flourish. The movements operated within Marxist-Leninist discourse and responded to state violence with political violence that saw itself as anti-imperialist. Convergence between nationalisms and the new social movements did not get off the ground until after the death of the dictator in 1975. Some violent nationalisms survived well into, and subsequent to, the transition to democracy (especially Basque nationalism).

In France, by contrast, both emerged during the same era; the nationalisms built up in the 1960s from the same sources as the social movements (except for the imitation effect upon nationalisms brought on by the end of the French Empire in Africa and Asia). Because of this, with the exception of Corsica - which had a different set of problems to the mainland - new nationalisms weakened at the same pace as the new social movements.

In Brittany, after the Second World War, the development of Breton culture and language moved over to the Parisian mediums of Breton emigrants through newspapers like *Brittany in Paris* or cultural associations like *Ar Pilhaouer*. However the most interesting development of these years is the creation in Brittany in 1949 of the *Committee of Studies and Links with Breton Interests* (CELIB), an association of modernising "notables" that strived to obtain the best advantage possible from what they saw as the inevitable breakdown in traditional society. The CELIB was constituted as a pressure group before the French parliament, in defence of a territorial ordination that would assist Breton regional development.[XVII] The movement obtained notable success at the beginning of the 1970s, achieving the establishment of a regional plan for Brittany that would be emulated in other regions.

But the CELIB did consider Bretons as French citizens whose situation should be equal to that of other peoples across the state. Notably the movement's major leaders adhered in 1962 to Gaullism, which produced leftist breakaways and gave depth for nationalisms to denounce the "betrayal of the notables".

By the fifties and sixties the tendencies that were emerging were both nationalist and leftist. Such tendencies strongly coincided with those that subsequently inspired Basque nationalism, and were similar to those motivating contemporaneously the Occitan and Corsican movements. From 1957, the Movement for the Organisation of Brittany (MOB) proposed the long-term independence of Brittany in the context of European federalism.

The Movement's refusal to declare its socio-economic political colours, reflected in its declaration that it was neither "red nor white" i.e. neither in favour of capitalism nor socialism, produced, in 1963, a split that led to the formation of the Breton Democratic Union (UDB). This socialist and internationalist inspired group took on the inter-war French communist Yann Sohier's definition of "double exploitation of Breton worker as Breton and as worker." However, the UDB expressively distanced itself from calls for independence and collaborated with the French parties in the *Union de la Gauche*. Throughout their history they have, however, supported Breton political prisoners (including those tried for violent offences).

The UDB, still operational, has split many times forming new left and right wing groups. The best known of these is the Brittany Liberation Front formed in 1966, which created the Revolutionary Breton Army (ARB), with the aim of attacking the symbols of "French occupation" (i.e. police stations, barracks, government buildings.)

In Occitania, a new generation of Occitans broke away from the *Felibres* and united alongside the wartime French Resistance. In 1945 they founded the Institute of Occitan Studies. Moreover, they broke away from a nationalist stance by defining the culture of the Oc language as part of French culture, which won over the sympathies of the French left, including the communists. From 1952, Robert Lafont, who wished to transcend the cultural boundaries of the movement, steered the movement towards the idea of regionalism.

However, the nationalist flame was relit by the Algerian war. In 1959, François Fontan founded the Occitan Nationalist Party, who sympathised with the Algerian FLN. The party became the (albeit reduced) permanent reference for independence and anti-imperialism. However, its support for De Gaulle, on the grounds that he was de-colonising, provoked a crisis in the party which caused the resurgence of Lafont's Institute of Occitan Studies. The Institute swung to the left in the 1970s as a result of its convergence with the struggle of the vine growers - in 1961 the Vine Grower Action Committees were founded with the aim of preventing the importation of Algerian and Italian wines - in the same year it supported the struggle of the Decazeville miners, which unified the linguistic struggle to the economic struggle. (French economic development had most benefited the north, the Isle of France and Lyon region, discriminating against Brittany and the Occitan south).

It was at this point that Lafont, who resigned from the Institute and formed the Occitan Committee for Action and Study (COEA), began to use the expression "internal colonialism" which he had borrowed from Mexican sociologists. The COEA, which regarded Occitania as a cultural denomination of a number of regions, became ever closer to the trade unions. The Committee called for federalism in the context of a "Europe of the Proletarian Regions" but distanced itself from anti-colonialism and rejected calls for independence. This allowed it to converge with the French left in the 1964 re-composition.

The events in Paris in May 1968 brought about the fusion of Occitan national sentiment with a number of revolutionary ideologies. There was a cultural explosion characterised by the new Occitan song, the Theatre of Carriera, etc.; and all of the broad range of leftist groups, of Occitan intellectuals, vine growers, Maoists, Trotskyites, and anarchists, were brought together in the Occitan Action Committees to denounce the "genocide" of rural communities. In 1971, the COEA was dissolved and the Lutte Occitane group was created. The group believed that capitalism was obviously in crisis and that nation states would dissolve. Over ten thousand people gathered in the Fair of Lutte Occitane in Montségur. In the campaign over Larzac, a town in which since 1973 several hundred farmers had been threatened by the extension of a military base, Occitan demands mixed with those of anti-militarist, anti-system ecologist and other alternative movements.

When in 1974 Lutte Occitane (Occitan Struggle) fell from grace, caused by the decomposition of "*gauchism*", Lafont inspired a new organisation "*Volem Viure Al País*" ('we want to live in our country') (VVAP) that took on the role of defending the territory; due to its convergence with the 'common programme', the French left accepted the organisation. The battles of the vine growers against the 'inhumane' Common Market became Occitan in nature, and in 1975 the VVAP and the Vine Growers Action Committees joined in the struggle. In March 1976, two (one from each side) died in a shootout between the vine growers and the CRS in Montrédon-Corbières.

The breaking up of the left-wing union in 1975 weakened the VVAP, which remained in crisis until 1981. The victory of the Socialists (supposed friends of the Occitans) generated hopes that, despite some cultural concessions, appeared to be unfounded. Although organised 'Occitanism' stagnated, its ideas gained ground. The regionalisation of what in 1982 was the most centralised state in Europe, France, whilst modest in results, has to be seen as the turning point from a technocratic objective of territorial planning from above to pressure for autonomy from below, especially in Occitania.

In Spain, the long duration of the Franco dictatorship meant the emergence of a new generation in the 1960s and 1970s that was accompanied by leftist radical nationalist formations, namely the Pro-African MPAIAC (the Movement for the Self-determination and Independence of the Canaries) in the Canary Islands; the Marxist-Leninist Unión do Pobo Galego (UPG) in Galicia; and the Partit Socialiste d'Alliberament Nacional, PSAN, in Catalonia. On occasion the armed activities of some of these groups continued (having split with their parent organisations) after the death of Franco. Currently this only continues in Galicia, although none of these cases reached the importance and social impact of that of the Basques.

Peripheral regionalisms/nationalisms grew in strength in the post-Franco period, although their relation with the 'State of Autonomies' established in the 1978 Constitution did not express itself in a unitary fashion but rather oscillated between adhesion, forced cooperation and open opposition.

Democratic Convergence, founded by Pujol in 1975, a political force supported by Catalan economic interests, was in fact the heir of pre-republican 'Catalanism'. The party has been hegemonic within the autonomous institutions from 1980 and later aspired to make Catalonia the Mediterranean's leading region within the European Union. Its dominant position is only threatened by Catalans who follow the historic Esquerra Republicana.

State-wide parties dominated politics in the post-Franco transition period in Galicia, with nationalist parties coming second in elections. However, the creation by the UPG of the Bloque Nacionalista do Pobo Galego, added to the UPG's *aggiornamiento,* made, by the end of the 1980s, the minority-supported yet dynamic bloc the unifying element of Galician nationalism, within progressive tendencies that evoke the spirit of Castelao.

Notes

[I] Diani (1992) defines social movements as "informal interactive networks between a plurality of individuals and groups who, sharing a collective identity, commit themselves to the resolution, for themselves and for pre-determined groups, to a cultural and political conflict".

[II] Tilly's theory of collective action salvages the gaps present in McCarthy's (1977) pragmatic tendency and in Oberschall's (1973) theory of resource mobilisation. From the point of view of the pragmatic tendency, organisational version of Olson's individualist model of rational choice, social movements initiate mechanisms of development and adaptation to the environment through which the individuals affected organise themselves in order to defend their threatened interests. In this approach social movements are hardly distinguishable from interest groups; for this reason it is not particularly useful in the study of nationalisms. If the culturalist tendency of the seventies and early eighties (Touraine, 1978; Offe, 1988) emphasised, in opposition to the pragmatic tendency which was predominant until then, what was new about the "new" social movements as against the "old", such as working class movements, in contrast, contemporary unitarian thinking (Melucci, Diani, Eder, 1993; Tarrow, 1991) analyses the dynamic and constructive aspects of collective identities forged by social movements through their discourse.

[III] Theories of resource mobilisation and the structure of political opportunity (which analyse the ability of social movements to evaluate opportunities for action present in the political context in which they operate) study the instrumental-rational external flow of these movements.

[IV] National movements only mobilise the entire range of resources available in their political phase. According to the three phase scheme of development, A, B, C, devised by Hroch (1985), phase A refers to the work directed toward knowledge of history and culture carried out by the ethnic elite; phase B opens the way to the patriotic incitement which aims to produce a national consciousness; while phase C represents the passage from an elite movement to a mass movement, which diversifies the peripheral, at the outset homogeneous, reaction, and facilitates the appearance of new groups and parties. This scheme is, in any event, an ideal type; in reality the three phases overlap in the development of national movements.

V   According to Melucci (1985), collective identities carry out three interrelated functions: they formulate cognitive frameworks which provide a rationale for collective actions; they stimulate relationships among members who act, communicate and take decisions; and they make emotional investments which facilitate a reciprocal acknowledgement on the part of the individual.

VI Myths are collective representations of social reality which resolve contradictions in a poetic way; the myth produces utopia and brings about a convergence. The collective psyche ("l'imaginaire collectif", in French) feeds off the myth and utopia; but it produces a culture of escapism, dream and unreality. While the myth represents an attempt to resolve social contradictions, the collectif psyche avoids them. Myth and the collective psyche appear indissociably linked (the former provides the cement which unifies the distinctive social strata of the group, while the latter provides the gratifying self-evaluating images) in the genealogy of the clannish psyche which all pre-modern ethnic groups develop, whether they are dominated or dominant. For this reason, the fact that these mythic or imaginary constructions, which are found in the arcane of the historic memory of all peripheral groups, give rise to feelings of superiority among the intellectual elites of the centre, when they are not openly scoffed at, turns out to be triply baseless. Firstly, because such reconstructions appear in all pre-modern societies, in those which have been involved in constructing the state as well as those who have not. Secondly, because they are functional in the sense that they lay the foundations of the relationship of human beings with a certain continuity in time in the long term and with a given inhabited space. Thirdly, because, like folklore, mass literature and art form part of the bitter culture of these peoples, being structured by categories of aesthetic perception on which the specificity of each one of these peoples is based.

VII Nativism and the millenarian utopia of ethnic origin have not been widely accepted. According to Muhlmann (1968), beyond ideological nationalism there is a psycho-sociological stratus which is much deeper and is apparent in "nativist" movements. These consist of a process of collective action which aims to restore a consciousness of group endangered by a superior foreign culture, thanks to the overwhelming evidence of a particular cultural nature, that is, the restitution in an archetypal mythical means of figures, symbols and themes characteristic of the group in question. Each of these movements possess, Muhlmann argues, common features: universalism and integrism, the logos of messianic suffering, egalitarianism and anarchism; and their mythical structures revolve around the archetype of the restoration of the original state of purity, the recovery of the earthly paradise, or the "Golden Age; the" Saviour's hope means the conquer of the "Millennial Kingdom," in which, through purging, the last shall become the first. Muhlmann believes utopianism and millenarianism to be related. According to the author, they are closed and imaginary systems, isolated from the external world. For this reason they have in common the "themes of purging and exclusion (sometimes compulsory) of all that is foreign, heterogeneous and incompatible". On the other hand, Bastide (1975) argues that millenarian movements are strategies in a search for dignity in the development of a new ethnic identity being carried out by colonised or ex-colonised peoples. These movements appear in periods of change in society and culture, change which casts doubt on the old values and balance of status, whether it be as a result of external causes, such as decolonisation, or internal causes, such as a transformation of castes in mobile social classes, the movement representing a guarantee of the dignity of the pariah group with respect to the privileged group.

VIII Political ideologies are the segment of belief systems, core elements of political cultures, which have a triple nature: dynamism, proselytism and the fact of referring to developed theories and political philosophies (Ansart, 1975; Schemeil, 1985). Macro-ideologies, that is, the most important historical perspectives such as liberalism, socialism, conservatism, anarchism, etc., are the mould of a number of continuously widening field of micro-ideologies which are distinguished by the heterogeneity of the "noble" material with which they develop. In this respect there is a similarity between the development of mythical or primitive thinking and political ideology. Primitive thinking (Levi-Strauss, 1962) serves as a repertory of heterogeneous tools of very diverse origin creating devices out of them that present the internal coherence of their utility. Ideologies are similarly inventive; with practical objectives as their goal they reorganise heterogeneous elements originating from the processes of the collapse of ideologies, political theories, earlier philosophies and even religious beliefs.

IX Caminal (1996) argues in favour of a tri-partite, and not a dual, scheme which differentiates between (rightly, in agreement with Seiler) juridical, political and cultural nations. Seiler defines Caminal's juridical nation as a political nation, and his political nation, as an elective nation. According to Seiler, the political nation, an objective concept, gives expression to the bond between the state, territory and citizens; the cultural nation is the community created over the long-term; the elective nation incorporates a subjective dimension based on the feeling of belonging and refers to the nation and not the state. In order to designate the same meanings, the terms of political-judicial nation, cultural nation and political-elective nation are used here.

X It is worth pointing out that the same applies to some state nationalisms which emerged out of a foreign occupation. The Spanish liberals of the Court of Cadiz confronted the Napoleonic troops as an invading force; but to a certain extent both camps shared a similar ideological opposition against the Old Regime.

XI According to Tiryakian (1994), the first generation of those who studied nationalism, Hayes (1966), Khon (1949), Shafer (1964), Hobsbawn (1983), were influenced by the fascist experiences of Germany and Italy; it was out of this that the "black legend" arose. They had argued that nationalism had been a progressive force in the formation of nation-states, but after this defining moment it became a reactionary force which threatened, from within, the formation of nation-states, jeopardising civil society, excluding the participation of certain groups and heightening the risk of conflict with neighbouring states. Liberal and orthodox Marxists agree with this view of nationalism as an anti-modern force which prioritises idiosyncrasy over universalism. In the seventies and eighties new theoretical tendencies related nationalism to rationality and modernity. These tendencies viewed agents of nationalist movements, with regard to their behaviour and organisational activities, as rational beings which sought to redress situations of weakness and show solidarity, while the movements themselves are seen as part of the modernising process. In contrast to the Kautskyian view that economic development tends to form units which are increasingly large and integrated, thereby reducing the national identity, the new generation of authors view nationalism as a functional factor in the protection of modernity (according to Gellner (1988) for example, nationalism is a consequence of social organisation based on cultures which are internalised and dependent on education systems, each of them being protected by their own political form of self-government).

XII Engels, a German, felt that this definition matched the "multiple small national groups that inhabit the south east of Europe." He felt they were stuck in a patriarchal or feudal state and were therefore contrary to progress.

XIII Anarchism has been a macro-ideology resistant to nationalism. Bakunin condemned state and patriotism equally as "this supreme state virtue, this expression of the soul of the state and of its force." The anarchist anathema also covered centralist and peripheral nationalisms, Anarchists were the only grouping during the Spanish Civil War that refused to take part in the Basque national government presided over by the Basque nationalist José Antonio Aguirre. However, such incompatibility weakened after the Second World War, and some peripheral movements have used certain developments from the classics of anarchism. Proudhon proposed a social organisation based on a federative contract of communes and self-managed worker societies: Europe would end up as the federation of its natural peoples. This conception has fuelled European Federalist tendencies that led to the idea of a Europe of the peoples. In this sense the concept of worker self-management (inspired by anarchism and syndicalism) converged, along with other influences, in May 1968 in France and broadened the understanding of peripheral nationalisms with respect to their base movements.

XIV The Basques are a good example of the huge ideological diversity of the system of nationalisms within an ethnic group. The ideologies that inspired the PNV towards the end of the 19th century were those of the principle of nationalities within the school of historic rights, traditionalism, neo-Thomism, and the social doctrine of the church. By contrast the ideologies that have guided ETA since the middle of the 1960s have been distinctly Third Worldist with a viewpoint of legitimising violence against foreign domination. The writings of Mao, Fanon, Guevara, and Lenin's theory of the revolutionary vanguard justified armed organisation.

XV Revolutionary ideologies return power to ideology by denouncing the oppression and defects that the conservative ideologies attempt to hide. Additionally, they initiate the demythologising of conventional legitimating discourse. Only revolutionary ideologies are capable of creating from nothing ideological groups based on the adhesion to new principles (Lenin's theory on how revolutionary theory creates revolutionary movements is applicable in these cases). For those individuals that suffer domination without consenting to it, revolutionary ideology that devalues in the idealist sense the forces of repression will restore their dignity through a Messianic inversion in the order of their values. On the other hand, the uplifting legitimation will contribute to the process of exoneration necessary for the acceptance of the violence implicit in revolutionary action, which is why such ideologies are profusely utilised by radical nationalisms. Conservative ideologies search for conformity with the established order and power structure, with whom they establish a faithful relation. They are systems of inclusion and exclusion, and the internalisation of their codes reinforces the conformity of their behaviour into institutions. Distance between the governed and those who govern is justified and the confidence of the former in the latter is organised. Taken to an extreme they make possible an ideological terror which allows only the legitimate holders of power to decide what is just or unjust (Ansart, 1975).

[XVI] The "end of ideologies" theory, ideological in itself, brings together neo-liberalism and neo-conservatism, especially in the Anglo-Saxon societies (Pastor, 1994). This ideology is fuelled by the conviction of the generation of the 1930s and 1940s that political ideology (identified with Nazism and Stalinism) is a dangerous mirage (Vincent, 1992). It believes that in industrial societies based on consensus and the diminution of social differences, ideologies play no more than a decorative role. For example, both left and right accepted in 1960s the logic of the welfare state (Lipset, 1969), and supported a state that is both omnipotent and a benefactor, disregarding on grounds of inefficiency *laissez-faire* style self-regulated capitalism. There is also a general belief that the majority of social and moral problems have been resolved and therefore controversial ideologies are out of place (Eccleshall et al., 1993).

[XVII]Brittany has a lucrative primary sector that contributes 40 % of French fish stocks and is the state's leading region for dairy farming, pork and beef, but industrially is very weak, with notable shortages of prime materials and electricity. This means the territory is a clear example of a colonial economy, dominated by French and multinational companies that buy agricultural produce at low prices to develop them outside Brittany.

5 Peripheral Nationalisms (2): Organisation and International Context

Introduction

After an initial phase of linguistic-cultural demands, peripheral movements assert their claim for institutions that take into account their differences from the centre. In this phase, when conflict with the centre is based around either access to state institutions or the substitution of such institutions by their own, the movements are guided by a rational-instrumental logic. Some of the institutions that the movement seeks to control will be related to the ethno-cultural characteristics of the group (educational etc.), others should give the ethnic group access to economic and political power, since without the second set of institutions the survival of the first set will be problematic. Demands of ethnic political institutions often end up competing with the constitution of the existing state, that therefore becomes a "map of ethnic power", being such a "map" the focus for controversy, and eventually the battleground for policies designed to modify it (Roessing, 1991). On occasion, when the ethno-cultural groups are not territorially based, other non-territorial based formulas are tried out.

The identity-ideological dimension does not disappear in this phase. It is this dimension that conditions the moderate or radical nature of the movement's demands, and that determines whether the movement keeps its unity or if it fragments. The identity-ideological logic itself is dependent on the rational-instrumental dimension. The lack of success in achieving its stated aims can weaken the movement; which should not be equated with the group's ethnicity disappearing. Contrary to the relatively recent argument of a number of authors, ethno-national identities have produced more than enough examples of their capacity to survive. As soon as they are proclaimed they tend to establish themselves in a stable fashion in the psychology of group members. But continual failure can also make the identity-ideological *gestalt* more radical. The movement becomes then more isolated and communicates only within itself, which can lead to the use of political violence.

Having their demands granted usually strengthens movements. However, what can also happen, as it does with new social movements, is that on becoming institutionalised the identity dimension of the movement is weakened, which reduces the threat it poses to the centre. This situation can be due to intra-state factors, e.g. the privileged situation of the Swedish minority in Finland, but is more frequently caused by transformations in centre-periphery conflict induced by supra-state processes.

Having become institutionalised, national movements organise either into a pressure group, a one-party system, a multi-party system, or on occasion an armed group. Party organisation becomes clear after universal suffrage is brought into place, in the phase of mass political participation. This reaches its full potential in composite states, as a means of accessing the representative bodies of the territory.   To serve its institutional objectives a movement organised in this way will need to mobilise the same resources it did in its phase as a social movement: the support of financers, the number of activists it can count on, the competence of its leaders, information and peer networks, its ability to troubleshoot, the movements image etc.

It is during this rational-instrumental phase when many ethno-national movements gain interest in transcending their own ethnic limits by including in their demands all of the citizens in their territory, even though not all of them share the same ethnic attributes. This is even more so the case if they employ universal or class-based ideologies. Peripheral nationalisms are, therefore, not merely cultural; in common with the nationalisms of the centre, they can also have an electoral base. The client group of cultural nationalisms is their private ethnic group; the client group of electoral nationalisms is national-territorial. In societies that are hosts to immigrants, peripheral nationalisms can either develop ethnocentric or even racist reactions, or, alternatively, accept the immigrants in case they adhere to the goals of the movement; they can also (above all where there are institutions of territorial power), consider them co-nationals independent of their subjective national identification. States have at their disposition mechanisms with which they can accommodate ethnic identities within their logic; similarly the rational-instrumental dimension of the ethno-national movements allows them to win over groups that are outside the ethnic group to the movements objectives.

In terms of the institutional structures, the fundamental distinction is if the movement is systemic or anti-system.[1] In western societies which have horizontal political participation, the question of whether a nationalist movement is pro-system or not depends on a number of factors: its identity-ideological composition (that at the same time determines the degree of radicalism of their demands, which can range from autonomy and federalism to self-determination and independence), its social-ethnic capital, and the attitude of the system bloc towards it. Moderate nationalisms are, in principle, systemic, while radical nationalisms are anti-system.

An ethno-national movement which accumulates strong social capital (Mota, 1996) will tend to be systemic. If social capital insufficiency is replaced by ideology, the movement will tend to be anti-system. Social capital that generates the voluntary cooperation between the members of a group emerges from mutual confidence and the norms of reciprocity that govern the group's relations, which are associated with dense networks of personal communication, social interchange, civic commitment, and horizontal interaction between neighbourhood associations, aid organisations, etc. Every society mixes in its own idiosyncratic way the political, social and cultural elements of its social capital. In the peripheries, when this is overwhelmingly the case, the dislocation produced by the passage between traditional and modern society (that is, from ethnic personality to ethnic identity) is diluted. Social capital exercises a moderate influence over collective identity, and through that, over the instrumental dimension of the nationalist movement.

This explains the differences between Catalan and Basque nationalism. In Catalonia the strong social capital, controlled from the top down by economic forces, structured a nationally self-organised society, but one that was conciliatory to the idea of the Spanish state. In the Basque country the Carlist wars were, as well as a reaction against the centre, a civil war that turned the passage between traditional society and modernity into a traumatic one, dislocating the former social capital whose deficiencies were later overtaken by the ideological dimension, which drove Basque nationalism to a less systemic stance than that of the Catalans.

The institutionalisation of an ethno-national movement creates a contradiction that is difficult to overcome: it is highly improbable that a peripheral movement, naturally reactive, can redirect all its activity towards the system, whose logic in all nation states is that of the centre.

In the vertical societies of the Third World the contradiction can push the entire periphery onto the anti-system side of the fence. In the West this is often resolved through the peripheral movement breaking up into systemic and anti-system factions. This fragments the initial unity of the movement and precipitates the emergence of distinct organisations.

The issues raised by anti-system nationalisms controvert the actual structure of political power, which (with some exceptions) the system is not prepared to allow. This makes the contradiction impossible to breach and a system of conflict predominates. Anti-system nationalism becomes therefore (whether it uses violence or not) the "internal enemy" of the system, or the "Foreigner at Home". Because of this the response of the systemic bloc to these nationalisms can be studied using approaches that analyse the international politics of states: theories of lack of security, imperialism and modernisation, the decision-making processes etc.

The lack of security generated by conflicts between elites can lead a section of them creating a broad consensus (thereby neutralising anti-elites) through a war against an external enemy that will increase social cohesion. Such a war carries lower risk and greater probabilities of success if the enemy is not an external force but an anti-system nationalism. The pressure to detect the "internal enemy" can also stem from an increase in political participation of the masses, who give as much support to the persecution of this nationalism as they did to the imperialist expansion of their own states towards the end of the 19th century. It can also come from the tensions caused by economic modernisation, when the prosperity of the state is not accompanied by the international status it believes it deserves; this can direct them to external conflict or to harass their peripheral "internal enemy".

However, the centre is not homogeneous in its attitude towards an anti-system nationalism; as in the international sphere, in these conflicts (especially in composite states in which the centre has to align itself with the peripheral moderates), there is intervention by many actors who have diverse interests and fluctuating commitments. The constant arguments and changes in line and policy of the two system blocs of the Madrid and Ajuria-Enea Pacts (drawn-up both in Spain and in the Basque Autonomous Community to unify attitudes against the ETA-KAS-Herri Batasuna anti-system bloc), and the disputes, break-ups and reconciliations of its member parties are a good example of the intervention of diverse agents with changing objectives.

Nationalisms and Pressure Groups

The organisation of peripheral nationalisms in their instrumental-rational phase into political parties and pressure groups produces frequent conflicts with their underlying cultural-identity based nature.

Pressure groups that defend interests of particular sectors often based on economic status, seem to be, in principle, incompatible with nationalisms, whose base is almost always territorial. The fact that such groups do not question the legitimacy of the governments they are trying to influence also distances them from nationalisms. Their respective client groups are also distinct. Those of an interest group are either large latent collective groups such as workers (whose affiliation should be won by the offer of material incentives) or groups that are numerically small but are endowed with a strong organisational potential, as with corporations. The client groups of the nationalisms are the inhabitants of the ethnic territory, who the peripheral elites should win over to the causes of their national movements, offering them incentives that are identity-ideologically based rather than of a material nature.

However, the incompatibility is not total. Groups that promote altruistic causes sometimes have borders that blend in with those of new social movements which themselves have much in common with nationalisms. There can be de-territorialized ethnic groups with specific interests (e.g. immigrant worker collectives) that promote associations that defend amongst other things the groups professional, cultural and recreational activities. In developed peripheries, economic forces can create pressure groups (such as *Fomento Catalán*, that is over two centuries old) with the objective of gaining concessions favourable to their interests from the central state (e.g. the protectionist measures that Catalan industry demanded and were granted in the 19th century). Although the strategy of these peripheral economic groups is not dictated by national objectives, on occasion they can even precipitate the organisation of nationalist parties.

There are pressure groups that are formed with national objectives, and which disappear as soon as they have obtained them. The organisations that emerged in Swiss Jura to win over the local political parties to the cause of forming a new canton are an example. On occasion low intensity sporadic nationalist violence can have the same effects as a pressure group (Seiler, 1982). This was the case in Wales in the 1930s, and was subsequently so in Scotland. It also emerged in Flanders and Brittany before the sixties, in the Italian Tyrol, the Alto Adigio, in the seventies, and in Corsica before the emergence of the FLNC (Corsican National Liberation Front.)

Political associations of a territorial-institutional nature can emerge in peripheries that already have nationalist parties. The existing parties reject insertion into these groups as they regard their objectives as being too moderate, or because they see the newcomers as rivals to the party's grip on nationalist hegemony. The Scottish Constitutional Convention is an example of this. The Convention, a devolutionist platform created in 1987 to support the "Campaign for a Scottish Assembly", was supported by local institutions, social movements such as feminists, Scottish Trades Union Congress affiliated unions and almost all political parties (except the Conservative Party). Several months after its foundation, the Scottish National Party (SNP) distanced itself from the Convention; the SNP feared Labour Party hegemony in the Convention and felt the campaign, in terms of its range of political options, was limited to constitutional reform of the United Kingdom that would make Scottish autonomy possible, excluding other options such as "Independence in Europe" which was called for by the party at that time (Moreno, 1995).

Trade unionism, both in its early stage of defending the political rights of workers and in its current stage of neo-corporative negotiation with management and the state, has historically mistrusted the inter-class nature of peripheral nationalisms and feared that the workers' movement would weaken with the fragmentation of state-wide union action. However, nationalisms themselves can create (or use those that previously existed) workers' trade unions that take on board the national objectives of the movement. This requires a number of pre-conditions: the national movement must have a large base of potentially unionised workers at its disposal; the existence of a peripheral political party that can utilize the union is needed; and also required is the appearance in the nationalist ethos of an instrumental-rational logic that allows the defence of the interests of specific sectors without risking fragmenting the movement as a whole.

Peripheral trade unions have always provoked at their foundation serious hostility from state based trade unions. If they were founded in times of moderate nationalism they were dismissed as being in league with the bosses. (The Spanish Socialist Syndicate UGT accused, in the 1920s and 1930s, Basque Workers Solidarity (ELA) of being an instrument of Basque employers). If the unions are founded in times of radical nationalism, their actions are denounced as, at best, being anti-union and marginal, and at worst, as being openly pro-terrorist. Until recently other unions described the Basque Syndicate *Langile Abertzaleen Bartzordeak* (Patriotic Workers Commission) created by the *Aberzale Socialist Coordinator* (KAS) in 1976 as such.

Yet peripheral unions can channel worker discontent against decisions taken by the state which are considered damaging to their territory, and which were taken with or without consultation with state based trade unions. An example of this was the decision by Spain to cut back iron, steel and naval production that led to government closures of firms in the peripheries; that coincided with the peak of the inter-union confederation in Galicia (CIGA) and the ascension of ELA-STV to the status of dominant union in the Basque country.[II]

Nationalist Political Parties

The principal organisational form of peripheral movements is through political parties. This occurs to such an extent that the genre of national movements tends to be confused with the type of its parties. A large proportion of Third World political parties are peripheral in nature, yet the following analysis refers to those from the West, especially those of Western Europe.

Leaving aside the parties that arose out of the disintegration of the Central Empires, the emergence, sooner or later, of these parties is related to the strength of peripheral movements and to the weakness of the centre. It is no coincidence that the first time such parties emerged was in the south of the Pyrenees (the *Lliga Catalana* and the PNV emerged at the end of the 19th century; the latter is now the oldest surviving nationalist party), and that the Scottish SNP and Welsh Plaid Cymru had to wait for the end of the hegemony of the British Empire after the First World War before coming to light. Nor is it a coincidence that peripheral parties emerged in France in the seventies (with some exceptions), at both the end of the French Empire and the peak of the country's social movements. In the same sense in Italy (with the exception of Sardinia) the emergence of peripheral parties in the Alps and the isles followed the collapse of fascism, and their resurgence in the North followed closely after the clientelist break-up of the welfare state.

Seiler (1990) relates the phases of such development to the phases of mobilisations reactive to the centre. Shortly after the democratisation of European institutions, many such peripheral parties were to be found within the imperial German and Austro Hungarian Parliaments. They were also present in Ireland, Finland, Norway and Spain. This phenomenon intensified after the First World War: not just in the states that emerged or were broken up by the peace treaties, but also in the political systems within which they already existed (as in the Spanish Republic). Parties even appeared in regions that were not included in the treaties (for example, Flemish nationalism in Belgium, or Frisian nationalism in the Low Countries).

By contrast, in the years that followed the Second World War, the movements entered into decline. Exhausted after years of struggle against fascism, some nationalisms were reduced to exile or secrecy (the case with Iberian nationalisms under Franco), some were regarded with suspicion for their collusion with the Nazis (e.g. the Bretons, Flemish and Frisians), while others were regarded as just hangovers from tribal folklore. Italy is an exception to the general rule, with post-war liberation leading to the formation of two ethnic parties with firm roots in the Aosta Valley and in Tyrol.

However, nationalist parties elsewhere re-emerged in strength during the sixties and seventies. They not only made a recovery (diversifying themselves into party systems) in historically strong nationalities like Catalonia, the Basque Country, Flanders, Northern Ireland, and Quebec, but nationalists parties also rose up in peripheral areas that had been on the margins of such processes, such as the Canary Islands, Andalusia,

Scotland, Wales, in French speaking Brussels, Wallony, Corsica, Sardinia and Veneto.

Whilst the movements did enter into decline again in the eighties, the fall-out was not as strong as it was following World War II, and did not affect the nationalist movements in Northern Ireland and Spain. Some states (e.g. Spain, Belgium) modified their structures. In Corsica, the Southern Basque Country and Northern Ireland the peripheral political systems included parties involved in political violence. The state that currently has the greatest number of peripheral parties is Spain, with the Basques and Catalans being the most developed. Whilst such parties have diminished in the United Kingdom and Belgium (caused here by the division of state parties into Walloon and Flemish wings), they have increased in Italy, especially in the Po Valley.

All political parties are vectors of a social conflict, and will take up a position with one of the sides in the conflict (Seiler, 1982). Nationalist parties are born out of centre-periphery conflict aggregating the demands of the latter. However, there is also a deficit between the structure of nationalist movements and their organisation into political parties. Parties are instruments of conquering political power for sides in conflict, yet peripheral movements do not wish to take power in the existing state, but to organise their own structure of political power. In addition, existent norms in all states dictate that parties do not act like factions, but see themselves as part of a whole, participating in a certain consensus on unity. However, peripheral parties emerge from national movements that conceive themselves as a whole in their own right.

Moreover, a number of the functions that are normally undertaken by political parties have been taken upon by the movement itself; other functions are not congruent with a movement's peripheral nature. In effect, movements in their identity-ideological phase have undertaken the global political project, the development of which should be the responsibility of political parties; they have also initiated an alternative political socialisation to that of the centre (although the party now carries this out with even greater intensity than before). The mobilisation of people and resources was also previous to the party's foundation; it will continue this from where the movement left off with even greater efforts, although the party is often criticised by those who oppose mobilisation through electoral methods.

Sartori's (1980) systemic definition of a party is a political group that stands in elections and through such elections can put its candidates into positions of public authority, therefore naturally accepting the electoral system and its rules. This definition weakens when it is placed in relation to the aims of many nationalist parties. For anti-system parties electoral campaigns are only occasions for mobilisations and to assess the social support they enjoy, that is why elections do not tend to be followed by taking up the parliamentary seats that they have gained. Even the systemic nationalist parties regard the seats they have gained with extreme caution (especially those within the central parliament), frequently excluding those in such positions from the governing body of the party. (The PNV has done this since its foundation).

However, the institutionalisation into parties of the nationalist movement strengthens its systemic tendencies, leading to the possibility of conflict between its instrumental-rational dimension and its identity dimension. In addition, the change from the cultural leaders of the initial phase of the movement to the leaders of party apparatus is susceptible to internal tensions (even divisions) that argue for a revival of the lost purity of their identity as a way of "returning to the roots".

The intensity brought on by mobilisation and socialisation within peripheral parties gives them a similar structure to that of the mass parties of a century ago, which themselves are currently undergoing a inverse process in their transformation into electoral or "catch all" parties.[III]

The base units of peripheral parties, councils and assemblies, evoke territorial units of the former mass parties. The political activity of these parties (especially those which are radical, but also up to a point those which are moderate) breaks the existent division between professional public political space and private space dedicated to consumption and leisure, causing the fusion of both. In the Basque Country, in a process contrary to that of the 100 year-old socialist (and communist) "People's Houses" the meeting points of nationalist political parties, the PNV's *batzokis* and Herri Batasuna's *herriko-tabernas*, polarise the social life of their supporters and are the catalyst for a large part of political activity in the country.[IV]

Their strong emphasis on the identity dimension and their "National Alliance" programme prevent such parties structuring themselves along class lines, and makes the existence of organised ideological tendencies problematic, despite the possible presence of a strong, and even charismatic, personal or collective leadership. (In radical nationalisms an armed group, anonymous by definition, can also take on this role). The emergence of "Barons" within the movement often heralds an imminent spilt.

The level of territorial de-centralisation, on the other hand, is quite varied. Radical nationalisms tend to be unitary. However, if the organisational ensemble is directed by an armed group there can be an extremely complex structure in which the latter takes on the decision-making functions, therefore relegating the civil party to a subordinated role consisting only of image making and electoral functions. Alternatively some nationalist parties maintain in their organisation the layout of ancient diversity of the historical territories that were later encompassed in the project of a national community. This is so with the PNV, whose territorial structure has been confederate since the times of Sabino Arana.

Nationalist parties are formed with the aim of politically unifying the ethnic group. When a national movement diversifies in to a subsystem of parties, relations between them are always problematic. Plurality has always been the fruit of divisions born out of the appearance of movements that have reformed the former identity markers in a different and often antagonistic way. This provokes splits within the collective (and hence personal) identity of supporters. Each party within the nationalist parties' sub-system also occupies a different place in the scale of legitimation/de-legitimation of state power, which causes reciprocal accusations of "betrayal" between the groups. (However, the existence of elected bodies in the territory can oblige their coexistence).

The ideologies of these parties are those of the national movement from which they came from. Their successive emergence coincides in Western Europe with phases of reaction of the periphery against the centre. These are - leaving aside the legitimist ideologies that belong to a previous phase to the appearance of parties - the populist in defence of ethno-territorial self-government, the progressive of the sixties and seventies and the neo-centralist, which arose out of the differences in economic status between the centre and the periphery (Seiler, 1990). Populism was, until the sixties, the central core of Western peripheral nationalisms, and in the present day continues to define many of them, often in combination with neo-centralism.[V]

Peripheral parties can be systemic or anti-system, depending on the greater or lesser extent of the radicalism in their strategy. Categorising systemic parties Seiler distinguishes between government parties and tribunes; the former participates in governmental coalitions and the latter does not. The author in his 1982 work employs a double ideological and strategic model to produce a general classification of peripheral parties.

According to their strategy the anti-system parties would be Herri Batasuna in the Basque Country and Sinn Féin in Northern Ireland. The tribune parties would be the Vlaamse Nationale (Flanders), the Ressemblement Walloon (Belgium), the SNP (Scotland), Plaid Cymru (Wales), the Andalusian Socialist Party (Andalusia) and the Valdotaine Union (in the Aosta Valley, Italy). Government parties would be Svenska Folk Partiet (the Swedish party in Finland), Sud-Tyroler Volkspartei (Tyrol, Italy), the Social Democratic and Labour Party-SDLP (Northern Ireland), the Party Québecois (Québec), Volksunie (Flemish), the Front Démocratique de Francophones (Walloon), Euskadiko Ezkerra and the PNV in the Basque Country, Esquerra de Catalunya and Convergencia i Unió (Catalonia) and the Partido Aragonés Regionalista (Aragón).

According to Seiler, classification in accordance with ideology would be:

1. Populist: VNP in Flanders, SNP in Scotland, SFP of the Swedish minority in Finland and the SVP in Tyrol, Italy.
2. Progressive: PAR in Aragon, ER in Catalonia, the PSA of Andalusia, HB and EE in the Basque Country, SF and the SDLP in Northern Ireland, PC in Wales, UV in Italy, the Walloon RW and the Flemish VU.
3. Neo-Centralists: CIU in Catalonia; PNV in the Basque Country, FDD in Belgium and the PQ in Quebec.

The existence of legal peripheral parties means that the system of state parties is competitive (it is most unlikely that a non-competitive system would allow the political representation of territorial plurality) and multi-party, with the presence of parties that emerged from at least two of the cleavages (or fracture lines) outlined by Rokkan and Seiler which are: centre-periphery, rural-urban, religion-laicism and work-capital. (However, during open war between the communities within a state the representation of each community may well be concentrated in one party: as happened in the Bosnian Islamic, Bosnian Serb and Bosnian Croat communities of the Republic of Bosnia.

The threshold of relevance criteria is different for state parties than for peripheral ones.[VI] Owing to this, legislation sometimes provides opportunities for the formation of peripheral parliamentary groups in central parliaments.

The resulting system is only polarised (in the sense that there is a ideological gulf between the extremes) if it contains anti-system nationalist parties, and not if the opposite is the case. This is because, paradoxically, the systemic peripheral parties represent a huge potential for coalition with state parties, as they operate along separate cleavages: both canvass their votes from specific separate electorates, which means competition is not antagonistic, but consotiational. Nationalist parties of the same periphery, even if they are systemic, are more in competition with each other, which explains the coexistence of the PNV and the Basque Socialist party in successive Basque autonomous governments, contrasting with the difficulties of coexistence difficulties between the PNV and Eusko Alkartasuna, which broke away from the former in 1986.

The relevance of peripheral parties to the state party system depends on various factors: the unitary or composite nature of the state, the importance of the population of the peripheral ethnic groups, and naturally the number of votes that these parties obtain in their respective territories. In Western states it is rare that all of these factors are concurrent; the

Spanish party system, which incorporates a number of relevant peripheral parties, is an exception to this. The presence of peripheral parties in central governmental coalitions is influenced by the possible existence of other peripheral parties, less systemic than the formers, or clearly anti-system, therefore inclined to denounce participation in a coalition as a betrayal of national principles.

As a result, and despite their potential for forming a coalition, such participation is rare. In the United Kingdom it has been impeded by the British two-party system, added to the unitary character of the state. In France only a small percentage of votes go to peripheral parties. In Belgium the division of the state parties in the sixties and seventies into Walloon and Flemish wings prevented the electoral rise of nationalist parties in the two communities. In Spain, since the PSOE lost its status of dominant party in the nineties, conditions for the participation of CIU and even the PNV into central governments has been concurrent. However, the existence of more radical nationalist parties in the respective peripheries (EA, and especially HB in the Basque Country, and Esquerra in Catalonia) and the fear of criticism from them, has contributed to avoiding this eventuality. Anyway, the CIU has end up holding the balance of power in Spanish bi-partisanship, determining the programmes and policies of successive PSOE and PP governments under threat (that was followed through on one occasion), of fresh elections by a switching of allegiance in Parliament.

Being systemic in nature and the lack of competition appear to be the best way for a peripheral party to gain access to central governments. This is shown by the SFP that has used such conditions to uphold the status quo of the Swedish minority disseminated across Finnish territory. [VII]

Although an analysis of the participation of peripheral parties in central government makes almost all appear to be *tribune* in nature, in composite states and in respect to territorial power almost all want to be *governmental*. Managing their own territory is in effect the privileged objective of systemic peripheral parties. Anti-system parties on the contrary, who don't recognise the legitimacy of the territorial institutionalisation of the state, can adopt the same stance of non-participation within parliaments in their territory (if such institutions exist) as they do to central parliaments.

In these cases the system of peripheral parties becomes highly polarised, displacing the focus of conflict to the acceptance or rejection of the territorial apparatus. This focus has a parallel, in territories where political violence is present, with acceptance or rejection of such violence. The systemic nationalist parties in both cases take the same line as state parties, whilst the anti-system parties take the opposite position.

In the Basque Autonomous Community, the anti-system Herri Batasuna polarises in a hostile manner the rest of the parties (whether they are Basque nationalist or state-wide parties) with respect to the use of violence and participation in autonomous institutions. The Madrid and Ajuria-Enea Pact blocs have been in existence since 1988. However, the Basque nationalist parties are now expressing reservations about them, the less systemic these parties become the greater the reservations. The PNV has voiced frequent concern over the Pacts, which intensify in correlation with their differences with the governing party in Madrid. Eusko Alkartasuna had doubts in adhering to the Ajuria-Enea Pact and did not participate in the Madrid Pact. The greatest defenders of the Pacts are the two main state parties, the PSOE and PP, who fear that in the Basque Country the divide will centre on Basque nationalist parties opposing the state parties.

In Northern Ireland there has been no local parliament since the 1970s. The centralist Unionist parties do not have a state role in the territory, their role is that of a Northern Irish pro-British nature. Because of this the Irish SDLP, although it rejects the use of political violence, has not been able to form a stable alignment with the Unionist parties against IRA-Sinn Féin. (However, to re-establish Home Rule in Northern Ireland, the SDLP did participate in a non-sectarian executive in 1973-4 that was headed by the Unionist Faulkner. This decision, described as "treason" by Sinn Féin, broke relations between the two Irish nationalist forces for almost twenty years.)

In Corsica, following the application of a Statute in 1982, the Corsican Peoples Union (Union du Peuple Corse, UPC) participated in an autonomous coalition government presided over by left wing radicals who were integrated into the French left. The withdrawal of the UPC from the coalition in 1984, added to the electoral weakness of Corsican nationalism as a whole, and the recent foundation of its distinct forces, has impeded the polarisation between moderate and radical Corsican nationalists. The UPC stood in the 1986 elections in coalition with the Corsican Self-Determination Movement (a party aligned to the armed FLCN group.)

A systemic party can become an anti-system party without using political violence (and even specifically rejecting it) from the very moment that its instrumental program becomes unacceptable to the constitutional order of the centre. Its potential for coalition disappears overnight, and it becomes an "internal enemy" to the state. The Northern League, a relevant party from 1992, aligned in 1994 with the Berlusconi's right-wing coalition. Having broken away from *Forza Italia* the left had reservations about bringing them over to their camp, but the secessionist proclamation of the Republic of Padania in 1996 made the group into an "anti-system" force that is rejected just as much by the Italian left as it is by the right.

International Context

The international context is a crucial factor in the origin and evolution of the centre-periphery conflict. The initial phase of the construction of the states, which determined which ethnic groups were going to be central and which were going to be peripheral, took place in an context dominated by the wars, treaties, expansions and retractions of Monarchies and Empires. It is international economic, political and military factors that have determined, in the era of the construction of nation-states, the successive phases of the above conflicts and their diversification over geo-political areas in the West, Eastern Europe, the Third World, Islamic countries, Latin America, etc. Finally, it is the international context that almost always determines the viability of peripheral demands when demands for self-determination or independence are made.

Such claims have seldom been resolved in the west through the recognition of the independence of the periphery by the centre, stemming from a self-determination plebiscite or not. Norway, a Kingdom in its own right since 1815, although subject to the ultimate authority of the Swedish King, gained its independence in 1905 after a plebiscite summoned by the Storting was ratified by Sweden. Iceland, also a Kingdom from 1918 in a regime of personal unity with Denmark, also gained independence in 1944.

Politically integrated western states almost always oppose any demand that attacks their territorial integrity. Therefore, conflict is difficult to resolve in the state sphere, regardless of whether the nationalist movement is pacifist or resorts to violence.

This means that demands for self-determination and independence only stand a chance of succeeding if they correspond with the orientations of the dominant bloc of states at world level. For this to happen certain circumstances are needed: firstly, that the state in conflict with the periphery is not a member of the dominant hegemonic bloc and, secondly, that the dominant bloc feels that the new nation's independence would be beneficial for its geo-strategic interests, the new nation aligning itself as a linkage group to one of the foreign states integrated in the bloc. (A similar situation arises when the power aligned with the nationalist movement, whilst not belonging to the dominant bloc, has sufficient power to change the will of the state from which the peripheral group demands independence).

Affinity between the peripheral movement and the foreign power can take on a number of different forms. It can be an ethnic (or civilisation) affinity; it can be related to the foreign state as a whole, or to an ethnic minority within it that is in a good position to influence the will of its state in favour of its linkage group (Zolberg, 1985). The Jewish-Palestinian conflict has become a major international problem because the Palestinians share an ethnic belonging with the neighbouring Arab states and a religious creed with Islamic states. The fact that the Republic of Ireland is contiguous to Northern Ireland, and that thirty million US citizens claim Irish origin, are factors that exert a constant pressure in favour of a solution to the problem which has the interests of the Catholic community at its core. The Tyrolese of the Italian Alto Adigio speak the same language (German) as the Tyrolese in Austria. The three million Tamils in Sri Lanka share an ethnic belonging with the hundred million inhabitants of Tamil Nadu Drávida in the south of India, and the Hindu religion with the majority of the inhabitants of this enormous country.[VIII]

The access of groups of states to independence tends to come in waves, in times of realignment of forces within the world system of states. This tends to coincide with, or follow, major wars, or the disintegration of a political regime that had universal validity. Chronologically such phenomena start with the wave produced in western Europe in mid-nineteenth century on behalf of the "Principle of Nationalities". This was followed by the wave of World War One in central and eastern Europe against the central empires, under the auspices of US President Wilson; later on by the process of de-colonisation that followed the Second World War against the small European imperial powers in the Third World, in the name of self-determination, which was helped by the United States (with some reservations) and by the Soviet Union. Finally, the final wave is the reconstruction of the national-state map in central and eastern Europe caused by the disintegration of Soviet and Yugoslavian federalisms.

In the West such waves have coincided with the theoretical elaboration of the right to self-determination. The theory emerged at the beginning of the nineteenth century under the "Principle of Nationalities" formula and up until the First World War was considered to be a subversive principle. In the mid-nineteenth century the Italian Mancini gave this principle its classic definition, founding it in French Rousseau style vision and German cultural romantic conception. According to Mancini "a nationality is a common thought, a common principle, a common objective [...] it is the succession of all men grouped by their languages, by geography or by history, that recognise the same origin and march under the authority of a right they share in unity to the conquest of a definite aim."

The Principle of Nationalities, that remained as the political objective (albeit not recognised in law) of nationalist movements until the early twentieth century, was reformulated in central and eastern Europe as a right to self-determination by the II International, which could depend on reflections of Marx and Engels for theoretical support. Under Kautsky's influence the II International Congress of 1896 formulated the right to self-determination, which was written into an official document for the first time as "the right of all nations to govern themselves" (Gallissot, 1980). This definition does not differentiate between the concepts of autonomy and self-determination, and was interpreted in two opposing ways. Whilst the Brünn Congress of 1899 vindicated autonomy, distinguishing it from self-determination, Lenin's Bolshevik Social Democratic Party supported in 1903, in the pragmatic manner that was witnessed, the defence of self-determination of all peoples in the Russian Empire.

After the Russian Revolution the Bolshevik Government solemnly proclaimed this right in 1918, which, coupled with the anti-imperialist struggle, became in 1919 one of the basis for the Third International. However, despite declarations that the Union of Soviet Socialist Republics (USSR) formed in 1922 was a voluntary union of republics, based on the right to self-determination, in practice the route to independence was blocked and substituted for linguistic and cultural autonomy.

The Allied victorious in the First World War supported the Principle of Nationalities, which became the basis of a programme by President Wilson, set down in the Fourteen Points for Peace as "the right of all peoples to govern themselves" (deviating therefore from the Soviet formula). This is how the principle appears in international law, recognised in the Peace Treaties of 1919 and 1920. However, the application of this law was a continuation through juridical means of the Allied military victory over the central Empires. Because of this, even on these occasions when Treaties involved consultation of the populations that would be affected by the configuration of the state maps in central and eastern Europe, their rights were waived when they were on a collision course with secret treaties, economic interests or strategic arguments that benefited the winning states, and rejected for national minorities within the winning powers and neutral states. Additionally, the principle was not considered applicable to the colonial territories of the winning powers, which continued to be subject to colonial ties under the euphemistic denomination of "protected territories".

The desire of the two superpowers which emerged at the end of World War II (the United States and the Soviet Union) to get rid of the colonial empires of the European minor powers, gave rise to the introduction of the right of peoples to govern themselves in the founding of the United Nations Charter in October 1945. This right is explicitly declared in the first Article, and is indirectly declared in Articles 63 and 66 that are about protectorate territories. The subsequent actions of the United Nations reinforced the identification between self-determination and de-colonisation. The key text in this area is the General Assembly Resolution of December 1960 on the right of colonised peoples to self-government, which states that the only path to de-colonisation for colonised peoples is that of independence. This declaration represents the closest point of convergence between Third World peoples and the Soviet Union.[IX]

However, although the UN has defended the application of this right to "protectorate" territories, it has not extended this defence to non-colonial territories. The UN has created a concept of peoples that only allows them to become holders of this right when the three features of geographical separation, ethnic difference and political subordination from the metropolis are present. This concept, that excludes the peoples of the industrialised world from the category of groups protected by this right, has been ironically defined as the "salt water theory" (Obieta, 1980).

The vulnerability of Third World states and their fear of secession emptied of any meaning the right to self-determination, until 1989, for all those in situations unrelated to decolonisation. Because of this the UN sustained radical opposition to secessionist movements (such as: Biafra in Nigeria, Katanga in the Congo, those present in Pakistan and in Kurdistan, Iraq), which was later extended to the petitions eventually made by ethnic minorities in the industrialised world.

In 1975, the first European Conference on Security and Cooperation linked the exercise of the right of self-determination to the maintenance of peaceful relations between the states of the two opposing regimes. Following from this, any internal political upheaval in a state in Europe, especially if this involved changing borders, came to be considered as a threat to European security.

This concept of self-determination was dominant in the 1945-1989 period of bi-polar World Order that, according to Hoffmann (1985), was characterised by: the stability of the western centre through the deterrent effect of the nuclear threat; the displacement of armed conflicts to the periphery, that is, the Third World; a plurality of supra-state political systems, some of which were imperial in nature (the Soviet Empire, and US neo-imperialism) while others were of a federal type (the European Union); and the survival of the nation state as the fundamental community of the international system (although weakened by the trans-national character of military alliances and economic trends).

The disintegration of Soviet imperialism has brought a new world order that is characterised by the existence of one hegemonic superpower, the United States, which has given itself a more easily defeated enemy: Islamic fundamentalism. For central and eastern Europe, the new order has erased the concept of self-determination as a colonial right and the intangibility of European borders as instruments of world peace. It has also increased the importance of a federal-style supra-state system, the European Union, expanding towards the east of Europe, which has direct relations with the institutionalised regions of its member states. This new situation has brought about a transformation in centre-periphery conflicts in Western Europe.

Notes

[I]   Peripheral nationalisms are systemic when they formulate their demands (inputs) through the normal instances in an accepted way, therefore determining the decisions (outputs) of the central or territorial authorities. The authorities' decisions in turn provoke three types of reaction; satisfaction (support for the system), dissatisfaction (rejection), or indifference (apathy), with which the systemic cycle reinitiates. (Cotarelo, 1988.)

[II]   ELA reinforced its class-based nature in this process, which distanced it as much from the PNV as from the Basque Autonomous Government. LAB, which was marginalized for a great deal of time, has achieved in recent years union membership that is equivalent or even higher than the number of votes received by its political counterpart Herri Batasuna, although in this case a clear line was not drawn between what is union business and what is a political matter. It is significant that LAB has left KAS.

[III]   Electoral or "catch all" parties (Kirchheimer, 1985) are the result of the adaptation of mass parties to a more mobile and less conflictive society, and to a mixed economy which rewards pragmatic policy. Whilst this has happened both on the right and on the left, the latter's cases have been more noteworthy. Such parties have tried to go beyond the limits of the old class grounds, directing themselves at the majority of the electorate. The information revolution makes possible the direct communication between leaders and the electorate through the mass media. Party figures can learn the electorate's desires through surveys, which allows them to prescind of the party apparatus. Through this the party abandons the organisation and socialisation of its activists and gives priority to specialised professionals and political showmanship (Martínez Sospedra, 1996). However this evolution towards "catch all" parties, common in the West, has also affected peripheral parties: especially the neo-centralist parties and those who assume governmental responsibilities at a territorial or even state level. This evolution is frequently denounced by the defenders of the initial identity dominated phases of in the movement, aggravating the existing tensions between cultural and political leaders.

[IV]   The *batzokis* are the headquarters of the PNV's municipal councils. Since the collapse of Franco's regime, such places have functioned as recreation rooms, with bars and restaurants for use by militants and staffed by members themselves on a rotary basis. At the same time they are also the centres of political and economic business of the municipality: nationalist councillors and often mayors (many of whom are PNV members) meet in the *batzokis* to discuss municipal problems. The *batzokis* are also where the local assemblies of the PNV meet to discuss political questions. The money comes from council members' solidarity or from the revenue generated by the bars, restaurants and breweries. Herri Batasuna's calls for social mobilisation, its lack of confidence in institutional participation and its utopian stance meant that in the first few years of the post-Franco transition (1977-9) it was the refuge for radical discontent within society. Because of this the ecological, anti-nuclear, anti-Army, feminist etc. groups that emerged at that time tended to set themselves up in the municipal spaces under

Herri Batasuna's control, the *herriko-tabernas* becoming, in addition to entertainment centres, the "territorial boarding houses" of the alternative movements. The social activities of the "People's Houses" run by the Spanish left (the PSOE, and the PC) have diminished in the post-Franco period. Their activists rarely socialise together, with the exception of some parts of the Bilbao estuary.

V Populist nationalism, according to Seiler (1982) is "a union of progressive and conservative attributes expressed through an emotive sentiment for the reconciliation of classes in the ethos of a general (or in this case national) fraternity. Populist nationalism is extremely traditionalist, in this sense it is the legitimate heir of the legitimist reaction. The community's traditional values, including language and religion are of up most importance to populist nationalism. Norwegian, Irish, Basque and Flemish nationalists have proposed the restatement of an official ethnic language [...] At the same time Populism consists of a struggle for popular and national emancipation, which rejects old concepts of legitimacy, and which should be taken as a break from the legitimist reaction". "Peripheral nationalism is also democratic. In a bourgeois state, when centralist nationalism has to take sides in class conflicts, it always sides with the right; populist nationalisms, in similar circumstances are always travelling companions of the democratic movement. This can be easily seen in Ireland and the Spanish state. The Spanish Civil War gives us an excellent example: the PNV, despite being staunchly catholic, preferred an alliance with anti-clerical republicans [...] "Additionally, peripheral nationalism makes possible a non-state path to a cultural nation. In the bottom of his or her heart, the populist nationalist dreams of an independent state. However, in the majority of these parties, although not all of them, they officially propose to find "realist" solutions to the national question [...] Populist nationalism in any case, tends to postulate the community as a member of a federal state or confederation or as an active subject of a form of self-government."

VI For state parties it is 3% of state votes; for nationalist parties 1% of state votes can be considered relevant, as with the PNV.

VII There are other possible variants. The end of a bi-party system to one of a dominant party, caused by the falling from grace of one of the major parties, may turn a peripheral party into the state opposition. This happened in Canada, where the practical disappearance of the previously ruling Conservative party following the 1993 elections, turned the nationalist Bloc Québécois into the main opposition to the ruling Liberal party.

VIII However, when the dominant bloc of states is not in crisis, the international context is hostile to peripheral demands for self-determination or independence. The factor here is the universal solidarity of states in defence of their territorial integrity, which fear a domino effect seeing set off within its minorities as a result of the triumph of a secessionist movement. This hostility is at a maximum in the Third World whose recently decolonised states have very vulnerable borders, and also in the cases in which the national movement of an ethnic group has a trans-frontier character (e.g. the Basques and the Kurds.) Under these circumstances states tend to join forces to combat the respective movements of the group on both sides of their borders; although if the states themselves are in confrontation they can pragmatically use the minorities to serve their own interests. This happens in Kurdistan.

IX The USSR's interpretation in the sixties and seventies of the right to independence of colonised countries, and its proclamation that this could be achieved through referendum or through war, was the origin of its discrepancies with the western bloc states over the international definition of "terrorism". For the western bloc the struggles for national liberation were "terrorist phenomena", for the USSR and the former colonies they almost always constituted an expression of the will to free disposition of the peoples (Freedman, 1976).

6 Centralist Nationalisms

The nation-state depends on nationalism in order to obtain the political loyalty of its citizens and their commitment to the two-fold task of the construction of the national society and community. These state nationalisms require the emergence of a national elite, and they activate the formation of a collective space-time identity. Both of these processes are later mirror-imaged by peripheral nationalisms. Alongside these "normalised" nationalisms there are reactive nationalisms, which are consequently centralist in nature. "Functional" reactive nationalisms emerge from the reaffirmation of the centre in response to opposition to its construction; or they may emerge from a reaction to a range of variables such as class struggle or the international status of the state. "Dysfunctional" state nationalisms are those which do not help in the construction of the centre; conservative rural movements and racists belong to this category. Peripheral nationalisms provoke centralist reactions in which state nationalism, in its turn, counter-imitates the methods used by them. These anti-peripheral nationalisms may operate within the sphere of the state, or they may be restricted to specific territories of the state in the form of local centralism. They may also emerge as mobilisations of the peripheries within the periphery against the latter, which acts as the centre.

Elite and Genealogical State Nationalisms

In effect, each nation-state develops nationalism as an essential instrument of legitimation. In horizontal societies of the West, state nationalism is not ethnic, even though it may have as its bastion the pre-modern symbolic capital of the dominant ethnic group. The process of construction of the national society is overseen by a state elite formed by statesmen, party leaders, the upper echelons of the civil service, high-ranking army officials, department of environment management, well-known artists, university heads-of-departments, ecclesiastical dignitaries, mass-media executives, leading journalists and revered television and radio presenters, famous sports people, all of whom are recruited from throughout the state, including its peripheries, and who identify with the centre. The elite, who personify the majority, take it upon themselves to represent values, monopolising the totality of humanity; they label the other groups, denying them total humanity and defining them as closed and, therefore, minorities. Members of these minorities are viewed as nothing more than a group or part of a group while, on the other hand, members of the dominant majority define themselves according to their individual status. This also explains how, in each nation-state, there is a historic-geographic centre which occupies the favoured place of the national psyche, appearing as the "non-region", while the regions appear as marginal and peripheral to the centre.

In every state society there is an "ideal group" (Geza Roheim, 1967), whose signs and symbols of power are located at the centre. This gives rise to a dichotomy between positive and negative identities, on the basis of which ethnic stereotypes are constructed. Members of the national elite who come from the periphery are those who react with greatest hostility to peripheral nationalisms (and even more so if they reside in the periphery): they reaffirm their values in contrast to the peripheral counter-values, and proclaim themselves to be national while not "feeling" nationalist (since they identify, to a certain degree, peripheral nationalisms with what is marginal and even pathological).

In their early stages, nationalisms in Western states (especially if there has been a rupture with the *ancien régime*) see themselves as a project of the present which looks toward the future and lacks a past. However, the construction of the community very soon (if it is not from the very start, in those state nationalisms which drew inspiration from Herder or Burke) requires the creation of a space-time collective identity. While the relation with space is unproblematic, since this coincides with state territory (unless there are irredentist claims with regard to territories located in foreign states), the development of a consciousness of continuity in time consists of a process of genealogical reconstruction in which, like peripheral nationalisms, historical memory fuses with myths and the imaginary.

These reconstructions, built in Western states throughout the 19th century, continue to have an effect in the 20th century. However much their assumptions have been overtaken by subsequent developments in history and archaeology, and even though organic intellectuals of the centre take greater pains to demythologise genealogies of the periphery than those of the centre, they continue to live on in the popular psyche and often too in children's school books.

In the middle of the 19th century Spanish national historians recast in a nationalist light the mythical stories of Numancia and Sagunto and the figures of Viriato and Sartorio propagated in the Golden Century (16th) by treatises such as *Crónica general* by Florían de Ocampo or *Historia de España* by Father Mariana. *Numancia* by Cervantes portrayed the Goths, supposed predecessors of the Spanish, as a people who had come from afar and who, on conquering the Romans had revenged the fall of Numancia; therein lies the identification of the stereotypical Spaniard with honour, bravery and religious values.

Modesto Lafuente propelled these myths in his *Historia General de España,* published from 1850 to 1867. The collective suicide of the inhabitants of Numantia and Sagunto, who refused to hand over their cities to the besieging Roman and Carthaginian armies, founded a Spanish nationalism of a heroic type. The figure of Viriato, who fought the Romans in Lusitania, a territory belonging to Portugal in the 19th century, facilitated the annexationist objectives of the Spanish nationalism of the era. This national imagery, represented by the historical works of the painters Ribera and Madrazo, continued to prevail in the chromolithographs of the 20th century.

This self-satisfied nationalism centred on Castile, which rested on a weak national construction (contrary to Paris, Madrid was a small capital in the 20th century), was unsettled from 1880 onwards by the fear of social revolution and the mobilisations in the Catalan and Basque peripheries. For this reason the centre did not develop more myths of the Golden Age, such as that of Tubal, supposed ancestor of the Basques, or that which claimed the Basques were the first Iberians (Díaz Andreu, 1994).

Even though France was, until World War II, an armoury of national ideologies for all of Europe, which were developed in tandem with national imagery for local consumption, the country was, however, sharply divided by a left-right conflict. The diverse range of "noble" political theories, which had been transformed into national ideologies by the 1789 Revolution, were exposed in the 19th century to different concepts of the French nation. Michelet argued that France was a person who had evolved from a central nervous system, the capital, extending outward to encompass the periphery. Renan claimed that France was a spiritual principle, based on a common stock of memories and the will to live together. For Vidal de la Blache and Siegfried, France was distinguished by its "physiognomy" and its "physical features" (Noiriel, 1994).

From 1880 onwards, the church-state conflict was responsible for the production of different school history text books in Christian schools and public schools. In the latter, Renan's contractual conception of the French nation predominated; tinged with culturalism it affirmed that France was, above all, the "land of our fathers". For Catholic schools, France was the "first-born daughter of the church" who was baptised in Reims together with the Frankish kings (Deloye, 1994).

There was however one figure above the left-right conflict: Joan of Arc, who came to occupy a place in the calendar of saints' day. Liberal-left historians rediscovered her in the times of the Restoration and produced documents proving that Joan of Arc's plan was to liberate France from the English and that this brought her up against Carlos VII and his Court, making her into a national symbol of freedom of the people in their struggle against the monarchy. Michelet rescued her from political prejudice. Joan of Arc was a heroine who embodied the nation and who was betrayed in part by the people themselves. She could therefore take her place in the pantheon of French history together with Clovis and other figures.

From 1870 until the Great War, Catholics and republicans battled over the use of Joan of Arc. The plan to create a national holiday in her memory was proposed by the left-wing against a church which they accuse of burning her. However, the new generation of Catholics began to view France without a king, which allowed them to revive Joan, and Bishop Dupanloup requested her canonisation. At the same time, the republicans lost interest in her as a symbol; since 1890 a national and Catholic Joan of Arc has prevailed over a left-wing Joan of Arc of the people. Nevertheless, it was the image of Joan as a person of the people which prevented the right-wing monopolising the myth in the 20th century. The attempt by Marshall Pétain to make her into a figurehead for his New Order was opposed by General de Gaulle, according to whom Joan's mission coincided with his own project for national liberation (Krumeich, 1994).

Centralist Reactions

State nationalisms become "centralist" when they defend the centre against threats. Given that these threats can be real or imaginary, the line between defensive and aggressive centralism is often blurred. Since these are not numerous, the strength of the centre often allows it to deal with them. However, the appearance of such phenomena reveals the presence of strong pockets of resistance to the construction of the centre (Seiler, 1990). For this reason they arise in the dense urban centre-European network; in France they emerged in the form of Bonapartism, with the objective of counteracting the opposition of the Church. State construction failed in Europe in those areas where the counter-reformation (Spain) had greatest influence or where this influence was combined with membership to the Europe of the city-states (Italy and Germany in part). Centralism would express itself at a later date in these areas in the paroxysmal form of fascism.

In these regimes the centre reacts violently against class struggle, which it considers to be the supreme anti-national threat, and against the international status of the state (particularly in Germany, angry after its defeat in World War I). They are, therefore, "functional" centralist nationalisms as they use the construction of the centre for their own ends (however depraved these might be ethically). These nationalisms activate different psychic mechanisms in their citizens: identification with the leader (*duce, führer, caudillo*) in the most self-destructive fashion; and aggressive over-compensation (Adler, 1975), as a means of overcoming, through expansionist state policy, the sense of inferiority rooted in infancy.

The intersection of these state nationalisms with cleavages different to the centre-periphery conflict leads to a reaction on their part against certain consequences of industrial development, such as the immigration of groups of migrant workers, or to take sides in the rural-urban conflict in support of the former. Given that these positions impede the construction of the centre, such nationalisms may be defined as "dysfunctional".

"Rural" nationalisms emerged in the second half of the 19th century as part of the massive opposition of the "notables" to the Industrial Revolution's role in ending the traditional way of life. They reacted against the appearance of capitalist modernising elites and the consolidation of working class organisations, both of which signalled the end of their hegemony. For these reasons the "notables" often sided with the church in the conflict which brought it into opposition with the state.

The idealisation of rural life, a mythic idealisation, since it is based on poor knowledge of the reality of peasant life, is from the point of view of a "return to the origins" the most direct consequence of the rupture of the liberal mould (inherited from the Enlightenment), based on the idea of linear and indefinite progress (Poulantzas, 1978). This rupture translates into the imaginary recovery of a society, the pre-capitalist peasant society, where cycles are repeated forever and in which the past is made present and origins are renewed. Given that this return to the past is conscious and laborious, operating not through an impossible re-creation of pre-national societies, but through the nation, ruralising nationalisms may also be defined as "essentialists" since they seek to detect in each nation-state that which is the national "essence", the "genealogical mould" which produced the nation.

This imagined rurality, which is never the work of peasants and rarely that of intellectuals who come from the countryside, is produced by urban ideologues or rural "notables". It appears as a depository of the national "essence" and embraces different denominations, *France des profondeurs* (Deep France) and the Spain of the noble Castilian lands. Given that the construction of the centre rests on the interrelation of national and industrial revolutions, these nationalisms, which aim to bring about opposition between the former and the latter, are dysfunctional. "Ruralism" may also pervade the nativist reactions of peripheral nationalisms in the early stages. (Sabino Arana praised the egalitarian nature of the "ante-iglesias", rural Basque communities, in contrast to the urban centres in the past and impiety and the marketing of the mines and factories of Biscay in his time). However, such an approach soon loses its privileged position, on account of its dysfunctional nature.

Sudden contact between ethnic groups in hostile situations generates ethnocentric reactions. When ethnocentrism becomes ideological it turns into racism, which is used as a weapon by the native population of a state against outsiders. Racism, which was already present in the Middle and Modern Ages in the treatment meted out to Jews by Christians for religious motives, acquired new characteristics as it gathered social strength in the second half of the 19th century. The interplay of a range of factors during this period facilitated the rise of racism: the philosophy of the Enlightenment, which proclaimed the diversity of cultures within the human species; the evolution of biology and anthropology in the 19th century, and proletarianisation and colonisation brought about by the Industrial Revolution.

An explanatory theory of these social facts was put forward in 1853 by the Frenchman Gobineau, who in his *Essay on the Inequality of the Human Races* proclaimed the superiority of the white race, and within it, the Aryans. He argued that this latter group, descendants of Japeth, were the aristocracy, and that the mix of races degenerate and kill off peoples. These genealogical theories are ancient, and some have their origins in the Middle Ages (the Castilian chronicles which portrayed Tubal, the son of Japeth, as the mythical founder of the Iberian Basques, goes as far back as the 13th century). What is entirely new is the portrayal of racism by Gobineau as a difference of nature among human groups and as the perfect identity between somatic and cultural potential.

"Race" distinguishes a human group with an imagined biological marker.[1] Those who belong to that framework are always minority groups which are, in sociological terms, in a position of dependency or inferiority. In this way, nature seems to copper fasten, in an irreversible way, the peculiarities of each group, thereby giving rise to the belief that there is a wall between them which makes it impossible to move from one culture to another. The majority view the biological difference, which is physical, and the cultural, which is social, on the same level. In relation to itself, the majority sees no framework or marker, believing therefore that it belongs to no race (Guillaumin, 1972). In the last instance, the majority syndrome may come to resemble the racist syndrome.

Racist nationalisms straddle the borders between the functional and dysfunctional. They are dysfunctional in the sphere of internal politics as they question a consequence of every industrialisation process, that of migrant workers. For this reason they cannot access to the status of central nationalism of the state (except in extreme cases of unequal vertical societies such as South Africa until recently). In the international sphere, racist ideologies such as those put forward by Chamberlain or Rhodes, often mixed with cultural myths such as the "white man's burden" as espoused by the British writer, Rudyard Kipling, or the universal civilising work of France, have functionally collaborated in the building of Western imperialism in the Third World. They may be tolerated more as states of mind than structured nationalisms when they help to sustain the discrimination against specific ethnic groups in the labour market. Nazi anti-Semitic (and anti-gypsy) racism constitutes an exceptional case which has been hugely ideologised and not linked to the processes of industrial migration. The identification and subsequent annihilation of these supposed "races" as the "internal enemy" of the "Aryans" forged the national alliance of Germans as the nucleus of a frenzied plan to create a universal racial Empire.

Centralism and racism have re-emerged in the contemporary European Union as a result of the combined effects of economic crisis and the immigration of workers who belong to non-European cultures. The situation has given rise to the appearance of xenophobic political parties in the more industrialised European states such as Germany, Great Britain, France, etc. (Seiler, 1990).

Ethnocentric and racist reactions also occur in peripheries. When the immigrants who are the victims of these reactions are foreigners these reactions are no different to those which occur in the rest of the state. On the contrary, when they come from a less well-developed state centre, ethnocentrism is structured in ideological discourses which are integrated into peripheral nationalism. However, these peripheral racisms almost always lack the virulence of the racism which emerges in the centre. If racism condenses the situation of social superiority of the racist group over the outsiders, the socio-economic superiority of the peripheral racists in relation to the immigrants from the centre becomes inferiority in what restricted access to the cultural institutions of the state is concerned. Besides, the upper-class of the periphery who need the immigrants quickly establish an alliance with them, particularly with the "immigrant elites", against peripheral ethno-centrism. (This alliance will continue to exist even when the nationalism in question is no longer racist in nature).

Finally, racism restricts peripheral nationalisms to the limits of the ethnic group. Therefore, once they organise themselves into political parties and develop a general-territorial logic (which seeks to win over to its objectives all the inhabitants of the ethnic territory, regardless of their origin) they erase all elements of racism. For this reason, for the past half century, there has been no trace of racism within moderate Basque and Catalan nationalisms. In the case of radical nationalisms, almost all of which emerged subsequent to the Nazi aberration, class ideologies transmitted a type of anti-racist discourse, an example of which was the claim by Basque radicals in the sixties, "Basques are those who sell their labour in the Basque Country". In this way, immigrant membership of these movements often became a means of speeding up the process of integration into the new nationality.

The nation-state frequently employs the use of stigma and stereotypes when dealing with peripheral mobilisations. If they show signs of becoming a threat, it may instigate the emergence of centralist nationalisms adopting various forms. Sometimes they are authoritarian, or even military, in nature. In 1917 the "Catalan question" gave rise to the centralist corporatism of the *Juntas de Defensa* (Army defence militias) which were the forerunners to the dictatorship of Primo de Riviera. Also, Catalan and Basque demands provided a basis for the arguments of Spanish proto-fascism in the 1930s. José Primo de Riviera, son of the dictator and founder of the *Falange*, defined Spain as a "unit of destiny within that which is universal", while Calvo Sotelo preferred a "Red Spain to a broken Spain".

On other occasions the disintegration of a state, and often an empire, will bring about a re-composition of the elements of the old nationalism and the appearance of a new centralism which sets out to bring together the remnants of the wreck. The surge of Eurasianism in post-Soviet Russia constitutes a case in point.

The territory which the old Soviet Union (and the present-day Russian Federation, although to a lesser extent) encompasses is the largest and most mixed ethnic jigsaw puzzle on earth. There are one hundred and thirty different peoples living there who speak a range of languages: Indo-European (Slav, such as Russian, Germanic and Latin), Altaic (Turkish peoples), Uralic (Estonians), Caucasian (Georgians and Chechnyans), etc. Of these peoples, twenty-three of the nationalities have populations of more than one million, and sixty-eight of the small groups number have less than ten thousand members.

Soviet federalism was counteracted by the development of an opposing tendency that began with Stalin and culminated with Kruschev, Brezhnev and Gorbachov, and which claimed that in the USSR a Soviet people had emerged which had overcome nationalist divisions and had created a new type of man, the soviet man. There were four aspects to this theory which were, in practice, official: the compulsory intercommunication among all peoples in Russian, "the second language of each citizen"; the mobility of labour throughout Soviet territory; intellectual homogenisation through Soviet education and common culture, a result of the combination of Marxism-Leninism, and of a system which, it was said, guaranteed welfare for all its members; the promotion of Soviet patriotism which, in practice, was confused with Russian patriotism, and which was expressed, above all, through institutions such as the army.

However, the elements which formed Soviet-Russian patriotism started to fall part at the end of the eighties. The hegemony of the Russian language was contested in the Baltic countries and the Caucasus. At the same time, the mix of peoples was rejected by Turkestan and the Baltic countries in protests which were often incited by the old *aparatchiki* of the Communist Party, who viewed themselves as the newly emerging "national class". Growing awareness of the widespread economic disaster (precipitated by, among other things, the draconian conditions imposed by the West) unleashed disputes between wealthy and poor republics, which set the scene for the emergence of a new Russian isolationism. The Russophile tendency of the army was denounced everywhere, and the Ukraine began to demand a national army (Carrere D'Encausse, 1991).

The end of the Soviet regime, brought about by the failed coup d'état in August 1991, paradoxically put an end to the doctrine to which Gorbachov had remained loyal until the end, that of the "soviet people" as an active subject of patriotism. The Minsk Accord of December 1991 was the basis for the creation of a Community of Independent States (CIS). It was signed, significantly, by the three Slav Republics of Russia, the Ukraine and Belarus, which were intent on salvaging from the demands for a split, a space for economic and military relations among the former republics of the extinct Soviet Union.

Nevertheless, the Russians' feelings of humiliation which were born out of many factors fostered a Russian nationalism of a new type. These factors were: the comparative sense of economic injustice with the West; the repatriation, from 1990 onwards, of almost one million soldiers of the Soviet army who were quartered until then in the states of the Warsaw Pact and their impossible integration into Russian society; the move to the status of second class citizen suffered by many Russians in the newly independent states, especially in the Baltic Countries where the ethnic Russians formed the majority of unqualified workers; and the unanimously anti-Serbian and pro-Croat position of Western powers in the Balkan war.

Although the most well-known expressions of this new nationalism are aggressive and caricatured in nature, such as those aspects associated with the political movement led by Zirinovksi, the "serious" ideological basis of this new nationalism is to be found in the theory of Eurasianism. This is not a new theory as it goes back to the teachings of Prince Trubetzkoi in 1920s, but it has spread massively in the present day through both political and academic spheres.

With regard to the controversy between the pro-Westerners and the Slavophiles which has divided the Russian intelligentsia since the times of Peter the Great, Eurasianism takes the side, although from a critical standpoint, of the latter. Although this theory was fed at the outset by the anti-Bolshevism of its founders, all of whom were obliged to go into exile, in the present day it only rejects those aspects of Russian Marxism based on Western communism and cultural internationalism; in no way its elements of Russian Messianism. In fact, Eurasianism vindicates, in its own way, Genghis Khan and Ivan the Terrible as well as Stalin.

This theory puts forward a new paradigm for political and cultural development which takes into account the Euro-Asiatic nature of the Russian state. In this enormous territory which reaches from the Sea of Japan to the Black and Baltic Seas there would have been a civilisation made up of large and small ethnic groups and numerous religions, synthetic in nature, although its means of communication, it is claimed, would have been the Russian language and culture. Reconciliation between the Asiatic east and the European west of this enormous terrain and the fusion of its different parts could give rise to the development and blossoming of multi-coloured cultures which are nourished by community values, presumably of Russian origin, and opposed to the mass European and American pop cultures. Such cultural reconciliation would be incompatible with "hyper-ethnicism" (which is linked to Western pop civilisation); in fact it must surpass the harmful consequence of hyper-ethnicisms, which would be national egoism. The right to self-determination may only be granted in exceptional cases, the most deserving of solutions being a federalism based on territorial autonomy and full recognition of the cultural expression of ethnic groups. In practice, Eurasianism provides the ideological cement which favours Russian predominance in the CIS, helping, above all, to maintain the territorial integrity of the contemporary Russian Federation (Titarenko, 1994).

Centralist nationalisms may emerge in those territories of the state which are subjected to most peripheral pressure, and are restricted to their own region. Pro-British nationalism in the Protestant community of Northern Ireland has been organised into unionist political parties which have dislodged Labour and the Conservatives from the local political scene. When these local centralisms perceive themselves to be abandoned by the state, they may drift toward white terrorism, as was the case of the French colonists who were against the independence of Algeria, and out of whose protests emerged the OAS.

The periphery may, in its turn, be polycentric, either because of its location in areas of dense urban networks, such as the Po Valley, or on account of its insularity, as in the case of the Canary Islands. Like the state, the peripheries also have their centre and periphery, born out of economic, historical or cultural-linguistic differences which often arise in their extremes (as it happens, for example, in the Basque Country, Quebec and Scotland). These peripheries of the periphery may align themselves with the state centre against their common enemy, particularly when the periphery is institutionalised. This perspective facilitates an understanding of the appearance of parties such as the *Unión del Pueblo Navarro* at the end of the seventies, founded to prevent the integration of Navarre into the Basque autonomous Community, and the *Unión Alavesa* in 1990, which threatened to push for the segregation of Alava from the Basque Country if the Basque Autonomous Community initiated a process of self-determination.

Note

[1]   Races do not exist as biological categories, they are only collective representations. For this reason the ethnic group should not be confused with the racial group. The former has a social reality, while the latter does not. The racial group is the group which is socially defined as racial and, in this way, it may embrace several ethnic groups, or none (Simon, 1976).

7 The Transformation of the Conflict (1): Western Nationalisms and the Welfare State

Two processes, one at the global level (the generalisation of the welfare state) and the other unique to Western Europe (the construction of Europe), have, in the last fifty years, altered the terms of the centre-periphery conflict, shifting it from bipolarity to multi-polarity. This has modified the identifying features of the peripheral movements, their ideological content and their instrumental/organisational dimension (which has to act within a field of many competing players) as well as the international context, which has become one more structural factor in the conflict.

Ignoring the difference between state and civil society of the previous liberal phase, the welfare state was initially an agent of centralisation which encouraged a disregard towards the periphery. Later, however, the functional overburden of the centre, subsequently provoked the forced transfer of functions to lower echelons. This was accompanied by the strengthening of local and autonomous institutions which gave rise to a new political factor - the region. This ambiguous player can take on many forms due to the varied modes of administrative organisation and territorial division of power displayed by the respective states.

This new political entity gave rise to new types of regionalist movements. Regionalism and regionalisation are not, however, the same: regionalisation comes from above as a geographical frame for state intervention and as one layer of territorial regulation, whilst regionalism comes from below in the form of demands for self-government. Ultimately, these claims may at times reinforce, at others diverge from, those of the peripheral nationalisms, because frequently the territory of the region does not coincide with the historical/imagined space of the nationalism. All of this creates a conflictive field of play in which regions compete not only with the nation-state but also amongst themselves - the state acting as arbiter, and at times as third player aiding some players to the detriment of others.

The triple objective of the European construction (eliminating the possibility of a new conflagration stemming from Franco-German disputes, reinforcing the ideological separation from Eastern Europe and turning Europe into an economic power capable of competing with other world powers - primarily the US but also Japan) lay entirely outside the centre-periphery conflict. The creation of this super-state actor, however, has added to inter-state processes such as international commerce and peace-time military pacts in eroding the sovereignty of the nation-state. Since the late 1970s this erosion has been reinforced by the effects of regionalisations brought about by the surcharge of the welfare state. That is why, although European Community regional policy developed since that time has been somewhat timid and too complacent towards the states, it has made possible a multidimensional field of play in which regions, bypassing the frontiers of respective states, maintain inter-state relations - whether of a cross-border nature or not - and associate in groups at a European level. At times they have also acted in collaboration with the state to put pressure on European authorities, while at other times - especially since the end of the 1980s - they have worked together with those authorities to face their respective states.

This new orientation of the conflicts has not done away with the logic of the previous phase but has followed on from it. The territories, which have a new meaning in this situation, display a maximum of social capital where a strong nationalism organised into a system of political parties exists. (It has been especially historical nationalist parties which have taken on a particular imaginary conception of the EU as a 'Europe of the Peoples' which ignores the states). In addition, although decentralisation, promoted as much by the late welfare state as by the construction of Europe, has channelled - or re-orientated - many of the contradictions, acute conflicts continue to confront nation-states with concrete peripheries. The disputes in the Basque Country, Northern Ireland and Corsica bear witness to the harsh reality of some of these.

The Welfare State and the Two Phases of Regionalisation

The liberal state, guarantor of public order in the face of internal threats and of the survival of the nation in the face of external ones (De Blas, 1994), was not passive. Its definition as 'minimal' state arose from its support for the free functioning of market forces. The exclusion of workers, not only from political life but also from the consumer economy, the consequent capitalist crisis of overproduction, and the pressure from the organised labour movement - which intensified with the introduction of universal suffrage - provoked the transition to the phase of redistribution of social wealth.

The concept of the welfare state, therefore, gradually emerged which aimed to overcome the dichotomy between state and civil society and to fine-tune capitalism by assuming the responsibility for the welfare of citizens through intervention in the market economy (Cazorla, 1983). Since this action was concentrated in a number of specific areas such as health, education, housing and social security, it has also been referred to as the 'social state'.

After the experimental phase of the inter-war period, the welfare state achieved universal coverage in the west between the 1950s and the 1970s. Keynesianism, on which it was based, brought together the concerns of liberals for full employment and of socialists for the redistribution of wealth, and responded to the double premise that economic growth had to be achieved within the rules of the capitalist game, and that social welfare was necessary to guarantee social peace as well as maintain demand (Sanchez, 1996). This premise, which nourished the intellectual humus from which the theory of the end of ideologies was born, went almost unchallenged until the 1970s, criticism being inhibited by the real growth of western economies since the 1950s and by the extension of welfare to wider social strata.

The development of the welfare state forced central governments to rely on intermediate levels (the local sphere in the territorial hierarchy) in order to apply their policies (Meny, 1992), which led to the appearance of the concept of regional policy. According to Loughlin (1995), this constituted the territorial aspect of regional and city planning policies of the welfare state. Regional policy has two facets - economic and social. Its origin is found partly in economic planning, in the neo-Keynesian belief that the state's macroeconomic management will have beneficial effects in investment and will lead to job creation in less developed regions. It is also partly linked to the advanced social policy of the welfare state: what traditional social policy did for individuals or groups must be replicated by regional policy, with regard to social welfare and equality, for the less developed areas of the state.

At that time, however, attempts were being made to achieve these objectives through state intervention and centralisation. This explains the long neglect of the periphery from 1920 to 1970, and the negative view that the liberal conceptions, as well as the Marxist ones favourable to the welfare state, offered of it in those fifty years. This produced a new reaction in the periphery, the defence of traditional features of culture and identity in addition to the protest against marginalisation and, sometimes, against the prejudices that regional policy had provoked despite its stated aims.

The types of regional policy which were applied in the post-war period - some of a political type, the support that the state gave to underdeveloped regions, others economic, which tried to use otherwise inactive resources to increase state-wide production - were influenced by technocratic and depoliticised criteria (Keating, 1996). In the 1960s, the methods became more complex, growth poles were created and more detailed regional planning attempted to include local and regional elites in its administration.

Antagonistic viewpoints soon emerged regarding the priority of state as opposed to local planning. The state had increasing difficulty fulfilling its promises, international competition impeded its control over the territorial logic of the economy and increased the cost of the regional 'clientelist' system (the practice of securing votes through the promise of government posts or economic benefits). In a free market it is in the interests of the state to favour its most competitive sectors, but these are frequently in the centre, and always in historically developed zones. In addition, the geographical mobility of labour threatened local identity, and the spread of the media eroded ethnic cultures.[I]

At this time the term 'region' (which can have different meanings depending on whether it is interpreted by economists, anthropologists or geographers) became understood as a level of government and territory below the state level.[II] The terms 'regionalism' and 'regionalisation', however, which appeared to converge in the 1950s and 1960s, separated in the following decades. Regionalisation (the region considered from a state perspective) focused on the region according to a government definition within the context of its overall policies. It could lead towards centralisation and the exclusion of regional representatives, a policy predominant in western states between the 1950s and 1970s. The regions created by unitary states do not, from this perspective, respond to the logic of vertical devolution of power, but are rather territorial levels of administration lacking political power (Loughlin, 1995).

Regionalism, on the contrary, is an ideology and a political movement which advocates that the regions should control the centres of political, economic and social power within their territory, generally through the establishment of their own political, administrative and legislative institutions. In any one state, regionalism and regionalisation can diverge in one historical period - such as in France in the 1960s - or converge subsequently - France twenty years later. (The territory of one regionalism, however, may or may not coincide with that of an historic nationalism - even in this last case, both may reinforce or contradict each other).

At the beginning of the 1970s, however, the cycle of prosperity came to an end. The economic crisis (monetary and energy) became a crisis of the political model. Public spending increased uncontrollably, often as a consequence of extravagant policies which electoral campaigns and spectacle-based politics broughtabout, and it was found that spending and inflation were not mutually exclusive but that both could grow simultaneously.

The welfare state was subjected to a flood of criticism from different sources. Neo-liberals denounced the bureaucratic overburden and the consequent governability crisis (Crozier and Huntington, 1974), while the New Right criticised the inefficiency and low productivity of the model as well as the supposed denial of liberty and individual initiative that accompanied it (Hayek, 1960; Rawls, 1970). Other groups demanded a return to individualism and a minimal state (Nozick, 1974); neo-Marxists highlighted the fiscal crisis (O'Connor, 1973) and the crisis of legitimation (Offe, 1984; Habermas, 1978).

The nation-state was, therefore, forced to lighten the burden and transfer authority to lower levels of government. This not only modified its nature but gave a new meaning to both local power structures and those of sub-state level (regions, autonomous entities and federal states) created by the territorial redistribution of power. In effect, the state was subjected to a twofold pressure which eroded its sovereignty: parallel to the limiting effect of the consolidation of supra-state organisations - such as the European Union -, a delocalisation of power was created at lower levels which brought it closer to the citizens, increasing the possibility of bottom-up control. The devolution of authority over, for example, services, health and culture forced public policy to take account of territorial imbalances and pressures from below.

Post-industrial society (heralded by the appearance of new social movements) imposed flexibility, decentralised institutions and created a new definition of the individual; new inter-state actors, responding to challenges beyond the limits of national frontiers (international co-operation, the media, attacks against the environment), became obligatory interlocutors. National and local identities were developed, differences were revalued and it was agreed that "small is beautiful" and that "the sum of diversities enriches" (Pastor and Ribo, 1996).[III]

From the end of the 1960s to the mid-1980s, regionalist and nationalist movements proliferated. Was the nation-state of necessity weakened in this metamorphosis? It could also be strengthened, as the delegation of specific tasks to lower levels - especially responsibility for political failures - allowed it to dedicate itself with greater efficiency to the tasks it judged to be a priority. In addition, the transformation of the welfare state did not always translate into devolution of power and growth in grassroots democracy; in the 1980s in key western states (the UK, US and later France) a neo-liberal model was imposed in which a mercantilist model of public policy abandoned the social provisions of the previous period and created distrust of the public sector, developments compatible with a strong political authoritarianism (Loughlin, 1995).

Local Power and the Logic of Decentralisation

In this process of power transfer towards the base, the meaning of local power changed, becoming the foundation of participative democracy. Historically the nation-state has been built from a centre which penetrated into the periphery making it homogeneous (Rokkan, 1970); local power structures - although more in the continental European tradition than in the Anglo-Saxon one - were an instrument for the consolidation of central power. Marxist ideology as much as the liberal one assumed the centralist model of politics; local power was confused with the power of the state representatives (prefects, governors) who shared it with the "notables" (*caciques* in Spain) and set up clientelist relations with each other as in a multi-cellular hive. Although capable of exerting pressure on the centre, local power was characterised more by its external and marginal character (Mabileau, 1985).

The nature of local power was modified first through the election of local institutions by universal suffrage, later on by the consolidation of the welfare state. Its form was shaped by the conjunction of organised structures, some institutional, some political (local party organisations) and others social (unions, local interest groups) which participate in the relationship that local power maintains with the political regime in the form of channels of communication and mechanisms of exchange with the centre.

Their proximity to the interests of the population allow them to fulfil a systemic function of mediation between political and civil society; they articulate and aggregate horizontally the social demands of the individuals or groups taking part in local institutions (inputs) and direct these vertically towards government decision-making centres involved in the territorial distribution of resources (outputs). Thus, a division of labour operates in which the state takes charge of social investment while local institutions, acting as administrators of goods and services, take care of local consumption.

However, although dependence on the centre predominates at local level, since it is the first stage in the election of politicians and the channel of participation closest to the population, the possibility of adopting an attitude of resistance is also born there. The transfer of the state's responsibilities to local structures, indicative of the second phase of welfare state, reinforces the change of attitude. The delocalisation of politics which accompanies post-industrial society at the same time differentiates the local from the infra-local (neighbourhood groups, civil associations calling for an improvement in the quality of life or opposing environmentally unfriendly decisions by local or central authorities); this being the level where pressure is applied directly to local institutions or indirectly to central ones.

The regions created by the welfare state as a framework for territorial administration in their turn become an intermediate level between the local and the central; the systemic functioning of local structures is duplicated and diversified, and regional movements occasionally facilitate this.

The capacity for resistance at the local level grows in those areas where there are ethnic, linguistic or religious differences. An area with multiple belongings oscillating between conciliation and conflict thus arise; participative democracy and the displacement of power fluctuate between integration and opposition, between the systemic and the anti-systemic. In this way, as suggested by Meny (1992), "in almost all European nations decentralisation policies (at the regional, local and infra-local level) set out to reinforce the ability of local authorities to intervene. Thus, the evolution of western societies, far from tending towards greater centralisation (as predicted by Parsons (1968), Deutsch (1961) and the developmentalist school), is characterised by a distinct fragmentation or by a likewise increasing co-ordination.

The form of territorial distribution of state power has itself undergone profound change. This is not now linked to the origin of the nation-state (transformation of the absolute state into a unitary one, evolution of confederations towards federal states) nor is it derived from the solution sought to the crises of the unitary state in the form of federal or autonomous initiatives. The vertical control of power and the forms of territorial organisation which that initiates are a consequence of the transfer of state functions to sub-state units intended to reduce the burden of the former.

In the decentralisation phase of welfare, the unitary state has to create territorial units invested with their own political powers; the composite state fills the old territorial frameworks with a new content. These units then attain a regulatory (legal self-regulation), executive (self-government), and financial capacity (the latter being the real indicator of the territorial distribution of power, and also constituting the most efficient system of administrative function because of its closer relation to the costs it incurs, the greater capacity for innovation it allows and the better improvement of public services into which it is translated) (Ribo and Pastor, 1996).[IV]

The territorial distribution of power responds to democratic tendencies such as the proximity of the administration to the citizens and the will to establish vertical control over the functioning of the state. This facilitates the appearance of a pluralist culture which values diversity and difference and which focuses on shared and negotiated solutions to problems.[V]

In fact, however, the creation of - or transformation of these sub-state units (generically 'regions' as used in the community sphere) responds to three distinct logics which ultimately reflect the diverse relations of power between the centre and the periphery.

1. If the centre predominates, the overriding logic is that of the centralisation of the first phase of welfare. The regions are the supra-local sphere which sits above the local. They constitute, therefore, the framework for the regulation of territory decided at the centre and are the level which allows the state, by delegating to them the functions it can not deal with, to concentrate on its activities in the general sphere. The geographical boundaries of regionalisation are decided by the high state bureaucracy on the basis of technocratic criteria. The French regionalisation of 1972 maintained the lack of administrative personality for nationalities such as the Basque Country and Catalonia, fragmented others such as Brittany and Normandy and combined into one region historically heterogeneous territories; the creation of regions such as L'Ile de France and Nord-Pas-de-Calais led some writers to wonder sarcastically if one should speak of 'Francilian' or 'Nord-Pas-de-Calaisian' identities (Pourcher, 1992). Such a method of operation can arise in composite states - autonomous or even federal - when its territorial structures become mere formalities. In the centralising phase of welfare the central structures of the federation are strengthened to the detriment of those of the federated states, diminishing the role of the latter in the formation of the will of the centre; its activities become subject to intricate regulations, and calls for a greater degree of mutual responsibility between the federation and member states conceal in reality new forms of dependence where the technical and administrative role of the states exceeds the political.

2. The decentralising phase can lead to a pluralist logic of 'shared allegiances' which generates an attitude of respect towards diversity, grassroots democracy and joint decision-making, elements which define a 'federal culture'. This can even happen - albeit with greater difficulty - in non-federal composite nations. In these cases it is expressed in the organisation at federal level of political and social players from the state sphere - such as unions, associations and social movements. The existence of a party-system with a federal structure is the best guarantee of this sort of culture. When the step is made from unitary to composite state, regionalist parties and movements frequently appear in the new sub-state units which defend the interests of their territories and aspire to positions in the new institutions. When the Spanish "state of the autonomies" was established in the late 1970s, parties of this type (some as self-declared regionalists, others not) arose in the Autonomous Communities of Navarre, Aragon, Valencia, the Balearic Islands, Andalusia and the Canary Islands. These parties differ in many respects from the nationalist ones (in fact nothing disgusts the latter more than to hear themselves referred to as regionalist). They are systemic, assuming

explicitly that their region is part of the nation-state; they do not question the form of territorial organisation of the state (as many nationalist parties would) but are created to provide content within it in their respective regions. They also lack the long period of maturation of identity and culture of the nationalist movements (notwithstanding that new regional identities may develop, which to this end disinter historic memories that can show a past of equal consistency, or on occasion superior, to those of some of the historic nationalisms). Their unitary nature derives, then, from their predominantly instrumental dimension which is not conducive to divisions motivated by criteria of class or ideology.

3. The third logic, that of conflict, arises usually in those areas in which in the long or medium-term strong ethno-national movements have developed led by nationalist parties organised into their own party-systems. Regionalisation, in the sense of transfer of functions, may strengthen nationalism, but it can also enter into conflict with its dimension of identity. The state may superimpose its sub-state units on ethno-national territories; but it may also ignore the nationalist territorial aspirations, using to this end the demands of the peripheries of the periphery. The social cohesion maintained by nationalism, as well as the social capital it accumulates, imbue the formal powers of the territorial institutions with real content. It is rare, however, that nationalist parties are satisfied with the formula for the territorial distribution of state power. When they make demands that exceed that, the centre tends to rely on the institutions of other regions, as well as on regionalist parties and the regional sections of state-wide parties, to oppose them, encouraging the comparative grievances which such demands awake in others. Suspicion reaches its height when the nationalist parties express their dissatisfaction with the position of their nationality within the state, making demands for self-determination and, eventually, having them approved by representative territorial organs.[VI] The attitudes of systemic and anti-systemic nationalist parties are, however, different. The former tend to accommodate their programme to the legal possibilities of action which the territorial organisation of the state provides, and where they do not coincide they concentrate their activities in the part of the ethnic territory within the official region. (The contradictions generated may at times be resolved where a nationalist party has acceded to institutional power through a division of labour in which the nationalist executive accommodates itself to systemic logic whilst the party sets itself up as the guardian of the purity of identity). Anti-systemic parties, in contrast, express their rejection of regionalisation and add the accusation of regionalism to that of collaborationism and betrayal levied at the first group.

Regions in Europe

Regionalisation in the nation-states of Western Europe has assumed various forms. Historical territorial organisations superimposed themselves on those created by the welfare state, which in turn operated in two time zones: the slow period of the 1960s and 1970s and the faster period which ran from the end of the 1970s until the present day (coinciding with the expansion of EU Regional Policy). However, the formal structures of territorial organisation are not the whole story; similar structures may conceal very differing power relations among political actors, regionalist, nationalist, peripheral, centralist, both among themselves and between themselves and the state.

Federal Germany organised itself, by virtue of Article 20 of its 1949 law, into fourteen member states (*Länder*) and two cities. Its federal nature is mainly due to pressure on the part of the winning powers of World War II. A chamber of representatives of the *Länder*, the *Bundesrat*, was set up as the means whereby representatives participated in the central legislature and administration together with the organs of the federation, the government and the federal parliament (*Bundestag*). The *Länders* are responsible for all power not directly attributed to the federation by the constitution. Each *Land* has a parliament, an executive government and a leader of government.

The territorial organisation of Belgium (a state formed by the juxtaposition of the Walloon and Flemish communities) tends toward two-tiered federalisation, which is unique in Europe. This process has undergone three phases. In the first phase, between 1970 and 1971, three linguistic communities were established (Flemish, Walloon and German), together with three economic-administrative regions (Flanders, Wallonia and Brussels). In the second phase, between 1980 and 1983, communities and regions acquired normative, executive and financial powers. The third phase, of a federal nature, was initiated with the 1988 Reform, which set up direct elections to the councils of communities and regions, whose members had been, up until that point, representatives of the National Assembly. These were then granted broader powers, and external representation.

Spain is, by virtue of its 1978 Constitution, a unitary state, based on the "indivisible unity of the Spanish nation," and at the same time autonomous, conceding to its nationalities and regions the right to autonomy. The criteria which differentiate one from the other, having voted on Autonomy Statutes during the Second Republic - a status granted only to the Basque Country, Catalonia and Galicia - gave these nationalities a more direct route to autonomy. However, the criteria are not rigid; in its day Andalusia took part in the direct route, while other regions are in the process of becoming nationalities through stands taken by their own parliaments. In fact, when nationalities and regions organised themselves as Autonomous Communities the law set out that their powers would become equal within the space of five years (although, in reality, differences have continued to exist among them). The juridical basis of the seventeen Autonomous Communities is constituted in their Statutes of

Autonomy; each one has a legislative assembly, an autonomous government, a president elected by the assembly and its own judicial power, albeit integrated into the state justice system.

France, a model unitary state, experienced the first attempt at regional reform as a result of a proposal made by General De Gaulle in 1969, but it was rejected by plebiscite. The 1972 Pompidou law regulating the organisation of regions was based on centralist criteria of a techno-bureaucratic arrangement. Each region had a Council made up of the Deputies and Senators elected in the constituency and by local representatives; a Regional Prefect appointed by the government, and a social and economic committee of a consultative nature. The decentralising reforms carried out by Mitterand's government in 1982 made the twenty-six regions, three of which were categorised as special regimes, Corsica, the Isle of France and the Departments of *Outre-mer*, territorial collectives, each assigned their own powers. Each had an elected regional council without legislative power, but which had decision-making powers on regional budgets; except for Corsica, which by virtue of its particular statute had a parliament with normative functions, a president of the council, elected by its members, and an economic and social committee with restricted powers.

The 1947 Constitution made Italy into a model of a regional state. Two types of legally differentiated regions were constituted in the Italian state. Five of the regions have a Special Statute, four of which (Sicily, Sardinia, the Aosta Valley and Trentino-Upper Adige) were established in 1948 and the fifth (Friuli-Venice Giulia) since 1964. The remaining fifteen regions, assigned ordinary status, had to wait twenty years until their respective statutes were implemented, as elections to the regional councils did not take place until 1970. The state is the source of power of the regions; each of which has an elected regional council without legislative powers (except for the Special Statute regions) and a regional board elected by the council. The government commissary co-ordinates regional administration in conjunction with the centre. Present political instability has created expectations of reform which have yet to be realised.

The United Kingdom is a unitary state, albeit of a weak nature (which has allowed for the autonomous functioning of peripheral institutions of civil society). Nationalist electoral gains brought about the approval of laws in 1978 for the devolution of power to Scotland and Wales. By virtue of this legislation, Scotland would have had its own assembly, not a parliament, with specific powers, an executive appointed by the assembly and a leader of the executive; a Secretary of State would be responsible for co-ordination. In Wales, the assembly would have had consultative and executive, but not legislative, powers. Financial powers were not planned for either assembly. The legislation was not applied, as it was rejected in the referendums held in March 1979. The Scottish, Welsh and Northern Ireland (a territory governed by London since Direct Rule was instituted in 1972) Offices acted as government representatives, having been assigned multiple powers. Decentralised administration has not taken into account ethnic divisions: there are fifty-six regions with planning and economic powers that are made up of twenty-nine

English and three Welsh counties, twelve regions in Scotland (nine in Great Britain and three in the islands); and the province of Northern Ireland.

In Portugal, which is a mono-cultural state, the 1976 Constitution set out the creation of regions, but this mandate was only applied in the case of the Azores and Madeira. Both regions have a regional (non-legislative) assembly, a regional commission and a regional council. The other five regions draw the territorial limits of development plans.

Denmark, like Portugal, is a mono-ethnic state, although its centralising nature goes along with strong administrative structures in the fourteen territorial collectives (*Amstkomuner*), which have an elected assembly and a chief executive of the administration. Greenland and the Feroe Isles, which each have 50,000 inhabitants, have far-reaching autonomy which resembles quasi-independence. These territories do not belong the European Union.

In the 19th century, Holland unified its historically confederated structures. Although power has not been distributed on a territorial basis, local institutions are very powerful. From 1962 onwards, twelve provinces have been operating with an executive council and a commissary, as well as supra-local regions of a collective statute. Both these institutions have powers with regard to territorial arrangements and local management.

Ireland is one only region based on local organisation: twenty-six counties, each of which having an elected council and a manager responsible for co-ordinating local institutions, and regions set up to receive funds from the EU Regional policy.

Greece is the most centralised state in the European Union. Its regions are organs of decentralisation, without juridical nature, as their president is appointed by the state (Castro, 1992).

Nationalisms and Regionalisation

The decentralisation and regionalisation inherent in the welfare state have modified the nature of peripheral movements in terms of their identity, ideology, organisation, configuration of their international context and link with ethnic territory.

Identity reactions may assume three forms. If the centre is strong and the periphery is weak, the orientation and territorial boundaries of the regionalisation are decided by the state bureaucracy, and if the state has managed to integrate peripheral elites into the regionalisation process in the initial phase of the identity reaction, then the political relevance of the ethno-national identity may diminish with respect to the stability of the state. During the course of this process, which usually develops in tandem with the stagnation and formal "picturesqueness" of ethnic elements, ethno-national groups, or important sectors of these groups, come to accept the reality of their situation in the political regime. They are satisfied with the powers devolved and with the institutions granted to their region; they accept the state as their legitimate representative, and consider any process of hostile acculturation harmful to their economic, social and political development.

Such cases, which even in their most extreme versions do not lead to a loss of ethnicity, are uncommon. The norm is the survival of ethno-national identities which structure themselves, in their relations with the centre, along a continuum which ranges from conciliation to conflict. If the former predominates, a system of alliance will be formed; if the latter is the case, a system of conflict will prevail (although, if the ethno-national movement is strong and diverse in nature, its different tendencies will become absorbed into one or other system).

The system of alliance is fostered by the plural culture and federal mentality. Respect for the original sovereignty of the federated parts and their co-participation in the formation of the will of the centre corresponds, on the identity level, to a concentric identity, the different spheres of which reciprocally enrich each other. (It is rare to find this ideal-type in a pure state; it almost always coexists with situations of tension and often requires, in order to be able to maintain itself, the location of the state in supra-state constructions which counteract the bi-polarity inherent in centre-periphery relations.)

In effect, even though the welfare state makes alliances possible, it also multiplies situations of conflict. The construction of the national community by the centre does not inevitably lead to conflict with the periphery: state-national identity does not forcibly exclude sub-state identities, as it may accommodate the existence of dual or multi-dimensional identities; and the transfer of power on a regional scale facilitates the establishment of plural linguistic-cultural education systems.

On the other hand, the construction of society by the state is rife with new motives for conflict. The state does not now distance itself, as it did during its liberal phase, from the territorial dimension of economic development. On becoming a planner and manager, its own decisions can be considered responsible for the degree of development in ethnic territories. When the result of intervention, or of passivity, on the part of the state is the peripheralisation and marginalisation of ethnic groups, the latter will accuse the state of "internal colonialism." However, a high degree of development may also provoke hostile reactions when peripheral elites believe that they have not been given their due position in the leadership of the development process, or when the elites aspire to the formation of their own "techno-structure" which would be autonomous with respect to the state.

One case in particular that represents a deviation of the welfare state, as is the case in Italy, for example, is where the political centre, which does not coincide with the economic centre, transfers resources to less well-developed regions with the aim, not of promoting their development, but instead of creating clientelist networks which guarantee permanent electoral support for the parties of government. This eventually brings about mobilisations in the more developed regions, which consider themselves to be damaged by this policy.

The lack of coincidence between a planned territory and an ethno-national space may arise from the planning rationale of the centre, but also be due to a conscious effort to counter powerful nationalisms by breaking-up its historical-cultural territory. This gives rise to democratic reactions when the ethno-national group is the majority in the segregated territories, or to irredentist reactions when the group is a minority. Reactions are the more intense the more anti-system the nationalism is which drives them; while systemic nationalisms, on the other hand, tend to adapt themselves to the reality of new regional territories.

Likewise, the ideology of nationalist movements, or, to be even more precise, the *gestalt* that shapes together ideology and legitimating mechanisms, changes. In the liberal phase, the rational-type legitimation resembled monopoly of the centre, since it was only considered able to construct society. In the decentralising phase of welfare, the political ethno-national agents may carry out, if they control regional power, certain types of societal functions: political, the integration and cohesion of their society; economic, intervention in their own development; and cultural, the promotion of their own particular education system. A new ideology emerges in systemic nationalisms, "welfare nationalism", ideology cloaked in an identity cultural-historic base.

It is, above all, in the instrumental dimension where the most important changes take place, which tends to occupy the space that the ideological-identity dimension previously held. This dimension serves as a frame which mirror-images (with varying levels, according to the degree of regionalisation of each state) the nation-state's work of construction (albeit in a subordinate and not integral mode). For this reason, the creation of a system of systemic and anti-system peripheral nationalisms is only possible during the welfare phase; for the former, the instrumental dimension is their main raison d'être, while the latter intensify their ideological-identity dimension.

The international context determines the feasibility of demands such as self-determination or independence. However, although systemic nationalisms do not pragmatically reject these aspirations, almost all no longer consider them to be viable in western Europe, at least in the short or medium term. Declarations made in territorial parliaments proclaiming the right to self-determination do not call for their application.

However, the international context has now become "intra-national." In the welfare state decentralisation is not restricted to historic nationalities, as was the case during the Second Spanish Republic, instead they include the territory of the state in its entirety. Relations between the centre and ethnic peripheries are mediated by the other regions, which, on occasion, assume the role which agents in an international context would have. In theory, relations among regions should be presided by the principle of solidarity; state constitutions give this principle a financial scope which is rooted in the establishment of Funds for Inter-territorial Compensation. Rarely is it the case anyway that the economic solidarity required springs forth spontaneously from within the respective states (notwithstanding the fact that many demands for inter-regional economic solidarity ignore the obvious fact that such solidarity requires a resource capacity of the regions, without which it is meaningless (Ribó and Pastor, 1996). In this way, the state becomes the arbitrator of numerous inter-regional conflicts (caused by many different reasons, such as the passage of water deposits from one region to another); which then places restrictions on the self-government of all of them.

Being very diverse in nature, from simple frameworks of administrative delegation to historical political nationalities, the great differences among regions generate a range of needs, which, irrespective of the way in which power has been territorially devolved, requires asymmetric decentralisation. Conflicts between the state and systemic nationalisms which are in regional power are set out in this phase in these terms. The centre responds to the asymmetrical demands of those nationalisms by defending the logic of equal distribution of authority to regions which are legally equal. When the ethnic periphery replies by pointing out the unjust nature of the formal egalitarianism of "coffee for all", the centre mobilises in its favour the feeling of "comparative injustice" which the demands provoke in the remaining regions.

Nationalisms and the Welfare State: Case Studies

Progression from the centralising phase of welfare to its decentralising phase modifies, in its turn, centre-periphery conflicts. The state territorial intervention of the first phase, which was controlling and bureaucratic, gave rise to reactions among some peripheral elites, who demanded participation as fully-fledged partners in the development of their territories (although these demands were not expressed as nationalist in nature). It was in such a context that in 1949 the Committee for Studies and Links of Breton Interests (CELIB) was established; likewise the 29th November Movement was formed in Corsica in 1959 in response to Gaullist attempts to create a centre for nuclear tests on the coast and to close down the only railway on the island.

During the 1960s and 1970s, the welfare state aroused in other peripheries the desire to be owners of their resources. In Scotland, plans to nationalise oil in the North Sea fostered a rise in nationalism. At the same time, in Québec, nationalism was impelled by the determination of a techno-structure formed by politicians and technicians to manage their own welfare in a regime of sovereign-association with the Canadian state.

The restrictions in social services, brought about by criticisms of the New Right of the welfare state, have radicalised the protests of some peripheries, which considered cutbacks as an attack directed against their development. It is worth noting again that in Scotland widespread opposition to the neo-liberal and authoritarian politics of the Thatcher government drew into the camp of nationalist demands political forces which were not nationalist, such as the Scottish Labour Party.

Paradoxically, some peripheral reactions have aligned themselves with neo-liberal logic. This is the case in Northern Italy, where rejection of the clientelist corruption of the welfare state has led the Northern League to cloak its demands for self-government in anti-corruption terms.[VII]

In Scotland, the expansion of British authority from the end of the 19th century onward (in 1885 the post of Secretary of State for Scotland was created, which was directly dependent on London, being transformed after World War I into the Scottish Office) was the cause of growing protest. The situation was exacerbated by the fact that after the Great War, and once again after 1950, Scottish industry went into decline which raised the rate of unemployment above the British figure. Although the relative difference with respect to the economic decline of the United Kingdom on the whole was not dramatic (the average income per capita varied between 1920 and 1990 from 86 % to 96 % of the British) it did, however, act as the catalyst for a series of campaigns in favour of greater self-government and constitutional change which have continued right up until the present. During the inter-war period a Scottish nationalism emerged led, from 1930 onwards, by the Scottish National Party (SNP), which defended in the early years a nostalgic and parochial image of Scotland free from corruptive influences.

After the Second World War the British state became the source of subventions for Scotland and Wales and established agencies for development in both countries with a view to addressing the dissatisfaction. This, however, was not enough to prevent numerous Scots having doubts about the advantages of the Union, especially when the discovery of oil in the North Sea gave them the hope of leaving behind their "poor British" image becoming "wealthy Scots." Consequently, the demands for self-government became increasingly urgent in the sixties and seventies, in tandem with the electoral successes of the SNP (which obtained 30 % of the votes in the 1974 election). The Labour Party, the hegemonic party in Scotland, where it has the electoral support of a nationalist sector, devised a plan to prevent constitutional reforms, the Devolution of Power, which would re-establish the Scottish and Welsh parliaments, assigning legislative powers to the former. The furious opposition of the Thatcherites and of a section of the Labour Party in England made it necessary to hold a referendum in 1979, which was defeated in Wales but approved in Scotland by 53 % of the voters (this figure did not, however, reach the pre-established majority of 40 % of the electorate). Callaghan's Labour government therefore refused to grant the Devolution of Power to Scotland, for which reason, among others, Labour was defeated by the Conservatives at the end of 1979.

If calls for self-government began to fade at the beginning of the 1980s, reflected in the electoral decline of the SNP, then the hyper-centralist and anti-Scottish policy of the Margaret Thatcher era (for whom the Devolution of Powers represented a step toward independence) helped recover lost ground and even led to the Scottish Labour Party assuming a strong nationalist position. Conservatives related the underdevelopment of peripheral nations to collectivising attitudes of the post-war years, which demanded a response of greater privatisation and cutbacks in the public sector; this policy led to the loss of one fifth of workplaces in Scotland. Moreover, Conservative attacks on the autonomous institutions of civil society and local authorities were particularly resented in Scotland, where a *de facto* Home Rule operated. For this reason, Scottish identity became much stronger (opinion polls in 1991 indicated that 40 % of those interviewed felt themselves to be only Scottish, and 29 % felt more Scottish than British). The population, including Labour supporters, began to question the right of the Conservatives to govern in Scotland, where they had obtained the support of a small minority (about 20 % in the 1980s and 1990s). From 1987 onwards a Scottish National Convention, led by members of the Labour Party, campaigned for the enactment of the Devolution of Powers by Westminster. In the 1992 general election, the bloc of parties which supported the campaign obtained fifty-eight of seventy-two Scottish seats; in September 1997, the constitution of the Scottish parliament was passed.

This explains why growing national consciousness in Scotland has been reflected more in the strengthening of Labour Party hegemony than in an increase in SNP votes which, in the current decade, remain at about 20 %. In any case, Scottish nationalism has changed its nature with respect to its origins. It is now more open and progressive, and its ideology is left-wing socialism and a modernising factor (Keating, 1994 and 1996; Penrose, 1994; Moreno, 1995).

In Quebec the new nationalism (which substitutes the idea of the French-speaking Catholic nation for that of the sovereign *Québecois* nation-state) is the result of the "quiet revolution" which initiated in the 1970s the entry of Québec in the phase of welfare. The governments of National Union had supported, until 1960, the idea of an agricultural, Catholic and liberal Quebec as against the interventionism of the Canadian social state. However, a rapid process of industrialisation and monopolisation brought about profound social change. Local and regional businesses were absorbed by large Canadian enterprises; new requirements produced new classes of French-speaking technicians, administrators and professionals who were nevertheless marginalized, both in the private sector and public administration, by English-speaking staff. When liberalism entered into crisis, state intervention was perceived to be a solution, but not managed from Ottawa, rather it was to be directed by sovereign structures of self-government. In addition, the 1964 reforms supported trade union negotiation of a Keynesian type with public authorities; the trade unions wanted to establish in Quebec, but not in Canada, their own intermediaries and framework for labour relations.

Likewise, the intellectual hegemony of the church began to fragment, under a two-tiered offensive. On the one hand, catholic morality found itself under attack by the new hedonism and secularisation, in tandem with a cultural explosion; on the other, social conservatism came under attack by socialist and national liberation ideologies, those being the years of the Cuban and Algerian revolutions and of anti-colonialist advances.

The economic base which sustained the movement was the nationalisation (in Quebec) of enterprises such as electricity, as well as the creation of public companies (the Credit Unions set up by the Desjardin Movement, the Industrial Development Society of Quebec, the General Finance Society of Quebec) which supported self-financing. This process led to the establishment of links between state-owned business, small and medium-sized privately owned firms, bank capital and trade unions. Owing to this, even though the nationalist parties which had emerged since 1960 reflected the tensions rooted in their varied and opposing elements (independence, anti-imperialism and interventionism), their hegemonic nucleus was dominated by technocrats trained in the economic sectors of the administration in Quebec. Their spokespersons, originating in the modernist wing of liberalism, were René Levesque, minister architect behind the nationalisation of energy and father of the "Quiet Revolution", and Jacques Parizeau, economist of the bourgeoisie and ministerial consultant. His stance in favour of the sovereignty of Quebec was, for this reason, comparable to his defence of economic association with Canada (with its dollar, its banking system and, above all, its market).

The sixties decade saw the coming to maturity of the *Parti Québecois* (PQ) through the synthesis of pre-existing various nationalist forces. Some of those which arose in 1960 converged in 1966 in the *Ralliement Nacional* (RN), a traditional and Christian-Democratic party. The *Rassemblement pour l'Indépendence Nationale* (RIN) which supported mass mobilisations, was the closest to the doctrines of socialism and decolonisation and to the independentist project (although it did not exclude association with Canada once sovereignty had been won). Out of some of its tendencies the Quebec Liberation Front (FLQ) arose. In 1970, this armed organisation kidnapped a British diplomat and killed a regional minister; after the military occupation of Québec, ordered by the government of Trudeau, the FLQ effectively disappeared.

The group led by Levesque, which tried but failed to impose on the Liberal Party the idea of a common-market between Canada and Quebec, became more radical in 1968 when the liberal Pierre Trudeau took power in Ottawa; Levesque founded the Movement for Sovereignty-Association (MSA). Months later, out of the merger of the MSA and RN, the *Parti Québecois* (PQ) appeared, with Levesque as its president. The RIN, whose leftist leanings Levesque did not trust, chose to disband itself a few days after the merger and its leadership asked its 14,000 activists to give their personal loyalty to the PQ.

The *Parti Québecois*, with a membership of 50,000 in 1980, signalled the end of the pluri-party approach of the 1970s and the substitution of past mass mobilisations for electoral and parliamentary political battles. Its manifesto, which supported the creation of a de-centralised and democratic system that would be favourable to the workers, advocated the sovereignty of Quebec and association with the rest of Canada. At the same time its slogan "build-up Quebec" envisaged the centralisation for Quebec state apparatus and the promotion of a national bourgeoisie, fully supporting the PME and local corporatism. The party had excellent relations with the Union of Agricultural Workers and the main working-class trade union in the country, the Federation of Workers of Quebec, as well as the network of Desjardin Credit Unions. In spite of efforts made to win over the PME, the relationship between the party and employers remained cool. Its social base was French-speaking and mainly young, with strong representation from the education and university sectors. The PQ was not, however, immune from tensions between the technocrats and small party cliques, between moderate and left-wingers and between gradualists and independentists. The party won the 1976 elections and repeated its victory in 1981.

The politics of the liberal Canadian Prime Minister, Pierre Trudeau, who was of Québecois origin, and held power in Ottawa, with a brief interval, from 1967 to 1984, were defined in the relationship with Levesque's PQ, in the same way that the PQ's politics were defined in relation to him. Trudeau, of a French and English-speaking family, had the achievement of autonomy of Canada from the United Kingdom as his political goal (although during his long years of power he only managed to substitute the British for the American way of life). His dream, on a cultural level, of a coast-to-coast bi-lingual Canada, was in conformity with his image and likeness. His bi-lingual policy, initiated in 1969, which supported the French-speaking minorities in Canada and the English speakers in Quebec, opposed the "PQist" idea of Quebec as the "main focus" of French-speakers. His recognition in 1971 of multi-culturalism and of the equality of the different provinces contested PQ's idea of the "two founding peoples" (which the "First Nation" Indian peoples also opposed).

However, his liberal perspective of linguistic rights as the individual rights of citizens contradicted the experience of Quebec: the individualist laissez-faire policy applied to a French-speaking small island in the middle of an ocean of English-speaking mass-media and huge populations in Canada and the United States could not but otherwise lead to the assimilation of the French-speaking minority, as it had happened in Louisiana. Because of this the protection of the French language demanded its recognition as the collective right of a different society, that of Quebec.

Therefore, after the PQ assumed power, the government of Quebec decreed in 1977, among other laws, a Charter of the French language which declared it the only official language with the aim of addressing two situations: the systematic tendency of the children of immigrants to Quebec to opt for education through English; and also that in Montreal, a city of three million inhabitants, all signs and public notices were written in English.

The referendum on sovereignty-association held by the nationalist government of Quebec in May 1980 was lost, with only 40 % of votes in favour. Although Trudeau had promised reforms favourable to Quebec if this result emerged, the reality was that those passed were the opposite to what was promised. His government repatriated the Canadian Constitution (amendments to it depended until then on British approval); the Charter of Rights and Freedom applied in 1982 defended the rights of individuals belonging to minorities (French-speakers in Canada and English-speakers in Quebec), as well as the right to be free from discrimination and to study in one's native tongue. However, the Charter was used to block legislation in Quebec favourable to the French language (the upkeep of which forced its government to invoke the "Derogation Clause", but with the approval of the High Court and for a maximum period of five years). Quebec symbolically refused to sign the new 1982 Constitution to which it was, nevertheless, subject.

In this way two opposing concepts crystallised; on the one hand, there was Trudeau's vision, based on official bi-lingualism, the Charter of Rights, multi-culturalism, absolute equality among the provinces and the strengthening of state institutions, a concept inherited by Canadian conservatives. The other concept rested on the distinct character of Quebec, on bi-culturalism and asymmetrical powers; this concept emerged out of the heart of Quebecois nationalism and penetrated all those political parties with a base in Quebec, including the Liberal Party. It subsequently, or simultaneously, inspired the campaign for a separate statute for Quebec, for recognition of Quebec as a different society or calls for sovereignty.

The 1982 Charter of Rights, in contradiction with its individualist and egalitarian ethos, recognised in Article 35 the collective political rights of indigenous peoples. Quebec, which claimed to recognise the existence of "distinct" indigenous peoples and their right to self-government, pointed out in any event the contradiction between accepting this fact and perceiving of Quebec however as a collection of individuals. Provincial and federal conferences which, from 1983 to 1987, attempted to define the extent of recognised rights of native peoples, failed in their objective; because, among other reasons, native peoples believed that if Quebec's demands were met, then their own would be indefinitely postponed.

In the mid-eighties the world economic crisis weakened Keynesian economic policies. Mulroney's Conservative Party won the 1984 Canadian election thanks to votes in Quebec. During their nine years of government, the conservatives, at the same time as they dismantled the structures of social state, proposed the reintegration of Québec into the constitutional sphere. It was then in 1985 that Bourassa's liberals defeated in Québec Levesque's PQ, which led to the latter's resignation. Bourassa followed up on Mulroney's proposal and put forward five conditions for an agreement with Ottawa, among which was the recognition of Quebec as "a distinct society", thereby guaranteeing the benevolent neutrality of the nationalists.

In 1987, the Prime Ministers of the Canadian Provinces reached an agreement on a proposal for constitutional reform, the Lake Meech Agreement, which proclaimed Quebec as "a distinct society within Canada," its differences consisting in having a French-speaking majority and an English-speaking minority. The proposal had to be submitted to federal and provincial governments for unanimous approval. After facing numerous obstacles, it was finally dismissed in June 1990 when it was defeated in votes in Manitoba, a western province, and Newfoundland, a coastal province. The outcome caused deep disappointment in Quebec.

There were three main reasons for the failure of the initiative, the roots of which go back to the birth of the state of Canada: indigenous demands for self-government, regionalism of the non-central provinces and, finally, Canadian nationalism. The leader of the "no campaign" in Manitoba, Elijah Harper, was a chief of the Indian Cree people. His opposition was unanimously supported by half a million indigenous people who lived on Canadian territory and who refused to approve any agreement which helped Quebec while their own demands for self-government of First Nations remained unmet.

The "no" result in Newfoundland symbolised the tension between the centre and periphery. The economic crisis had given rise to regionalist demands in these provinces, particularly in those of the west, but also in the coastal regions which accused Ottawa of bias toward Ontario and Québec. During the 1980s these provinces demanded reform of the Senate which would give them an equal number of representatives; from this standpoint they opposed any constitutional reform aimed only at Québec. Finally, Canadian Anglophile nationalism, ever prevalent, which had already expressed itself through opposition to the legislation in Quebec on the French language, also rejected the initiative.

Nevertheless, the approval of Quebecois nationalism of the terms of the agreement was reversible; as would again happen in the 1990s. The term "distinct society," clearly avoiding the use of the word nation, which would have conferred on Quebec the right to self-determination in the international sphere, did not grant additional rights to already existing ones. Given that this definition arose out of recognition of the existence, unusual in Canada, of a French-speaking majority and an English-speaking minority, it may even be interpreted as being, in agreement with the 1982 Charter, against the linguistic interests of the French-speaking Quebecois population.

From this, the new and definitive failure of the conservatives of Ottawa and liberals of Quebec (who had repeated their electoral triumph of 1989) to reach a new agreement in 1992, that of Charlottetown. Its terms had been worked out by the federal and provincial prime ministers and aboriginal chiefs on the basis of three principles: recognition of the right to self-government of the First Nations, which should be negotiated within each native community; recognition of Québec as a distinct society within Canadian federalism, but without granting it new powers; and a willingness to reform the Senate with regard to giving equal representation to the provinces.

The agreement, rejected by Quebecois nationalists and indigenous communities, was the subject of a referendum on 26th October 1992, and defeated in a vote of 54 % of Canadian citizens; the result was clearly negative in Quebec and in the four provinces of the West and Nova Scotia.

This failure, in addition to the ultra liberal policy of the Conservatives, with its consequences reflected in an increased rate of unemployment, economic decline and the closure of businesses, radically impacted on the Canadian political scene and re-launched, particularly in Quebec, the idea of a social state, which had been associated from its beginnings with Quebecois nationalism.

In the federal elections of 1993 the Conservatives were wiped off the political map; they moved from holding the majority of seats to winning only two. Chrétien's liberals (who, like Trudeau, was of Quebecois origin) won; and Bouchard's nationalist Quebecois Bloc became the second largest party in Canada, with fifty-four representatives, ahead of the ultra right-wing Reform Party, which held fifty-two.

In the September 1994 elections in Quebec, under the leadership of the historic Jacques Parizeau, the Quebecois Party was triumphant (although only by a slim majority over the liberals). Once again, the economic-social alliance which had given its support to the PQ during the 1970s brought together state-owned business, trades unions and small and medium-sized firms. This alliance, which was now more solid than ever, presented as a viable and achievable option the goal of full sovereignty of Quebec, as against the proven failure of Canadian federalism to obtain a satisfactory arrangement.

In fulfilling its electoral promises, the PQ government scheduled a referendum on sovereignty for the 31st October 1995. With the aim of convincing the undecided, an alliance was agreed among the PQ, the Quebecois Bloc and a small former federalist party, the Democratic Alliance of Quebec, which once again introduced a gradualist perspective, provoking some disquiet among the pro-independence sectors of the PQ. Quebec proposed an association pact with Canada with a view to consolidating the existing economic sphere; if, within the period of one year, agreement was not reached the National Assembly of Quebec would then unilaterally proclaim its sovereignty. The referendum was lost by scarcely one % of the votes (Seiler, 1982; Bourque, 1995).

Italy was, together with Germany, the last of the great nation-states to establish itself. The popular nature of the struggle against the causes of its underdevelopment (the break-up of the city-states, Genoa, Florence, Venice, etc.; Italian control by the royal European Houses, the Austrian empire in the north, the Spanish Austrians, later succeeded by a lateral branch of the Bourbons, in the south; the remains of the medieval dream of a politically unified Christianity which were, above all, the Vatican states) explains the consensus behind national unity for an entire century. Nevertheless, Italy is, of all today's states in the European Union, the country under greatest pressure to initiate constitutional change of a federalising nature.

In the mid 19th century, the drive towards unification in Italy, culminated in 1870, originated in the north, in the kingdom of Sardinia-Piedmont; meanwhile Garibaldi's troops aligned themselves with the elites in the south of the country in order to put an end to the kingdom of the Two Sicilies (Naples and Palermo). Support for national unity among Italian intellectuals (including Gramsci), inheritors of Machiavelli in this respect, had overshadowed the extreme polarisation of the new state. The three geographic zones of Italy, the fertile plains of Po in the north, at the foothills of the Alps and Dolomites, the long boot-shaped peninsula spanned by the Apennines, and the two large Mediterranean islands of Sardinia and Sicily, do not coincide, except in the north, with the three large historical-political spaces of Italy which are: the continental Italy of Milan, Genoa and Turin, industrially developed and united by the Lotharingian axis to the spine of Europe; the Tuscan centre and Lazio, birthplace of the Italian language, polarised by the bureaucracy in Rome; and the south (the Mezoggiorno, Sicily and Sardinia), under-developed and dominated by large land-owning families, with clan systems of social organisation which have tended toward, on account of stagnation, illegal activities (the Neapolitan "Camorra" and Sicilian Mafia).

The ethno-cultural map is even more varied. In the mountainous valleys of the north there are two linguistic minorities: French-speaking in the Valley of Aosta (200,000 inhabitants), brought to Italy by the Kingdom of Piedmont; and the German-speaking south of the Tyrol, a territory under dispute with Austria and annexed by Italy in 1919, included, together with the Italian-speaking group from Trent and the *ladina* ethnic minority, in the region of Trentino Upper Adige (1.5 million inhabitants). The population of Friuli, which is part of the Friuli-Venice-Giulia region (2 million inhabitants) speak their own language, Frulian; in Sardinia (with a population of 2.8 million) where nationalist movements have existed since the 1920s, the language spoken is Sardo. Finally, Sicily, a large historic island with five million inhabitants, experienced a strong but short-lasting independence movement after the second world war. Lombard, Veneto and Piedmont have occasionally claimed linguistic difference with regard to Italian.

The ideological division existing since the end of the 19th century between Catholic and socialist-communist sub-cultures had hidden the territorial dimension. As a result of the defeat of centralising fascism, the Italian Constitution set out a regional state with eight regions in the north, four in the centre and six in the south, in addition to the two islands. Five of these regions, border regions characterised by ethno-nationalist movements, were awarded Special Statutes.

In the first elections of 1949, nationalist organisations obtained strong support. The Sicilian Independence Movement (MSI) won 8.8 % of the votes, in Sardinia the *Partito Sardo d'Azione* obtained 10.5 %; in the south Tyrolean region of Bozen (Bolzano in Italian) the *Sud Tyroler-Volkspartei* (STV) obtained 67.6 %; in 1954 the Valdotaine Union won 29 % of the votes in the Aosta Valley. However, these movements, except for those in mountainous regions of the north, lost strength in the following thirty years. In Bozen, a budding segregationist movement which started in 1961 and carried out sporadic armed activities, disbanded in 1970 with the concession of Special Regime for South Tyrol in the heart of the region of Upper-Trentino Adige; the STV, the hegemonic south Tyrolean party is today a moderate force in a consociational agreement with Italian-speaking and *ladino* groups. The Valdotaine Union in the Aosta Valley, split between right and left, has become a catalyst for federalist ideas throughout Italy.

Meanwhile in the south, as has already been pointed out, Sicilian nationalism disappeared and Sardinian nationalism faded for decades. On the other hand, in the industrialised north during the 1980s strong de-integrating movements emerged which inverted the thesis of "internal colonialism."

From the inception of the Italian state, the political and economic elite in the north had ensured that there would be no political party representing the specific interests of the south. However, the establishment of the welfare state after the second world war dramatically displaced the geographical centre of hegemony. Successive Italian governments attempted to deal with growing differences in employment and production capability between the north and south through extraordinary measures of state intervention, which led to a high degree of bureaucratisation of the middle-classes in the south. This redistribution process guaranteed the markets of the south for large Italian industries but at the price of enormous fiscal demands, which led to protests by small businesses in the north. In tandem with this process, the growing presence of the south in the Italian political apparatus was gradually changing it into the political centre of the state; the state parties became in the southern regions clientelist forces which rewarded loyalty by directing the resources of the centre toward them. This explains in part both the clandestine links between large Italian political parties and the Mafia organisations of the south; also, the decline of southern pro-autonomy parties, which were increasingly unnecessary to obtain assistance.

At the same time, the evolution and growth of the Northern Leagues in the past decade took place due to the crisis of the "partycracy" of the centre and the south; its support base was made up of professionals, skilled workers and small and medium businesses who identified the welfare state with corruption and the "bottomless pit of the south" into which funds were being channelled. Ethnic mobilisations at the beginning of the 1980s in the north had common features such as the discovery of their own culture, protests against immigration from the south and anti-centralist campaigns. The Venetian League emerged in 1980 from a study group of the Venetian language; the Piedmont and Lombard Leagues emerged later, in 1985. While weak at the outset, the Lombard League, under the leadership of Bossi, grew inordinately toward the end of the 1980s (in the years when, at the same time, the regional policy credibility of the European Union was growing, thereby reinforcing the importance of the Lotharingian axis). The different Leagues (in Piedmont, Veneto, Liguria, Tuscany and Emilia-Romagna) merged in 1992 into the Northern League. In the elections of that year the League obtained 20.5 % of the votes in Lombardy and in the other regions between 10 % and 17 %, figures which have grown in subsequent years.

The ideology of the Northern League has passed through four phases. In the first phase emphasis was placed on ethnic elements, including language as the basis of identity, which would be under threat from the southerners engaged in "internal colonialism" due to their demographic predominance and support from the state bureaucracy (the racism implicit in this thesis was, in any case, challenged by the leadership of the League). In the second phase, that of electoral success, the ethnic discourse opened up into a position which prized national identity on a territorial basis of a "Padania" arising out of the federation of all the regions of the north, which encompassed all its inhabitants, including immigrants from the south. The third phase was initiated in 1994, the year in which the phenomenon of the "partycracy" broke up and in which the League formed an alliance with Força Italia, the Thatcherite Berlusconi party, who had been triumphant in the right-wing Freedom coalition, in opposition to the communist Progress coalition. The League proposed a federal structure for all of Italy which would be based on three states: Padania (in the north), Etruria (in the centre) and the south. Berlusconi's refusal, being handcuffed by the centralist National Alliance of the former fascists, to initiate any constitutional reform process with federal leanings, convinced the League, which had been electorally weakened, to break the alliance with Força Italia. This created a crisis which cost Berlusconi his position. The League, after making a provisional proposal in favour of an Italian Republic made up of nine states and twenty regions, in which all would enjoy wide-ranging fiscal autonomy, ended up proclaiming the independence of the Republic of Padania in 1996, which has made it into the main anti-system party in Italy (Strassoldo, 1992; Petrosino, 1994).

Notes

[I] A neo-Marxist current - Lafont (1976), Nairn (1979), Hechter (1975), Beiras (1995) - explains this situation based on the theory of 'internal colonialism'. Lafont proposes that it derives from a triple source: the disorder of capitalism, which takes into account competitive productivity over the interests of a population tied to a particular area; the deficiencies of the centralised state for which protectionism only makes sense when applied to the state in its entirety; and the passivity of peripheral representatives who fail adequately to defend their country. Hechter incorporates a fourth variable - dependent industrialisation, whereby profits generated in the periphery, instead of being redistributed in the region of production, are sent to the centre where decisions are made concerning the periphery.

[II] The diverse political and administrative situations of the states (especially within the EU) have necessitated definitions of the region which limit themselves to emphasising territorial continuity, homogeneity of the population and political representation of the regions. The Assembly of European Regions defines them as "entities situated at a level immediately below the central state, provided with political representation guaranteed by the existence of an elected Regional Council."

[III] These transformations of the welfare state generated new trends in political analysis. Some authors presented the state as one more collective player, involved with others in relations based on compromise, conciliation or competition (Schmitter, 1979). Others emphasised the reversal in relations between centre and periphery to the advantage of the latter (Mabileau, 1985) or indicated the new spread and horizontal reach of the political brought about by its extension to the masses (Sartori, 1992). The spread of power at the top brought about the establishment of a new branch of political science - public policy - to study the action of governments. Here, also, the catalyst for the appearance of this discipline was inflation and the feeling of a crisis of governability which accompanied the centralising aspect of the welfare state. Now there was no more talk of the technocratic revolution: the action of government became full of separate compartments, conflicts, compromises, impotent centres and autonomous peripheries. Thus, the study of public policy allowed an atomised outlook, highlighting the differences and peculiarities of a crumbling state (Thoenig, 1985).

IV According to these writers, the principle of financial distribution must not be confused with inter-territorial solidarity, which depends on a previous availability of resources to both parts. Neither should it be considered the principal instrument of social distribution of wealth which is independent of whether the state is unitary or composite.

V Pluralist culture may be taken on by the periphery and by the centre. From this second perspective, De Blas indicates that "the observance of shared allegiances will involve the form of vertical distribution of power and the acceptance of a liberal-democratic culture which fails to be dazzled by the charms of homogeneity nor falls for the temptation of exclusive allegiances."

VI Self-determination is subject to a political nationality. It has an external aspect, the right of a people to choose their own sovereignty, and an internal one, the free choice of their own social order and form of government. It is a non-prescriptive right which can thus be created through successive consultations and it displays very varied forms (as shown clearly in Quebec and Puerto Rico recently). Although its establishment always causes alarm in the centre, it is not identified with the creation of an independent state; but it may include this alternative. It may consist variously in giving a legal form to a process of decolonisation; in the separation from a state; in the regrouping of the state; in the recognition of indigenous groups; or in an increase in self-government (Ribo and Pastor, 1996).

VII The explanation of this process of the multi-ethnic nature of the Italian state was hidden until recently by the general enthusiasm which gave rise to the Risorgimento and Italian unification in the middle of the 19th century.

8 The Transformation of the Conflict (2): European Construction

Until the Second World War the nation-state was in exclusive control of its sovereignty. The old model had the following features: clear national sovereignty exercised by parliament and an executive; borders which were considered impermeable and inviolable; monopoly control of the apparatus of legitimate violence (army, police, prisons, etc.); the existence of a single focus of legitimation, in accordance with which only citizens of the state had the right to vote; political representation expressed exclusively through state institutions; and the exclusive authority of the state-national government in the political, economic and social spheres (Loughlin, 1993).

Factors inherent in the decentralising phase of the welfare state have eroded this type of state. In the upper echelons, economic developments of a political and military nature have proven that (within the areas of Western Europe and the Atlantic) this model is no longer feasible. From the 1950s onwards, trans-national businesses have made state borders into permeable membranes overwhelmed by economic flux, the impact of which is not only economic, but also cultural and ideological. On the other hand, NATO, when it de-nationalised armies, dissolved the basis of the old nation-state: territory, population and sovereignty.

Of the oldest supra-national macro powers which emerged after the Second World War, that which carried out wide-ranging changes to the centre-periphery conflicts within its sphere was that which had the project of European construction as its catalyst. Its specific nature becomes apparent if it is compared with organisations such as the North American Free Trade Agreement, an exclusively inter-governmental organisation without any supra-national institution engaged in policy-making (Keating, 1996).

European Construction and Regional Policy

Pro-European projects arose out of the era of a Europe destroyed by two devastating world wars and the determination to prevent a new war between France and Germany. They materialised in 1949, the year in which the Cold War began, by conferring ideological meaning on the border separating Eastern and Western Europe (Anderson, 1993).

The theoretical roots of these projects are to be found in European federalism of the inter-war period (which resulted in, among other things, the idea of a European Union, put forward by the French statesman, Aristide Briand). This federalism was fostered by, in addition to the remote influence of Proudhon's Europe of the Peoples, Mounier's personalism in the 1930s and the doctrinal *corpus* of Christian-Democracy.[I] Federal and functionalist theses entered into competition for a period of time,[II] even though they did have certain features in common. According to federalist thinking, the most renowned adherent of which was Altieri Spinelli, Europe would become integrated via a series of measured political decisions which would set up a real federation. According to neo-functionalists, whose best known leader was Jean Monnet, the most important step was starting the process since, once it was underway, even though it might only be by sectors (for example, in the economic domain), it would create its own irreversible dynamic of economic co-operation and then, later, political co-operation.

From 1948 onwards, when the European Congress was held in the Hague, pragmatic functionalism assumed the leading role over federalism; this revealed the lack of will to assign decision-making powers to European Community institutions and to make the regions into the political subjects of the institutions. Following the Schumann Plan of 1949, the result of a joint initiative he engaged in with Monnet, the Treaty which led to the establishment of the European Community of Coal and Steel was signed in 1951, establishing a common market for these raw materials. In 1957 the Treaty of the European Economic Community (EEC) was signed in Rome. It was also known as the "common market," a customs and economic union which allowed for the free circulation of people and goods within its terrain and also pursued a policy of closer links among its Member States. As a consequence of the 1965 Brussels Treaty the EEC established shared institutions: the Council of Ministers, the European Commission and the Courts of Justice and Accounts.

The intergovernmental model of European construction and the centralising nature of the first phase of the welfare state contributed to the overshadowing of the sub-state (or regional) entities during the fifties and sixties. Although the two funds established by the Treaty of Rome, the European Fund for Agricultural Guidance and Guarantees (EFOAG) and the European Social Fund (ESF), impacted on the regions, they were not perceived as an instrument of regional policy.

Although the Council of Europe, created in 1949, promoted associations among sub-state entities, the local aspects of this policy hid the regional aspects. Of significance was the name given to the association created in 1957 as the official consultative organ of the Council: the Conference of Local Authorities in Europe. This did not include regions until 1975, when it became the Conference of Local and Regional Authorities of Europe (CLRAE).

Two events modified this stand: the decentralising phase of the welfare state at the start of the seventies and the entrance in 1973 of Denmark, the United Kingdom and Ireland into the European Community. These states, particularly the latter two, were fearful that the common market would exacerbate the underdevelopment of some of their regions.

The first two structural funds, the EFOAG and the ESF, were reformed in 1972 to take the aims of regional policy into account; this was particularly the case in 1975 when the European Regional Development Fund (ERDF) was set up. However, paradoxically, the assignment of quotas for community budgets to states to allow them to carry out regional plans initially reinforced the states in relation to their respective regions. The strong representation of the different states in the ERDF (its committee is made up of civil servants appointed by the European Commission, but also by the states, and its budget is dedicated to projects previously chosen by member states) has been addressed but not eliminated in subsequent years. In 1979 a section outside of the quota was created, 5% of the total, which could be distributed by the Commission independently of the Member States. The 1984 reform represented the start of EEC Regional Policy. Funds distributed by the Commission exceeded 11% of the total ERDF, the budget of which had risen from 4.8 % in 1975 to 7.3 % in 1984.

The emergence of regions as a stage in the process of administrative decentralisation and horizontal distribution of power encouraged their interregional cooperation; this cooperation, promoted by the European Commission, pursued the direct object of reinforcing their situation with regard to Community Regional Policy, and the indirect object of modifying their relations of power with their respective states. These associations, which were of a wide-ranging nature (economic and political, or arising out of a specific shared situation, such as the cross-border nature of certain regions) proposed an exchange of information and the search for engagement with regard to Community demands; these were the real catalysts for the establishment of a third level of community power, the European infra-state power.

In 1973, about twenty regions set up the Conference of Peripheral and Coastal Regions of the EEC (they currently number sixty, many of which have been awarded Objective 1 status, indicating that they are the least developed regions according to Community Regional Policy). From 1989 onwards three large commissions have been in existence at the heart of this policy: the Inter-Mediterranean, the North Sea and the Atlantic Arc. The latter brings together twenty-five regions (five from the United Kingdom, one from Ireland, five from France, six from Spain - the entire Cantabrian cornice - and five from Portugal). The Association of Traditional Industrial Regions encompasses twenty regions suffering industrial decline; its objectives were to be taken into account in the eighties in order to form the content of Objective 2 status, applicable to those regions undergoing industrial decline.

This new framework of exchanges revived the European conceptual map of the Modern Age as outlined by Rokkan (1970), whose mosaic shape showed the Europe of the nation-states organised on a grid-system. Borders dissolved in the zone of the dense network of city-states of the "Lotharingian" central European axis, but remained on the extremes where they separated former imperial nations such as France and Spain or on new borders that emerged from bloody conflicts such that the one separating the two Irelands; in both cases against the will, and in spite of the resistance, of traditional ethnic groups divided by the borders. Therein lies the strength of Alpine cross-border associations which have emerged in recent years, in contrast to the weakness of their counterparts in the Pyrenees. In 1971, border regions of the Rhine basin founded the Association of European Border Regions (AEBR), which numbered fifty-two at the end of 1990. The AEBR addresses specific problems such as territorial policing, cross-border crime prevention and civil protection, regional policy, cross-border workers, etc. The EU INTERREG programme which promotes border regions, and the pilot project LACE, which the European Commission charged to set up a data base on those regions, have taken into account almost all the objectives established in the AEBR Charter.

The oldest Alpine associations are to be found among those grouped into mountainous regions: central (Alp), eastern (Alp-Adria), and western (Cotrao) Alpine regions. In 1983, the Community of Work of the Pyrenees was established, bringing together the seven regions of the French-Spanish Pyrenees, in addition to Andorra. Its achievements have been modest, mainly due to the opposition of states and to rivalry among member regions: while the Basque Country focused on the Atlantic Arc, Catalonia looked to the Mediterranean.

In October 1989, with the same objective of preventing the drift toward the Mediterranean, under the encouragement of the Atlantic Arc Commission, the Conference of Regions of the South of Atlantic Europe (SAE) was formed. The SAE attempted to deal with the difficulties arising from the peripheral nature of these regions consistent with their infra-structural deficit and weakness in research and development.

The Conference of Peripheral Coastal Regions and the Inter-regional Associations mentioned above promoted the foundation, in June 1985, of the Association of Regions of Europe (ARE), which was made up of some one hundred regions of the EEC and the Council of Europe (numbering more than two hundred in the present day). Regions of central and eastern Europe may participate in these organisations as Associate Members. The ARE, which is in competition with the CPLRE for the position of the most representative regional European association, aims to become the main representative of horizontal dialogue among regions and of vertical dialogue between regions and the Community, as well as the embryo of the future European Regional Senate.

The appearance of ARE coincided chronologically with the third stage, the political stage, of the European Community's Regional Policy. It was promoted by Jacques Delors, president of the European Commission from 1985 to 1994, Mounierist and Christian-Democrat in outlook; but also by the new member states, Greece in 1981 and Spain and Portugal in 1986, whose regional problems were more serious than those faced during the opening up of the EEC in 1973. The ambiguity of the final results was due to the contradictory attitude of the new member states, who wanted to deepen the economic aspects of European regional policy, while restricting the political consequences regarding the autonomy of their regions. A new playing field was therefore drawn which placed on one side regions, considered on an individual basis or through their associations, together with the European Commission and European Parliament, both supporters of Community autonomy; and on the other side, the Council of Ministers and member states.

The Delors plan sought the creation of a single European market: the Single European Act of 1987 aimed to bring his idea to fruition through the introduction in 1993 of the free movement of people, merchandises, services and capital. This economic unification had to be complimented by a strong social policy; regional policy was seen as the means whereby territorial economic inequalities could be addressed. The Single European Act therefore decisively transformed Community Regional Policy, which, with the goal of achieving the social and economic cohesion necessary for the establishment of a single market, established the reduction of inter-regional differences as its objective. It was consequently decided to focus the co-ordinated action of the three structural funds EFOAG, ERDF and ESF on five regional priority objectives. The first affected regions which were less well developed (those with a Gross National Product lower than 75 % of the Community average), the second referred to those regions undergoing industrial decline. A new emphasis was conferred on regions, advocating the negotiation of regional policy at three levels: supra-state (the Commission), the state and the region; support was given for regions to establish their own Plans for Development, to participate on the

Community Support Framework and to take part in the appropriate committees.

Reforms during the period 1985 to 1988 put an end to quotas and introduced a system of minimum and maximum financing. Regional projects could have a Community origin and be defined by the Commission; regional governments and private groups could collaborate in the development of their policies.

The new framework for Community intervention, even though it was based on economic criteria which ignored historical-cultural factors, offered, when it also became detached from politico-institutional conditions, some means of support to ethno-national movements of a trans-state nature. In effect, the economic logic of the European Common Market led Community institutions to make the distinction between three concentric territorial units, NUTs 1, 2, and 3 (Nomenclature of Territorial Units) which set out the different spheres of EEC economic intervention in less developed regions, in areas of industrial decline and in rural areas requiring support. The NUTs 2, which are conventional, refer to basic regions, such as the Autonomous Communities in Spain, in other words, the units of regional policy of member states. The NUTs 1 are, on the other hand, the large macro socio-economic regions into which several regions are grouped, the level at which analysis is carried out of the effects of processes such as the unification of customs control. The NUTs 3, refered to *Provincias* in Spain, *Kreise* in Germany and *Departements* in France, allow for rapid diagnoses of certain regional actions. The EEC includes 66 NUTs 1, 176 NUTs 2 and 829 NUTs 3. Moreover, cross-border co-operation promoted by the EEC planned a fourth, more extensive, level not officially contemplated, the trans-state "Euro-region" which would arise from the association across the borders of several NUTs 1 or NUTs 2 (Letamendia, Borja and Castro, 1994).

The Maastricht Treaty, signed in 1991, stated that the process of European Union would be completed in 1999. It aimed to achieve complete economic and monetary union, the harmonisation of foreign and defence policy of member states and the introduction of a common social policy. The directorates of Delors' team were based on the principles of partnership and subsidiarity (a term which means that superior units should only assume responsibility for those tasks which inferior ones are not able to carry out). However, does this principle apply only to the relationship between member states and the European Union, or does it also apply to the relationship between the states and their respective regions?

The Assembly of European Regions has fought for the application of the second interpretation, although its efforts have not been successful so far. Prior to Maastricht, the ARE set out the following demands which were approved in its 1990 Assembly in Strasbourg: application of the principle of subsidiarity, participation of its regional representatives in the EU decisions which affect the exclusive competences of the German *Länder*, of the regions, Italian, French or others, as well as of the Spanish Autonomous Communities; and the establishment of a regional organism of mixed and consultative nature which would also take part in community decisions. The Maastricht Treaty partially addressed these issues; Article 198 of the Treaty created a Committee of the Regions (but the representatives of regional and local entities which make up the Committee, appointed by the Council, are there at the proposal of each member state, although they are not subject to an imperative mandate). Besides, local entities must be equally represented on the Committee for the alleged reason that those states lacking regional organisation would not otherwise be represented. The Committee is obliged to state its position on matters such as education, youth and training, cultural policy, public health policy, trans-European networks, social and economic cohesion, structural and cohesion funds; it is also under obligation to act as an appropriate vehicle for the correct application of the principle of subsidiarity (ensuring that it encompasses the regions of member states). Nevertheless, the Committee cannot recur to the European Court of Justice, it does not have regional representation outside of local entities, and with regard to the Commission or the Council, its recommendations are not compulsory.

The old inter-governmental system, based on the sovereignty of the nation-state, is therefore rendered obsolete. A new political set-up emerges with new rules, new actors (regions and local authorities; nationalist movements and political parties also?) and a new framework which goes beyond that of the nation-state. New regional mobilisation takes place at two levels: at the regional level, the public and private sectors organise themselves within the core of each region in order to improve their negotiating position in relation to state governments, other regions and the European Union; at the inter-regional level, regions organise themselves into the aforementioned associations and engage in an exchange of information and technical knowledge (Loughlin, 1995).

This new political set-up influences the orientation of regional governments policy, as well as nationalist or regional parties, in those cases where they held power.

The Catalan *Generalitat*, led by the president of the nationalist coalition party *Convergencia i Unió*, Jordi Pujol, who the European Regional Assembly elected as its president in its sitting in Santiago de Compostela in 1991, holds outstanding influence over European inter-regional co-operation, which has made his nationalism even more pragmatic in nature. During the eighties the *Generalitat* focused more on co-operation with European regions than with national movements. From 1987 onwards, Catalonia has been a member of the group of "advanced regions" set up by the German *Länd* Baden-Württemberg and by the French region Rhône-Alpes, which were later joined by Lombardy, an Italian region (a group known as the "four motors"). This situation gave rise to the image of Catalonia, which was removed from its ethno-nationalist base, as a partner of the "wealthy regions."

The Community process was one of the main factors accounting for the rise at the end of the eighties of a strong regionalist movement in the Canary Isles. Following the defeat of radical nationalism in the elections of October 1982, the unique situation of the Canary Islands as the last European outpost before the African continent explains the appearance of this hegemonic regionalism. Its economic goal of free-exchange (due to its position as an enclave with respect to Japanese products and its commercial relations with sub-Saharan Africa, especially Senegal) do not fit in with the type of European integration proposed by the central government. Feeling immersed in a large European framework, the local bourgeoisie proposed a renegotiation of economic and fiscal relations between the Canary Isles and the EU, a process in which Spain played the role of the uncomfortable intermediary. Gari Hayek (1994) commented that, "in reality the new nationalism in the Canary Islands situated itself, objectively and subjectively, within the political tendency which is known as the Europe of the regions, or nations."

Inter-regional cross border relations increased the porous nature of state borders (which, as has been stated, is greatest in the old European zone of the Confederation of city-states and least in the old imperial states). Prior to its last extension in 1995, there were ten thousand kilometres of borders in the European community and ten % of the population lived in border areas. The European Parliament has approved two charters which protect groups and cultural heritage of a potentially cross-border nature: the Rights of Ethnic Minorities and the Charter of Regional Cultures and Languages. It has also approved, at the request of the CPLRE, the Framework Covenant of Cross-Border Co-operation. This ethos converges with that of the new social movements in which transnational and local orientation almost always takes priority over the state orientation, which facilitates co-operation on both sides of the border.

Nevertheless, in the case of those projects of cross-border co-operation where inter-regional and inter-governmental principles conflict, is the latter tendency which frequently wins. On occasion, pressure from the states in question has resulted in the marginalisation in these projects of regional and local authorities.[III] On other occasions, measures in support of under-developed regions are driven by a state logic which has had environmental consequences and which, according to Baker (1994), are not being adequately resolved.[IV] Finally, there are cases where the success of regional cross-border co-operation has encouraged states to copy the initiative, being these inter-governmental projects which have ended up by draining resources.^{V}

Transformation of the Centre and the Peripheries

Centre-periphery conflicts have been modified within the sphere of the European Union; state-national centres have become peripheral with respect to the EU, while peripheries operate on a variable geometry terrain in which there are already players besides the nation-state. Moreover, the intensification of interactions with a supra-national entity, distinct from the state centre, opens up the possibility of escaping from the game of mirrors which forces the peripheries to imitate that which had been up-to-date their only point of reference, the nation-state. For the first time they are in a position to engage elements not included in their counter-model, which allows them to place themselves ahead, and not to the rear, of the nation-state. In this way, the EU appears as a mosaic of territories, and not a hierarchical pyramid, which resembles pre-modern Europe (Keating, 1996). A hybrid system is thereby created which is neither the Europe of the regions nor the inter-governmental Europe; it is a confederation tending toward a federation in which there is multi-level governance (Loughlin, 1995).

However, not all changes weaken the state. Although European construction has modified the nation-state, it has not led to its dissolution for the following reasons:

1. The architecture of the EU gives priority to states in the Council of Ministers, which continues to be the key institution, more so than the Commission or the European Parliament.
2. Important decisions within the EU, such as the Single European Act or the Maastricht Treaty, arise out of engagements made by nation-states, which act on the basis of the advantages that may be gained from them.
3. Although borders have become permeable, those that pass through may be detained at customs posts if they are considered suspicious. Moreover, in those situations where there is a violent response to a specific nation-state, for example, in the Basque Country or Northern Ireland, state-national policy prioritises borders over EU economic policy, which may lead to their closure.
4. In almost all nation-states central governments have the final say on relations between regions themselves and between regions and Brussels.

Additionally, in the nation-states of the core of Europe only the elites appear to feel European; the working and middle class identify, instead, with traditional state-nationalism.

The lifting of internal EU borders has led to the strengthening of external borders, which, added to the deepening of the global economic crisis, has fuelled new state centralisms. From the outset the construction of the EU has been associated with the idea of an external border, which was, from the beginning, ideological: "free world" against "totalitarian communism". At the start of the nineties the external European border assumed a new significance as a protective element of political stability and a certain type of advanced society. The disappearance of the Soviet threat has given way to the fear that completely permeable borders would open the door to the "poor of the Earth" coming from the south and the east in search of a better way of life. From a conservative point of view this would lead to crime, subversion and drastic political changes. This fear of immigration, which feeds racism, is linked to the desire to create a large area of economic prosperity that would be protected from the demographic pressure from the south and east and from the competition of powerful and wealthy rivals such as Japan. "Fortress Europe" subsequently would reach an agreement with the North American Free Trade Area, "Fortress America," which would create the largest market in the world (Connor, 1994).

But "Fortress Europe", far from establishing the basis of a European pan-nationalism, has been the catalyst for the appearance of centralist state -nationalisms which are ethno-centric and isolationist in nature (especially in those states which are renowned hosts for immigrants). This resurgence of xenophobic nationalism has created internal psychological borders among European citizens, which explains the problems encountered during the process of ratification of the Maastrich Treaty.[VI]

The Single European Act stipulated, with the aim of establishing the Single European Market, the quadruple free movement of goods and services, capital and people by January 1993. It is the movement of people, which seemed to be inseparable from goods, which is the weakest aspect of the arrangement. The EU Council of Ministers decided upon, with the goal of dismantling internal border controls, the strengthening of external European borders, that is border areas with third countries, airports and big ports. This required a high degree of co-operation between police, customs and immigration officers which has not been achieved but nevertheless has reinforced the walls of "Fortress Europe." The influx of half a million people requesting asylum, most from Eastern Europe, has provoked demands on police controls and led to international agreements restricting the rights of exiles.[VII] The movement of students and executives has increased within the European Union; however, it is less and less common for EU citizens to reside in other member states (currently five million, as against eight million foreigners from outside the EU, excluding illegal immigrants). The reinforcement of external borders has led to a high price being paid by peripheral countries with extensive coastal border regions, and has damaged regions which shared economic and cultural ties with non EU countries. (Germany has tried to protect the fragile economy of its six new Länder by strengthening the Oder-Neisse border and convincing Poland to become a buffer state).

With regard to nationalist movements, the process of union has reinforced the European federal and Christian-Democrat tendency of some. The revival of the Europe of city-states should be considered as one of the factors responsible for the rise of the Northern League in Lombardy and other regions of the Po Valley, an area which has been historically integrated by the "Lotharingian axis." The ideology of a Europe of the Peoples (or regions) has been adopted by the movement in the Canary Isles, by John Hume's Social Democratic and Labour Party (SDLP) in Northern Ireland and by the Scottish National Party.

In Scotland during the 1970s and 1980s, public opinion was hostile toward the EEC. The Scots feared that the expansion of the European market would exacerbate the peripheral situation of their industries (steel, fishing, agricultural production, etc.) The British Labour Party, which was opposed to the EEC during the seventies and right through until the end of the eighties, was the most voted party in Scotland; and Labour Party Scottish supporters were even more hostile to the EEC than their English counterparts. In the 1975 referendum, held two years after the United Kingdom joined the EU, the no campaign was headed by the Conservative extreme right, the left-wing of the Labour Party and Scottish and Welsh nationalists. But in the eighties attitudes toward Europe changed and the construction of the European Community became associated with territorial self-government; in this process of change the rejection of English conservatism of the south-east and British centralism of Margaret Thatcher played a vital role. The defence of the European Social Charter led trades unions to change their position, being followed shortly afterwards by the Labour Party. In Scotland, the former Labour Party supporter, Jim Sillars, after setting up an ephemeral Scottish Labour Party, joined the main nationalist grouping, the Scottish National Party, converting it to a pro-European stance.

Nevertheless, collaboration between nationalist political parties is far behind that which has taken place between regions. This is due, besides to the inertia of the game-mirror fixation of these parties with their respective states, to their poor representation in the European Parliament. This could have opened up a channel of communication for peripheral parties, who, for obvious reasons, have always lacked Internationals within which to organise themselves. But the European Parliament was constituted late, in 1979; moreover, almost all states have set up single state constituencies for the European elections, which are to the great disadvantage of nationalist parties who have been forced to form coalitions which undermine their meaning and make them lose votes, drastically reducing their electoral representation.

This was the situation in successive European elections in 1979, 1984, 1989 and 1994. France pleaded unity of the homeland in arguing for the single constituency; Spain pleaded the same argument in 1989, in spite of being a state based on autonomous regions; only Germany and Belgium organised electoral constituencies on a plural basis.

Given the numerical and, often, ideological impossibility of establishing their own parliamentary group, the search for a specific space has been problematic for peripheral parties, since the European parliament is dominated by the two large coalitions of the European Socialist Group and the European Popular Party (Christian-Democrat), which leave behind the remaining groups, including communists.

> In the 1979 European elections, peripheral parties obtained five out of a total of 410 seats. One refers to the VU in Flanders, two to the French-speaking FDF in Belgium, one to the Scottish National Party (SNP) and one to the STVP in the Italian Tyrol. Except for the latter, which joined the Christian-Democrat group, the rest belonged to the Non-Aligned group.
>
> In the 1984-89 elections, the seats obtained by peripheral parties rose to ten. Four belonged to the Rainbow Group, made up of 20 parliamentarians (the rest belonged to the Greens and anti-EU Danish representatives). Two belonged to the VU (Flanders), one to the coalition of the Valdotaine Union - Sardinian Party of Action, and one to the Basque Party Eusko Alkartasuna (which headed a coalition). The remaining six belonged to the South Tyrolean Party, three to the Catalans, two CiU and one UD respectively, one to the Scottish National Party (SNP), and one to Herri Batasuna (HB) of the Non-Aligned Group.
>
> The 1989 polls saw peripheral parties win 16 seats. Ten of those elected joined the Rainbow Group: VU (Flemish) one, Europe of the Peoples, a coalition led by EA, together with the ERC from Catalonia and the PNG from Galicia, one;. *Avennue Corse*, Sardinian Party of Action, Andalusian Party, Scottish National Party and Fianna Fáil, one; the Lombard League with two seats. Of the other six seats, two belonged the Catalan CiU, one to the Nationalist Coalition, led by the PNV, one to the south-Tyrolean SVP (these four representatives joined the Christian-Democrat Group), one to *Izquierda de los Pueblos*, led by the Basque Party *Euskadiko Ezkerra*, and the other is HB, which joined the non-aligned group. Basques, therefore, obtained four seats, a quarter of all allocated to European peripheral parties.
>
> In the June 1994 elections, even though the number of Basque seats dropped, the total number of peripheral party representatives slightly

increased to 16, due to the gains of the Northern League. In Flanders, VU obtained one seat and the Flemish National Right-wing bloc won two seats. The Basque parties EA, HB and *Izquierda de los Pueblos*, as well as the Andalusian PS, lost their seats; CiU won three and the Nationalist Coalition, led by the PNV, won two. The Northern League won six seats, the SNP from Scotland won two and the northern Irish SDLP, one.

Although the stance of the European Party of the Peoples (EPP) is Christian-Democratic and European federalist, the presence of conservatives among its ranks (including the Spanish Popular Party) has softened its ideology. The Rainbow Coalition, which is ecologist in nature, has set as its goal the creation of an open Europe of Autonomous Regions and Nations. *Herri Batasuna*, vetoed by the Rainbow Coalition, forms part of the Confederation of Nations without a State, a grouping of left-wing nationalist parties of a non-parliamentary nature founded in 1984, which incorporates the Corsican *Cuncolta*, the Catalan *Crida* (until it dissolved), Sinn Féin and the *Bloque Galego*, etc.

The repercussions of European construction on peripheral movements, therefore, rests on other bases and, to be precise, on the opportunity which it opens up to them of becoming elements which reinforce their territories with regard to the new challenges that this process entails. It is the social capital which they accumulate that may become the factor that will counteract the combined destructuring effects of a global mass culture with an open trans-national market. They are best equipped to transform mere spaces subjected to the erosive effects of these forces in territories where the social, cultural and political dimensions are most coherent. The new regionalisms may also reinvent territory in order to adapt it to contemporary challenges; but they are at a disadvantage with respect to the accumulating factor of social capital carried out in the medium and long-term by historic nationalisms.

It is the strength of a specific national culture which gives rise, giving them cohesion, to those factors that determine the capacity of territories to adapt to new challenges. These factors include skilled labour, the existence of links with research and development centres, information and economic networks among businesses or the drive by regional governments to promote the ability of their territories to adapt to new economic challenges (Keating, 1996). It is of little relevance that the national culture, and the ethno-territorial language, are minorities in the state; what is important is the ability to forge the coherence of the whole.

Likewise, new social movements, which often converge in terms of goals and ethos with peripheral nationalisms, re-evaluate ethno-national territory. A large part of neighbourhood, feminist, ecologist and anti-nuclear movements have a local or peripheral-national base and a decentralised perspective. They oppose, on their terrain, the logic of the centralising, remote and bureaucratic apparatus of states and multi-nationals.

The new nationalisms which make this logic their own use the past, not to return to it, but rather as identity-formation capital, opening up their identity to plurality and debate; with respect to their instrumental-rational dimension, they are territorially inclusive and not ethnic-exclusive. They may articulate the different dimensions of their territories in several ways. In some cases they develop, through a process of extra-official and ongoing negotiations, the convergence of interests of different social classes in defence of their territory. In others, when problems appear to put this social pact into practice, economic, social and political groups, which are more interested in the territorial economic growth than in the norms guiding social relations, forge "pro-development coalitions" with the aim of making the territory more competitive in open market conditions.

Two variants are identifiable in this version, one conservative and the other progressive. The coalition may serve as the means of development which represents the interests of the property-owning classes; or it may represent wide-ranging social groups reflecting the interests of the working classes, non-economic groups and social movements. This second variant incorporates a social dimension as it attempts to correct the disadvantages inherent in strategies of development that favour the demands of the international market, biased toward more advanced territories and the owners of capital. In stateless nations, this social dimension is linked to the cultural dimension, as their ability to show solidarity and to integrate depends on the social capital accumulated by culture and identity (Keating, 1996).

On the level of nationalist aspirations, association in the European Union or in the Free Trade Treaty appears initially to reduce the economic cost of independence, as it removes questions on customs tariffs and commercial barriers. However, is the erosion of state sovereignty synonymous with autonomy or independence for those peoples without a state, or does it open up the possibility for a non-traumatic passage from one to the other? No answer can be forthcoming, given the emotional responses and, therefore, the predictable political costs that a simple declaration of self-determination provokes in central states.

One line of thinking claims that these peripheral nationalisms campaign for a type of self-government which makes independence superfluous, becoming heralds of a new phase in which the state has disappeared and power is scattered. However, in reality, such a claim is ideological. This phase does not exist because nation-states oppose it; it is sufficient to note their resistance, in western Europe, against the tranformation of the confederation of states which is the European Union in a federation ruled by a central power; or to something as simple as subsidiarity operating from below, in favour of regions.

The era of the nation-states has now passed, but it will take time to transform itself into something different; the old world has died but the new one has not yet been born. As long as peripheral nationalisms re-orientate their situation out of this crisis, they are the forerunners of a phase which does not yet exist but which will, becoming therefore a factor of progress with regard to obsolete states.[VIII]

Cross-Border Relations and National Conflicts

Given that the most well-known cases of violent nationalist conflicts in Western Europe have a cross-border dimension, the process of European construction, in so far as it offers a solution to these conflicts, has gained the consensus of all those involved, state nationalisms and peripheral nationalisms, and within these, pro and anti-system movements.

The scope of the solutions put into practice during the past decade, coinciding with the expansion of EU Regional Policy, has been limited; they have not resolved the Corsican, Irish or Basque conflicts, although in the case of Northern Ireland they have been an important factor in achieving movement in a dispute which is currently unresolved.

On the other hand, the process of European construction may modify, at least partially, the direction of the mobilisation of the peripheries of the periphery, leading to the substitution of their hostile position toward the latter and their support of the centre for one in which a complex territorial relation is accepted within a broader framework, with the intervention of other players. In this case the European Union assumes the role of reference which the nation-state had, up until that point, occupied. It is within this context that the split within the Navarre Union Party, UPN, which gave birth to the Convergence of Navarran Democrats (CDN) led in April 1995 by the ex-president of the *foral* government, should be understood. This party emerged as a consequence of co-operation between Navarre, the Basque Autonomous Community and the Region of Aquitaine, and its ideology explicitly embraces the European discourse.

> The process of European construction allowed Corsica to free itself from the economic tutelage of Marseilles and Nice, and confirm its position as a bridge between France, on the one hand, and Italy and north Africa, on the other. According to Dressler-Holohan (1995) the European dynamic has facilitated, since 1989, the renewal of Corsica's links with the Italian island of Sardinia, something which would have previously been impossible given the impact of Mussolini's fascism on Corsican irredentism (in Corsica, Sardinians were referred to derogatorily as *luchesi*). In May 1989, an INTERREG programme

of mixed development involving Corsica and Sardinia was unanimously approved in the European parliament. Given that more classical options such as building close links between the wealthy French region of Provence-Côte d'Azur would have deprived Corsica of EU funds destined for less developed regions, in the autumn of 1989 the Socialist Group placed a motion before the Corsican Assembly urging it to put the European programme into practice and at the same time to open up Corsica to the Mediterranean cultural world. However, co-operation with Sardinia was limited to southern Corsica, and the 1991 reform of Corsican Statute did not even mention it.

The borders between the two parts of Ireland emerged during the course of the struggle for Irish independence, which resulted from the war between the IRA, the British army and northern Irish Protestant loyalists between the years 1918-1921. Its boundaries were set out by Protestant unionists, allied to British Conservatives, who expanded the territory in the north to limits within which they would enjoy an undisputed majority. The three hundred roads which criss-cross the five hundred kilometres of border were difficult to defend from the IRA which used them in attacks against customs posts and British military garrisons. The British state, therefore, closed between a third and a half of them (O'Dowd, Corrigan and Moore, 1990).

The intensification of the conflict coincided chronologically with Ireland and the United Kingdom's entrance in the European Union in 1973; which represents a strong case exposing the contradiction between the goals of achieving permeability in EU borders and their closure for military and police reasons. Historically, economic co-operation between the two parts of Ireland has been weak; reciprocal imports and exports have not exceeded 5 % of their respective products. Nevertheless, since the end of the eighties, EU initiatives have improved the situation. The INTERREG programme embraces an area, all of Northern Ireland except Belfast, which is greater than that which is directly a border area. The 1985 Anglo-Irish Agreement set up the International Fund for Ireland (IFI), to which the EU has contributed 15 million euros. From 1989 onwards, after both parts of Ireland were awarded Objective 1 status, they have received about six hundred million pounds from the EU structural funds. These subsidies have financed cross-border initiatives in tourism, roads, health and energy. However, the absence of decentralisation in the Republic of Ireland and the United Kingdom has impacted negatively on the management of INTERREG: poor bureaucratic co-ordination in carrying out policies and a lack of institutional flexibility has made economic co-operation difficult.

In either case, co-operation between the Republic of Ireland and the north within the framework of EU initiatives is one of the few matters, if not the only, supported by practically all political parties in the two parts of Ireland. President of Ulster Bank George Quigley, representative of the business sector in Northern Ireland, commented that "Ireland, north and south, should form part of an 'insular integrated economy' in the context of a single European market, thereby facilitating the development of a Dublin-Belfast economic corridor" (Anderson and Goodman, 1993).

In the course of the 1992 political discussions between northern Irish political parties and the British and Irish governments, John Hume's SDLP put forward a proposal involving the devolution of political power in Northern Ireland to six commissioners which would have real authority; three were to be nominated by the political parties, two by British and Irish governments, and the sixth, by the EU; according to the media, some members of the EU Commission would have backed the proposal. However, Hume's initiative was based on a view of political power within the EU similar to that existing within the United Sates of America, which is in fact not realistic: the EU continues to be based on the sovereign will of member states. In effect, during a visit to Northern Ireland in November 1992, the president of the European Commission, Jacques Delors, commented to the media on Hume's proposal saying that, "I am aware of this alternative solution, but I do not think that the EU should interfere in the internal problems of a country, a province" (Hainsworth and Morrow, 1993).

In the Basque Country, as in the rest of the Pyrenees, cross-Pyrenean relations, which were well developed in the past, became restricted by absolutism and eventually frozen in the era of the nation-states. Until the start of the Modern Age, border Pyrenean valleys formed part of a federation based on *facerías,* treaties of alliances and peace and the use of pastures, which were exempt from all external political control. From the 16th century onwards, when the unity of the two large monarchies, French and Castilian, was consolidated, the valleys, under threat by far-off state conflicts, adapted the old pacts to the new situation by making them into meticulous treaties which established the boundaries of pastures, the use of water and forests, making provisions against the risk of war (Fernández de Casadevante, 1985).

The predominance of the French monarchy in relation to its Castilian counterparts was reflected in the first cross-border treaty of the Modern Age, the 1659 Treaty of the Pyrenees, and also in several boundary treaties drawn up during 1856 and 1866. Cross-border Pyrenean routes and bridges suffered the consequences of state suspicions. Until the end of the 18th century there were no roads, and even at the end of the 19th century there were only eleven; the main reasons put forward against the opening of more trans-Pyrenean roads were military. For this reason, the gauge of the first Spanish trans-Pyrenean railway was different to that common throughout Europe. Presently, French-Spanish change-overs only take place at the two extremes of the Pyrenean line. Two thirds take place on the link between Aquitaine and the Basque Country, where 51 % of merchandise passes; while 80 % of industrial products enter through the west, 77 % of agriculture enters by the east. Cross-border traffic is very low; only six itineraries resulted from French-Spanish negotiations.

There are two cross-border commissions in the Basque Country, although their output, which is dependent on the French and Spanish states, is very low. By government agreement an International Commission for the Pyrenees was set up in 1875 in which Spain was represented by all central government ministries and the army, but not by regional and local authorities. In tandem with this, another commission, of a local nature, was set up in 1978, the Joint Technical Commission of Bidasoa (CTMB), which was established as a consultative organisation of the CIP, noted for its inertia and bias to the detriment of local mayors.

Cross-border municipal authorities are strongly Basque-speaking, particularly in the areas of western Navarre and the northern Basque Country. From 1980 onwards, the promotion of the Basque language is a stated goal of all local governments in the Basque Autonomous Community and figures in the text of the Ainhoa Declaration subscribed to in January 1993 by the presidents of the Basque and Navarre Autonomous Communities and by the Regional Council of Aquitaine. Nevertheless, cross-border co-operation in the north with regard to the protection of the Basque language is difficult since it lacks the status of official language in Aquitaine and because France has refused to sign the European Charter of the Rights of Minority Cultures and Languages. There is, therefore, a legal vacuum in this state with regard to the protection of the Basque language.

The process of European integration is attenuating Spain and France's resistance to the external activities of its sub-state entities, mitigating France's long-standing centralist tradition (timidly broken by legislation in 1982 and 1992), and overcoming the resistance of the Spanish state, whose constitution attributes in its article 149 exclusive authority to the state in international relations.

Inter-regional co-operation between the Basque and Navarre Communities and the region of Aquitaine experienced a major drive from 1989 onwards, coinciding with the Reform of Structural Funds and the promotion of Community Regional Policy by the Single European Act. The presidents of the Basque Autonomous Community (CAV) and Aquitaine signed a Protocol of co-operation between the two regions in October 1989; co-operation became tri-partite after a new Protocol was signed in January 1992 by the president of the Navarre Community.

With differing degrees the main political forces which operate in the three border regions (CAV, Navarre and Aquitaine) agree on the goal of consolidating an Atlantic European macro-region which will increase commercial and cultural exchange and reinforce and complement economic exchanges. Basque nationalist parties, which have represented a stable majority of the CAV electorate since 1977, have shared, since the times of Sabino Arana, therefore for a century, a common vision of a Basque Country consisting of seven provinces, which embraces the four historical territories of the south, Biscay, Gipuzkoa, Alaba and Navarre and the three in the north, Laburdi, Lower Navarre and Zuberoa. Although the Basque Nationalist Party (PNV) and Eusko Alkartasuna (EA) support, albeit while being critical of its limitations and shortcomings, the process of European construction, contrary to HB, which defines the EU as "the Europe of multinationals and merchants," the three main nationalist groupings are in favour of, for different reasons, developing links on both sides of the border.

Right-wing French parties in the Basque Country, although they have diverging views on the EU (the Christian-Democrats argue for a federal

Europe and the distribution of responsibilities at local, regional, national and EU level; Gaullists are in favour of a Europe of the Homelands and reject any form of federal structure in the EU) support inter-regional co-operation. The new Basque nationalism in the northern part of the country, which emerged with *Enbata* in the sixties, supports it for patriotic reasons.

Cross-border co-operation is also backed by socialists within the Basque Autonomous Communities and Navarre and in the Aquitaine region, even though placing it within the framework of an Atlantic Euro-region. In this way, it represents the only point on which all the main political forces from both sides of the border (including radical Basque nationalism) agree.

This process is behind the evolution of the institutional definition of Iparralde (the French Basque Country). The joint opposition of state representatives and local Basque notables, who historically have played the double role of mediators between representatives of the state and peasant society, as well as that of inter-cultural translators between the "wise" culture in the French language and the popular culture in the Basque language (Euskera), has prevented since the time of the Revolution the creation of a Basque Department either on an ethnic or modernising basis (Chaussier, 1994).

New Basque nationalism in Iparralde led by *Enbata*, a modernising political factor in relation to notables, the hegemony of which has, nevertheless, remained unbroken, came to exercise an influence on Basque civil society in the seventies which was far beyond that of its electoral strength.

Radical Basque nationalism in the Spanish state which, because it lacks institutional responsibility, can openly defend the unity of the seven historical Basque territories of the north and south (in France and Spain), has become the obligatory point of reference for successive nationalist organisations in the north, first *Enbata*, later, from the mid-eighties onwards, EMA and Euskal Batasuna. This is even more the case when political refugees coming from the south suffered during the eighties under the French government policy of extraditions and political hand-overs, as well as attacks by Spanish police agents in Anti-Terrorist Groups of Liberation (GAL). It was these links and this presence, which the French government defined as the "foreign element", one of the reasons, if not the decisive reason, for the socialists' failure to fulfil their promise, made at the beginning of the eighties, to establish a Basque Department. However, at the same time, the anti-repression factor and solidarity with Basques from the south is one of the main reasons behind the increased nationalist vote in the French Basque Country: during the past decade it has risen from an average of 5% to 10 % of the votes. In this region there has been an armed organisation active since 1973, Iparretarrak (IK), "those from the north," but the frequency and scope of their attacks and their repercussions among the population are much less marked than those of ETA in the Spanish Basque Country.

However, the political set-up including four parts in the French Basque Country, representatives of the state, locally elected representatives, Basque nationalists and the economic, cultural and socio-professional sectors from civil society (to which a fifth element must be added, the Aquitaine Regional Council, established in the early nineties) has become currently more subtle. During the eighties, the crisis of the model of the welfare state eroded the French Jacobin structures. This led the Prefects, sub-Prefects and General Councillors in the Department of the Atlantic-Pyrenees to pay greater attention to the demands of Basque civil society; but their cultural and economic "live forces", if they do not strictly identify with Basque nationalism, they are under the influence of its initiative and plans. This nationalism, in its turn, becomes more positive and participatory; while locally elected representatives, caught between the two sides, have abandoned their role as mediating notables for that of modernising managers.

Decentralisation policy implemented in 1982 and aimed to be an instrument of modernisation within the French territorial administration, became, in fact, "the consecration of the big notables," who from then onwards were able to enter into direct contact with the population, without the mediation of the Prefects. However, presidents of regional Councils constituted within the framework of the Decentralising Law, lacking in consistent authority, were obliged to affirm their ability to act as territorial representatives through a policy of gestures which demonstrated, among other things, that their management was backed by decisions taken in Brussels on the distribution of EU funds which affected their regions. In this way, at the end of the eighties the Regional Council of Aquitaine put forward the idea of an "Atlantic Axis" which was to be the counterbalance to the two other existing European axes: the central European "Lotharingian axis" of Milan-Brussels-London and the Mediterranean axis of Lombardy-Marseilles-Barcelona with extensions to Madrid and Valencia. However, for Aquitaine, the compulsory partner in inter-regional co-operation which would put flesh on the bones of the Atlantic axis was the Basque Autonomous Community; therein lay its interest in signing the Protocol of Co-operation in 1989.

The Fund for Basque-Aquitaine Co-operation set up in 1990, vehicle of partnership between universities, economic and cultural organisations, becomes a factor, albeit modest, in the promotion of linguistic and cultural, as well as economic, co-operation between the northern and southern parts of Basque Country. However, the lack of territorial identity in the French Basque Country and the absence of institutional interlocutors are obstacles to this co-operation. Since the vehicle for cross-border co-operation already exists, it has become an element which objectively exercises pressure in favour of Basque institutionalisation of the north. The new situation brought about a convergence, unseen in the last thirty years, between the sphere of civil society driven by radical Basque nationalism and cross-border initiatives put forward by moderate or institutional nationalism of the south, a convergence with regard to which neither Herri Batasuna nor Basque nationalism of the north adopted a negative attitude in principle (Letamendia, 1994).

The combination of these new factors, which result from a change of attitude by all those involved, has created over recent years an opening in the institutional framework of the northern Basque Country. This opening expresses itself through a new commitment. Leaving to one side, for the moment, the matter of the Department of the Atlantic Pyrenees, an institutionalisation of a private nature is taking place in its Basque territories, which is expressed in the establishment of a "Council for the Development of the Basque Country" and in a "Council of Elected Representatives of the Basque Country". These Councils are perceived as new forums for debate between elected members and representatives of civil society, and in this sense, go against the general tendency in French decentralisation, which places the former in a position of privilege (Chaussier, 1994). As a sign of the change in attitude of state representatives, the initiative for the project, as well as the momentum for prospective working involving the project *Pays Basque 2010* (Basque Country 2010), came from the sub-Prefect in Bayonne.

The Council for the Development of the Basque Country put forward its statutes in July 1994 in the Chamber of Commerce and Industry in Bayonne. Its Plenary Assembly is made up of five sections: the "Elects" (15 members), "Social and Economic Activities" (30 members), "Teaching, Training and Culture" (20 members), "Administration" (14 members), and "Qualified People" (10 members). The Council of the Elects, which began its work in March 1995, is composed of 21 Basque General Councillors of the Department, 4 Members of Parliament and 36 mayors, delegates of rural or urban councils.

In October 1995, more than 130 Basque majors approved a motion requesting the establishment of their own Basque Department. In this way, cross-border co-operation, although it has left the socio-political set up in the south intact, has become a real factor in institutional change in the north.

Notes

[1] Mounier's personalism is a moral concept according to which the human being should overcome material demands and the de-humanising institutions which impose a materialist perspective of social life. Present-day Christian-Democracy, which is both reactionary and progressive in nature, wants to promote the establishment of an order neither monolithic nor authoritarian, that would be the result of the measured composition of multiple and hierarchical orders, born out of the natural sociability of human beings (Chatelet, 1986).

[II] After the Second World War Alexandre Marc created the European Union of Federalists, which supported a model of European integration in which the federated nation-states would devolve some authority to sub-state organs of government. This objective had many similarities with the Project for the Co-ordination of European Nationalities which was developed in 1943 by the exiled president of the Basque Country, José Antonio Aguirre, appointed as honorary president of the European Congress in 1949. European federalism inspired, in its most extreme aspects, Heraud's (1963) ethnicist theses, according to which Europe should be made up of a federation of ethnic nations, submerged up until that point in artificial states, states which should become federated before disappearing.

[III] In the multi-fund programme submitted to INTERREG by Spain and France for their border areas of Gipuzkoa, Navarre, Lerida and Gerona to the south of the Pyrenees and the French departments of the Atlantic Pyrenees, Upper Pyrenees,

Upper Garonne, Ariège and the north-eastern Pyrenees, approved for the period 1991-1993 with 50% funding financed by the EU, the regions and even the Community of Work of the Pyrenees were not included.

IV   These regions are often chosen as the sites for chemical plants or nuclear power stations, making them ecologically vulnerable. Other projects involve the construction of roads which are a threat to the environment (as is the case in Spain and Portugal). Ecologist groups have criticised INTERREG for not redressing the damage caused by those programmes financed by structural funds and for not confronting the states in question.

V   The Alp-Adria Community of Work emerged in 1978 as an initiative taken by the Italian regions of Veneto and Friuli-Venice Giulia, the Yugoslav republics of Slovenia and Croatia and the Austrian Länder of Carintia, Estiria and Upper Austria; they were later joined by the Italian regions of Trentino, Upper Adige and Lombardy and, in 1986, by the Hungarian regions of Györn-Sopron, Vas, Somogy and Zala. The success of this initiative convinced some states of the area (Austria, Hungary, Italy and Yugoslavia and, later, Czechoslovakia and Poland) to set up an organisation of inter-governmental co-operation called Central European Initiative. According to Delli Zotti (1995), the regional members of Alp-Adria must cover their own expenses, which has prevented them from carrying out large-scale projects. On the other hand, state financing has allowed the Central European Initiative to undertake projects of greater scope in areas of transport and infrastructure.

VI   By constitutional imperative, the Maastricht Treaty had to be submitted to referendum in Denmark for the first time in June 1992; in spite of the fact that the entire Danish elite, trades unions, political parties, mass media, asked for a yes vote, there were more no votes (50.7 %) than yes (49.3 %). According to Connor (1994), this result had an enormous psychological impact on the pro-European movement as it represented the first opportunity for the people to express their views on the Treaty. It underlined the gap between the European elite and their people; in fact, only three (France and Ireland, besides Denmark) of the twelve member states held referenda on the Treaty.

VII   On account of the 1985 and 1990 Schengen Agreements, Spain has been demanding since May of 1991, the year in which it signed the aforementioned agreements, entry visas from Moroccan citizens. This has encouraged illegal attempts to cross the Straits of Gibraltar, often with tragic consequences, and has affected numerous north African immigrants who travel each year to work in Andalusia or other regions of Spain.

VIII   In any case, the previous description of new nationalisms is also a definition of an ideal-type. It is often the case, when an historic national movement is fragmented, that its systemic parties lead pro-development coalitions of the ethnic territory; since anti-system parties are trapped in pockets of exclusion and social marginalisation, territorial development lacks cohesion. In extreme cases, where excluded anti-system movements opt for political violence, the internal fracture deepens and the competitiveness of the territory loses value in the international sphere. This undermines the confidence of foreign investors seeking territories where there is a secure return for their capital. In this case, the position of systemic parties in relation to international interlocutors is that of asserting their role in order to place a check on violent demonstrations, obtaining this way their support to assume key posts in the drive toward the development of their territories from a self-centred perspective.

9 Other Nationalisms: Break Up of States, Pan-Nationalisms, Non-Territorial and A-National Ethnic Movements

The previous chapters dealt with the peripheral national movements that emerged in horizontal nation-states as a reaction to state processes of political integration and national acculturation. These nationalisms assumed the form of social movements, which involved the re-elaboration of a specific collective identity and ideology by an elite: as the movement matures this process will eventually lead into mass mobilisations of the ethnic group. The nationalist movement passes from an ethnic and selective identity-based phase, mirror-imaging the community, to a rational instrumental phase characterised by the construction of society and demands for institutionalisation that are inclusive of all inhabitants of the territory and not just the members of the ethnic group. The decentralisation inherent in the latter phases of the welfare state and the emergence of a supra-state and trans-regional space associated to the European construction make this model more complex, but do not denaturalise it.

However, whilst this model fits well into the geopolitical area of Western Europe, its application on a global scale is uncertain. In practice, peripheral reactions have diverse origins, and alongside the nationalism/social movement model there are at least two further variants. One is encountered in the vertical multi-ethnic states of the Third World, and involves the reaction of self-organised tribal, clan-based (or ideological) societies against the centre's imposition of persecution or discrimination. A second variant is present in "imperial" federal states, which have disintegrated due to the decomposition of the Marxist-Leninist macro-ideology that held them together. In this case, political elites head state constructions that, at least in their early stages, are secessionist processes led from top to bottom, although this can later lead into violently ethnocentric mass mobilisations. Socialist states confront then peripheral nationalisms that appear to strengthen whilst the centre fragments (as is the case with the former Soviet Union and Yugoslavia) or that have historically opposed the centre (as in the People's Republic of China). Such nationalisms are by no means homogeneous; their differences depend on the existence or not of an imperial centre, and on the degree of consistence of the latter.

Further "deviated" cases include those in which the territorial dimension is either not well defined or non-existent. Such movements, which do not develop processes that mirror image the nation-state (since the state, after all, is the exercise of exclusive sovereignty over a population contained in a given territory) are based more on ethnicity (or civilisation) than national grounds. A number of different movements fall into this category, such as the 'Pan-Nationalisms': Pan-Slavism, Pan-Germanism, Pan-Americanism, Pan-Turkism, Hispanity, Pan-Africanism, Pan-Arabism, Pan-Islamism... Another category are the non-territorial ethnic (or ethno-national) movements of religious communities, as Jews, or of nomadic groups, as gypsies. A-national movements constitute a final variant, encompassing ethnic groups that have been left on the margins of the nation-state they inhabit: native Americans, Laplanders, Australian Aborigines... Those movements often arise within settler colonies (this is not always the case, as the Laplanders demonstrate), as a reaction to profound deculturation.

Break up of "Real Socialism" Federal States

It is within "State Socialist" federal states that contemporary processes of fragmentation are underway. In some senses comparisons can be drawn between some of these states, as former Soviet Union and People' Republic of China, and historical empires: both are multi-cultural states which draw together a plurality of peoples over a vast area, both mobilised resources and diverted energies into promoting universal projects - the socialist states, a strongly ideological project, involving the construction of state socialism before the final triumph of communism at a world level. Their pluri-ethnic structure is similar to many Third World states, although, unlike them they are not vertical, rather they possess a strong and institutionalised centre (in the Soviet and Chinese cases, but not in the case of Yugoslavia) concentrated in the ruling communist party, and another historical-social centre made up, respectively, of the Russian and Chinese 'Han' ethnic majorities: their multi-culturalism and ethnic pluralism are represented precisely by the federal nature of the state.

Such states are culturally held together by ideology. When this begins to decompose the relationship between centre and peripheries determines whether the crisis will provoke the territorial disintegration of the state, and whether subsequent relations will be of a conciliatory or conflictive nature; and, if this is the case, it determines also the orientation of the process of state construction in the new nation-states.

In China the imbalance between the giant "han" imperial majority and the "small [ethnic] brothers" of the border regions eliminated from the constitution (in contrast to the Soviet Union) any reference, even programmatic, to the right to self-determination. For the time being at least, this has made unviable any move to secession; national conflicts are concentrated on resistance to the Chinese centre by Tibet and the Islamic Xinjiang.

In the Soviet Union, which had transformed the Russian centre into a large Federal Republic and federalised the non-Slav and the non-Russian Slav groups, the emancipating movement had a double origin: first, nationalist movements, and second, the federal communist apparatus transformed into modernising elites, who petitioned for a state as their prerogative. The nationalist movements saw independence as the result of the mobilisation of a wide social base, while the federal communists perceived it as a process of political secession directed from above (and negotiated as much as possible with the Russian centre). The dominance of the recycled former elites has created a triangular playing field in which the new Russian Federation continues to intervene in the new nation-states, be this institutionally (through the Commonwealth of Independent States, CIS) or through the economic links and military commitments the former republics hold with the centre.

This duality of relations between the Russian centre and the new political entities on the one hand, and between the former-communist elites and nationalisms within each of these entities on the other, also occurs within the non-Russian territories which the USSR transformed into Autonomous Republics, because of their geographical position within Russia or because of the small numbers of their inhabitants. Relations between Russia and these republics range from rational negotiation to open warfare.

Civil society is weak when these federal territories become nation-states; experienced political elites overwhelmingly dominate their socio-economic counterparts, favouring authoritarian politics. Given that the new centre almost always finds itself submitted to threats from the old imperial centre and from new rival states, the new-born state itself becomes, as it happens always when the construction of the centre meets with deep resistance, intensely centralist. The formation of the community uses mechanisms of identification with power, which impede expressions of plurality; the construction of society is promoted through authoritarian means within which the economy serves political ends. If the former federal state did not possess an undisputed centre, but rather had a number of parts with similar strength, the new nation states may conclude that only the consolidation of a homogenous ethnic bastion will save them from internal and external risks. If the ethnic groups are spread out over the territory of the former state, as in the former-Yugoslavia, the minorities scattered in neighbouring states, on suffering increasing harassment, may reclaim their ethnic kinship with the mother state. If one of the new states lights the spark of ethnic homogenisation, the flames will spread to all the rest.

The community is produced in this increasingly harmful way through friend-enemy polarisation; a new ethno-centrism is born which generates a hostility that opposes largely secularly mixed human collectives. If the conflict is exacerbated by a clash of civilisations (as it is the case in the Balkans, where a new division between Latin and Orthodox civilisations has emerged, driven on both by the West and the new Russia), the catastrophe will be deeper, its depth correlating with the degree of ethnic and cultural plurality of the territory - Bosnia is a prime example of this.[1]

In the People's Republic of China there are currently 1,200 million people, 92 % of whom belong to the massive Han ethnic majority. There is also a mix of national minorities, which have been referred to as "small brothers" and who total 90 million people; they inhabit areas that occupy 90% of China's border regions and 60% of its total territory, including all of the west and parts of the northeast and south. These territories contain the majority of China's forests and cattle grazing lands, and a great deal of its mineral resources.

The "Han" ethnic group, the largest group on earth, was formed as a result of the fusion of three thousand peoples over the course of three millenniums (a fusion that is written into Chinese mythical history, in a legend that speaks of the union between the north eastern peoples and those of the Yellow river promoted by the Yellow emperor), which led to the emergence of the Hia, the ancestors of the Han A. Following their unification in the 3rd century b.c, the Han, who originated from the northern lands, continued to spread out and gain control over the non-Han peoples, who they looked down upon. The reciprocal relations between the peoples however were complex. Confucianism advocated the non-violent assimilation of non-Han peoples and two non-Han groups, the Mongols and Manchurians, governed the Chinese Empire for centuries. The final dynasty, in power between 1644 and 1911, itself Manchurian, violently repressed a number of non-Han rebellions. The process of forceful assimilation undertaken in the 20th century by the Kuomintang Chinese nationalist government did not succeed.

Although the Chinese Communists in their II Congress in 1922 adopted the Leninist principle of respecting national minorities within a unified state, in 1935 they reduced this commitment to a level below that of the Soviet Union, substituting federalism for autonomy and excluding the secessionist option of self-determination, considering it incompatible with the situation in China. The People's Republic organised along these lines in the 1950s as a multinational state with 41 autonomous areas (the territories inhabited by minorities) consisting of five regions, 31 prefectures and 105 sub-prefectures that covered 64% of the state's territory. 63 million of the 143 million inhabitants of these territories were members of non-Han groups. The 1984 Regional Autonomy of National Minorities law consecrated the principle of "a state of many nationalities" which was complemented by the principle that proclaims "political integration with cultural diversity". But the fact that autonomy legislation must subordinate itself to "democratic centralism", which is controlled by the central government, and the bonding of autonomy to territory (which excludes a third of non-Hans who live outside of these areas) have been sources of discontent.

Following the foundation of the People's Republic, 400 ethnic groups petitioned for minority status. This was granted to 54 of the groups in 1953 and was denied to the rest, who were seen as branches of other larger minority groups or of the Han. The size of the minorities ranges between the 15 million Zhuang (who live in the southern provinces which border Indochina and Burma) and the 2,300 members of the Lhoba people in the southeast of Tibet. On average minority groups have around 200,000 members, with 18 of them having over a million.

The large and depopulated western provinces are especially diverse in ethnic terms: Xinjiang, which borders Russian Siberia and former Soviet central Asia; Tibet, which borders the Indian sub-continent; and Qinghai, a province adjacent to Tibet and Xinjiang. Whilst the 2.2 million inhabitants of the province of Tibet are 90% Tibetan, the northern province of Inner Mongolia has less than 20% Mongols. The 17 million inhabitants of Xinjiang are 63% non-Han; the dominant groups within them are Turk-Altaic speaking Muslims, seven million Uygur, and one million Kazakhs, Kirghize, Mongolians, Manchus and Huis. Half the Qinghai population is made up of Mongols and Tibetans.

With the exception of the Hui and Manchu groups, who speak Han, the rest of the ethnic groups have their own languages, which belong to the Sino-Tibetan, Altaic and south Asian language groups. The Chinese communists promoted minority access to a written language (before 1949 only 20 groups had access to it). In the 1990s there were 91 universities in autonomous areas and 13 institutes of national minorities (3 of them in Peking) in which courses in these languages were available. However, within education Chinese Han has become stronger over the last few years at the expense of minority languages.

Many minorities possess strong religious traditions including Muslim, Buddhist and others; for some of them religion is their principle marker of identity (for example the Hui and Uigures, Islamic peoples of Xinjiang, and the Tibetans and Mongols, Lamaists). Although the constitution of the People's Republic of China guarantees religious freedom, which is sometimes viewed by the Chinese government as a useful aid in the struggle against illiteracy, on occasion it has been seen as an obstacle to progress or, at worst, as a pro-separatist and anti-communist agitation.

Chinese policy towards nationalities has had three distinct phases: first, initial flexibility, especially between 1949 and 1957 before the Great Leap Forward; second, extreme rigidity in the decade of the Cultural Revolution (1966-76) during which the existence of minorities in China was denied and the distinctive autonomous areas were dissolved; and, finally, the re-launch of autonomy in 1978. Development programmes in autonomous regions have occasionally provoked conflict due to their anti-ecological character and through the intense Han immigration they have produced- for example, the Han population in Xinjiang rose from 6% in 1952 to almost 40% in the 1970s. Despite the autonomy laws of 1984 which technically gave national groups decision-making capacity over their own development, the immigration of Han elites and the subordination of local authorities to the centre in Peking (where Hans are the majority) has depleted this capacity in important areas such as deciding the proportion of local people in top decision-making posts, the methods of resource exploitation, the level of Han immigration, and the deployment of nuclear arms within their border territories. In addition, the modernisation and the construction of a socialist market economy, encouraged from 1978, has privileged the heavily populated eastern coastal provinces at the expense of inland provinces where almost all minorities live.

Open national conflicts are concentrated in Tibet and Xinjiang, with Chinese presence in Tibet regarded as a foreign invasion. The Tibetan Turfan dynasty treated the Chinese Tang dynasty as equals between the 7th and 10th centuries. The Mongols who dominated China and Tibet in the 13th century granted the country to a Buddhist religious chief, which in reality created a theocracy governed by Lamaist monasteries. Since 1754, when the Ts'ing dynasty helped Tibet defend itself from the Ghurkha invasion from Nepal, Chinese influence established itself in the territories, producing a number of armed conflicts over the following two centuries.

Following Tibet's support of the nationalist Kuomintang government, the Chinese army invaded in October 1950 and declared the territory an autonomous region of the People's Republic. Although the central government made huge investments in roads, schools and medical care, it oppressed any Tibetan who favoured independence. This produced a rebellion over the period 1959-62, which was put down by Peking and led the political-spiritual head of Tibet, the Dalai Lama, into exile in the Indian region of Dharmsala. Following this, a Tibetan parliament was created in the area, elected by the 100,000 Tibetans who lived in exile. The moderate proposal put forward by the Dalai Lama in the European Parliament in Strasbourg, which would have given China control of the defence and foreign representation of Tibet in exchange for freedom of Tibetans to establish their own self-governing bodies, was ignored in Peking, and mobilisations and repression continue to the present day.

By 1962, 70,000 Kazakos had crossed the border into the Soviet Union from Chinese-held Xinjiang. In May 1989, shortly before the Tiananmen 4th of July movement, a book published in Shanghai about Muslim religion, that had offended the Islamic population, provoked an assault by hundreds of Muslims on the Party headquarters in the capital, Urumqi. In 1990, the Chinese authorities, fearing that Islamic contagion would travel over the border from the Soviet central Asian regions, began to spread rumours of "separatism." In the spring of 1990, in the region of Baren, which borders Afghanistan, there was a Muslim rebellion against the forces of order which reached the capital Urumqi and Kashga, which had to be closed down, and during which the leader of the movement "Freedom for Eastern Turkestan" died. The Secretary of the Chinese Communist Party, Jiang Zemin, promised help for Xinjiang at the same time as he threatened to oppress the "separatism". Tension persists in the region (Postiglione, 1995).

The collapse of Marxism-Leninism (defined by a single party and a planned economy) has broken the axis upon which the hierarchy of the peoples of the Soviet Empire rested, provoking differing social effects across its three rings: the outer ring, the former sovereign states of Eastern Europe; the middle ring, the former Federal Republics of the Soviet Union; and the inner ring, the Autonomous Republics of the current Russian Federation.

The middle ring consists of the former Soviet Federal Republics of the Baltic, Moldova, the non-Russian Slav republics (the Ukraine and Belarus), the three principal entities of the Caucasus, and the five Asiatic republics of Islamic Turkestan. Most, with some exceptions (Armenia and the Baltic countries), have undertaken a process of state formation within which the protagonist group - referred to by Marx as the "national class"- has been those who held senior posts in the Soviet administration, the *apparatchiki* of a party that had been structured on a federal basis (in the West, by the contrary, the industrial-finance bourgeoisie had fulfilled the role of national class).

The inner ring consists of the Autonomous Republics of the current Russian Federation. The interaction of the central Russian state with the Autonomous Republics has ranged from the brutality it exercises in the case of Chechnya to the establishment of an asymmetrical federalist relationship with Tatarstan and other Volga region republics, a similar relation to the sovereignty-association relations pursued by Quebec for years with Canada.

The federal constitution of the USSR facilitated the independence of the former Federal Republics in the 1990s and, therefore, greater autonomy for the current Autonomous Republics of the Russian Federation. The solutions put forward by Lenin and Stalin to the ethnic problem were territorial in nature, and whilst they did not open the door to national sovereignty for these peoples, they did facilitate the access of their members to cultural rights and nationality defined in federal terms. This solution led to the formation of a number of Federal Republics, and a greater number of Autonomous Republics, Provinces and Districts. As a result all citizens of the Soviet Union were also citizens of a Republic or Autonomous Province, and ethnic nationality was indicated (and continues to be so) in passports.

One month after the October Revolution, in November 1917, the *Declaration of the Rights of Peoples* in Russia proclaimed the right of nationalities to separate from the empire. Poland, Finland, the Baltic States, Georgia and the Ukraine became independent. In terms of internal territorial administration the Soviet's principle of power (Worker and Soldier's Councils) was imposed. Some nations, such as the Ukraine, Belarus, Azerbaijan and Armenia in the Caucasus and others in the Far East, organised as Soviet Republics.

However, the power of the Communist party, and hence centralism, clawed back ground from territorial autonomy from 1919 onwards. The Red Army conquered Georgia, which had been governed by Mensheviks, in 1922: the Bolsheviks who took power in December of the same year established a

Transcaucasian Republic consisting of Georgia, Armenia and Azerbaijan. On the 30th of December, the First Soviet Congress approved the treaty that created the Union of Soviet Socialist Republics (USSR) endorsed by the Russian Socialist Federative Soviet Republic, the Soviet Socialist Republics of Ukraine and Belarus, and the Federal Republic of Transcaucasia.

The Soviet Constitution of 1924, whilst safeguarding the linguistic-cultural rights of the ethnically diverse Federal Republics, conferred wide powers on the Central Federation. In 1936, when there were already eleven federal republics (the USSR would annex the Baltic States three years later following a no aggression pact with Hitler), the Second Constitution proclaimed the right to secession. It was, however, a purely verbal declaration condemned to be practically unviable. The constitution established the formal subordination of the periphery to the centre; the Presidium or the USSR Government could revoke or suspend the decisions of the governments of the Federal Republics. Real power, in the hands of the Communist Party, was also concentrated: the Communist Party of the Soviet Union (CPSU) was structured in the format of a pyramid, with its base being made up by the respective Communist Parties of the Republics totally subordinate to the CPSU. (Russia did not have its own Communist Party, because of the will to impede the emergence of an alternative focus of power to the CPSU).

However, although their mandate was not autonomous, the formal structures of the federal governments and communist parties of the republics generated a dual federal and communist elite that formed a single bloc of peripheral *apparatchiki*. Towards the end of the 1980s such elites, acting either alone, in unison with nationalist movements, or at times in opposition to them, led the process of independence of the Federal Republics and the reinforcement of power of the Autonomous Republics.

The territories of three of the areas of the former Soviet Union, the Baltic states, Transcauscasia and Central Asia, became independent in 1991 after the collapse of the USSR. Economic satellites and torn by inter-ethnic conflicts, the real extent of their subordination to Moscow has depended in each state on the interplay of three elements- the national movements, the local evolution of the old Soviet apparatus and the Russian centre.

In the Baltic, Estonia has 1.6 million inhabitants of whom 61% are Estonian and 30 % Russian; Latvia has 2.7 million of whom 52% are Latvian and 34% Russian; and Lithuania 3.7 million of whom 80% are Lithuanian and 9% Russian. Estonia and Latvia were occupied in the 13th century by the Germanics, who were displaced in the 16th century in Estonia by the Swedish and in Latvia by the Polish; both countries are religiously Lutheran. Lithuania, in contrast, maintained its independence, receiving Polish and Catholic influences. The great Dukedom of Lithuania arrived in the 16th century, before unification with Poland, from the Baltic to the Black Sea. These countries were occupied during the 18th century by the Russian Empire. In 1918-20, having defeated the Soviets, the three nations gained independence. Their regimes subsequently became right-wing dictatorships (in Lithuania in 1926, in Estonia and Latvia in 1934). In 1940, following the Molotov-Ribbentrop Pact, Stalin re-annexed all three, putting down mobilisations and undertaking mass deportations.

These circumstances made the three areas the most active focal point for anti-Soviet resistance, of which Lithuania with its active Catholic church and low proportion of Russians was the major protagonist. An intense historical-cultural renaissance took place in Lithuania in the 1970s within which the Lithuania-Helsinki Watch Committee denounced past and present repression, with intellectuals demanding the publication of the German-Soviet Pact of 1940.

Transcaucasia is a mountainous region that stretches from the Black sea to the Caspian Sea. Within it Georgia currently has a population of 5.3 million, of whom 70% are Georgian and 6% Russian; Armenia has 3.4 million, of whom 93% are Armenian and 2% Russian; and Azerbaijan 6.8 million, of whom 82% are Azerbaijanis and 6% Russian. Georgians and Armenians are the descendents of ancient indigenous peoples who converted to Christianity in the 4th century. Both within their own empires and whilst they were vassals of foreign empires, their national churches preserved their language and culture. The current Azerbaijanis originate from the Turk-Altaic speaking sheperd peoples of Muslim religion, who came from the northeast of Iran and who settled in the eastern plains of the Caucasus without developing their own form of state. The Russian Empire conquered Georgia at the end of the 18th century and absorbed Armenia and Azerbaijan at the beginning of the 19th. The Armenians were the most urbanised, their bourgeoisie dominated manufacturing in Tiblisi and the oil industry in Baku. Georgians were governed by their nobility, and the Azerbaijanis, whose identity was more Islamic than national, by their clerks.

In 1917, Georgians and Armenians each created nation-states. The Menshevik rulers of Georgia enjoyed the support of European socialists, whilst Armenians, from 1915 onwards, felt overwhelmed by their southern brothers who were fleeing Ottoman genocide; the Kemälist Turkish republic was perceived as supporting the Azerbaijanis. In 1920-22, the Bolsheviks unified the three republics into the Transcaucasia Federation, crushing rebellions by the Armenians in 1921 and Georgians in 1924. The Soviet Constitution of 1936 converted them into individual republics.

In the 1970s, the Georgia-Helsinki Watch Committee demanded official status for the Georgian language. Since the 1940s Armenia had petitioned for the region of Nagornyi-Karabakh to be added to its territory; the region, an enclave whose population is 90% Armenian, was ceded to Azerbaijan in 1921. In the 1960s, the Armenian Youth Union started to commemorate Armenian genocide at the hands of the Turkish. In the same decade the party of National Unification reclaimed territorial unity, as did the Armenia-Helsinki Watch Committee in the 1970s. Both were oppressed as a result.

Ex-Soviet central Asia is currently made up of five republics. Kazakhstan has a population of 16.5 million, 40% of whom are Kazakhs, and 38% Russian. Uzbekistan has 19 million people, 71% Uzbekistanis and 8% Russian. Kyrgizistan has 4.1 million, 52 % Kirguizistanis and 21.5% Russians. Turkmenistan has 3.3 million, 72% Turkmenistanis and 9.5% Russians. Finally,

Tajikistan has 4.8 million, 62% Tajik and 7.6% Russians. The farmers of the oasis and the nomadic shepherds of the steppes made up the historic population of the long corridor that runs between the Himalayan mountains and the Caspian Sea, bordering western China and Afghanistan and Iran to the south. The Arabs brought Islamic religion to the zone in the 7th and 8th centuries, and built up commerce in some of the region's cities. They were succeeded first by the Turks them by the Mongols, as Tamerlan made the area the centre of his empire. The area's vast prosperity fell from the 16th century onwards due to the discovery of new routes to the east, and the zone fell under Chinese and British control.

The Russians conquered the area in the middle of the 19th century, establishing a large-scale monoculture of cotton and setting up Russian and Slav colonies; they also introduced their own Cyrillic alphabet and discriminated against Islam. The rebellion that began in 1916 against famine and the political repression of Tsars, continued without lasting solution until 1923, directed by then against the Bolsheviks, who, whilst not Tsarists, were still European. The Basmachis, who were initially bandits, reinitiated the struggle towards the end of the 1920s, this time against the collectivisation of land.

Protest accelerated in the Baltic States in the Gorbachev era, as national and ecological demands united. In 1988 the Sajudis national movement was formed in Lithuania, on the basis of an anti-nuclear programme. The Sajudis defeated the Lithuanian Communist Party in the March 1988 elections, but despite this the LCP did adhere to the movement's independence programme. A human chain of a million people over the three Baltic countries was formed to demonstrate rejection of the 1940 pact, and in March 1990 Lithuania unilaterally proclaimed its independence. Gorbachev imposed economic sanctions, reinforced Soviet military presence and in January 1991 repressed mobilisations in all three countries. However, following the aborted coup of August 1991, the declarations of independence of the three Baltic States were recognised by the central government and supported by the West.

From this moment on tensions with Moscow were concentrated on economic, linguistic and military factors. The economic orientation of Estonia and Latvia towards the West allowed them to maintain centre-right governments; however, in Lithuania, where the economy is 90% dependent on Russia, the Sajudis suffered defeat at the ballot box in autumn 1992. Their Communist replacements promised better relations with Moscow. The Russian-speaking citizens of Estonia denounced as discriminatory laws on languages and citizenship. Opposition to the 200,000 Russian soldiers that remained in the territories forced their repatriation from Lithuania in 1993, and similarly from Estonia and Latvia in 1994.

Inter-ethnic conflicts in the countries of Transcaucasia have allowed Moscow to continue to intervene in their affairs. Georgia has a number of ethnically diverse groups including Abjazia, which demands independence, and Southern Ossetia which seeks unification with Northern Ossetia (and hence with the Russian Federation where the latter is situated). The Russian military strikes in 1989 against Georgian independence fighters strengthened the position of their leader, Gamsakhurdia, who took power in 1990; the attacks also provoked a more hard-line stance against Georgia's own minorities which led to clashes and deaths. In 1991, Georgia's minorities turned to Moscow following the independent state's rejection of affiliation to the Commonwealth of Independent States (CIS), which had been encouraged by Russia. The pro-Russian opposition managed to remove Gamsakhurdia at the beginning of 1992 and replaced him with the former-soviet leader Schevarnadze. After an initial phase of indecision from Schevarnadze in which Abjazians and Gamsakhurdia's followers gained ground, Schevarnadze changed his position and Georgia entered the CIS in October 1993. He had succeeded in putting down the rebellions of the two opposition groupings, but at the price of subordination to Moscow.

The Armenian protests of 1987-8 over Nagornyi-Karabakh and the damages to the region's ecology were ignored by Moscow. In 1990, the Communist party in Armenia was displaced by the National Pan-Armenian Movement of Ter Petrosian who, having been elected head of state, declared independence. In 1989 Azerbaijan saw the establishment of the Azerbaijan Popular Front, led by Elchibei. Following the massacre of Armenian residents in 1990, Gorbachev declared a state of emergency in the territory, but sided with the Azerbaijanis, and thousands of Armenians were displaced from border regions. In 1992, however, Russia switched sides, and Ter Petrosian's Armenia entered the CIS; whilst, in Azerbaijan Elchibei's nationalists won the June 1992 elections and requested military assistance from Turkey. The decision turned out to be disastrous for Azerbaijan; the Karabakh Armenians re-took the territory at the cost of a million Azerbaijani refugees. Elchibei was subsequently ousted in a mutiny of his own military and Moscow replaced him with the former communist leader Aliyev who, like Schevarnadze, joined the CIS and went on to defeat militarily Armenia and Karabakh.

In the Republics of central Asia, clashes between diverse indigenous ethnic groups, Kyrgyz, Uzbeks and in the Fargana Valley, Tajiks, began with the revolts of Alma-Ata in 1986 and continued to the 1990s. There were also clashes between these and European Slav colonies, and deported groups such as Armenians in Tajikistan. Ecological disasters in the region, such as the drying up of the Aral Sea, led to intense protests, and there was an emergence of National Fronts, including the Birlik in Uzbekistan, the Agzybirlik in Turkmenistan and the Kirguiz Democratic movement. However these Fronts developed at a later stage and with less protest towards the centre than their Baltic and Caucasus counterparts, and the five new states had no problem signing up to the CIS. In all of these countries the new government has evolved from the former communist parties. The able Kazakhstan leader Nazarbaiev obtained large amounts of aid from the West in exchange for disarming 1,400 nuclear warheads.

The greatest threat to Russia is derived from Islamic groups, especially in Tajikistan, a Persian-speaking enclave in a Turkish-speaking area, which is open to the influence of the Islamic Iranian revolution. Since 1992, the most serious war of the post-Soviet era has been fought in the area; a war in which a coalition of Islamic forces and reformists has fought against President

Rajmanov, head of the former communist elite, who has been supported by Russia and neighbouring states who fear the spread of fundamentalism from the south. The war has produced over 50,000 fatalities and 500,000 refugees, many of whom have fled to Afghanistan.

However, it quickly became apparent that the independence of the Federal Republics of the former Soviet Union in the second half of 1991 did not eliminate national problems in Russian state territory. Non-Russian peoples made up 46% of the former USSR; in the Russian Federation they continue to constitute 18% - around 30 million people. Most of these peoples live in autonomous republics, which are largely concentrated in three zones, the Northern Caucasus which encompasses eight republics (from east to west: Kalmykiya, Dagestan, Chechnya, Ingushetiya, Northern Ossetia, Kabardino-Balkar, Karachayevo-Cherkessiya and Adygeya) , the Volga-Ural region of the dominant Tartar ethnic group with six republics (from North to South: Mordovia, Mariel, Chuvashiya, Udmurtiya, Tatarstan and Bashkortstan), and the four Asian republics to the north of Mongolia, Gorni-Atai, Jakasia, Tyva and Buryatiya. Additionally there are also three climatically semi-polar republics, namely the Komi and Kareliya in Europe and the Shaka-Yakutiya in Siberia.

The transition from the totalitarian Soviet Union to the post-totalitarian Russian Federation generated the need for both a new distribution of territorial power within the new state and a change in the centre's relationship with federal subjects, especially with the Autonomous Republics, within which the non-Russian ethnic groups are concentrated. As with the Federal Republics this was characterised by a dual process, which involved the transformation of the former communist elites into representative elites (with the aim of maintaining power), and the appearance of nationalist parties that interacted with such elites.

Some of the Republics, Republics with economic resources and strong ethno-national personalities, such as Chechnya and Tatarstan, made unilateral declarations of independence in 1990.

Tatarstan is the leading republic in the creation of a new form of relations with the central Russian state. Six gas pipelines pass from Siberia through Tatar territory, and the republic possesses heavy industry and is rich in oil (producing 10% of the Russian total). Tatarstan has a current population of 3.6 million, of whom 48.5% are Tatar and 43% Russian, being the centre of a large ethnic group of 7.5 million Tatars who share Islamic faith and Turkish language and are spread across the Russian territory, with substantial pockets of population in the Ukrainian Crimea and Bashkotorstan. (Tatars are descendents of Genghis Khan and the Gold Hordes; the struggle against them led to the unity of the Russian Empire in the times of Ivan the Terrible).

Tatarstan's current leadership among non-russian groups has to be attributed to political, economic, cultural and religious factors- political, the consolidation of its governing elites and leadership of President Shaimiev; economic, the development of the petroleum industry and the production of military armament; culturally through the historic relevance of the Tatar peoples (who were the dominant group in medieval Russia); and religiously through the Islamic creed of Tatar. Additional factors include the easy interaction between the governing elite and the Ittifac, Azatlyk and Tatar Public Centre nationalist parties. The ex-communist elites and former Soviet administration, who are the leading group in Tatarstan, defend associated economic sovereignty, the territorial status of the Republic and the common citizenship within it of Russians and Tartars. The nationalist parties defend the political sovereignty and cultural rights of the Tatars (both those in the republic and those who are outside), but open conflict between the projects of both collectives has not broken out. This pioneering character has made Tatarstan the model for others to follow in the processes of constituting new federal subjects in the current Russian Federation.

When the risk became apparent that the disintegration of the middle ring would affect Russia internally, Yeltsin's government set to work on a Federal Treaty that would preserve the integrity of the Federation. However, ethnic and territorial-administrative principles conflicted within the structure of the treaty: its 88 subjects were not only the twenty ethnically differentiated republics, but also Russian provinces and territories (known as Krais, Oblast and Okrug), besides a number of important cities.

Just weeks before the 31 March 1992, the date the treaty was set to be signed, the Tatarstan government held a referendum over its earlier declaration of independence, which was passed with 61.4% in favour. Tatarstan and Chechnya were therefore the only republics that did not sign up to the Federal Treaty. However, from this moment on the processes within the new entities totally diverged. Tatarstan, as soon at it had proclaimed sovereignty, initiated negotiations with the Russian government that led, after the signing of 15 specific agreements over economic and social materials, to a co-federal-type Treaty signed on the 15th of February 1994 by Presidents Yeltsin and Shaimiev. Chechnya, on the other hand, demanded that Moscow integrally accept its independence, and ended up being subject to military invasion in December 1994 with the aim of its reintegration in the Russian state, a brutal aggression that came to an end temporarily in a ceasefire in September 1996.[II]

The Russian-Tatarstan Treaty granted Tatarstan asymmetric federalist powers. In its introduction the Treaty claims to be founded on the "right to self-determination universally recognised for all nationalities" and proclaims that the Republic of Tatarstan would unite with the Russian Federation on the basis of the Russian and Tatarstan constitutions and the Treaty in question. Exclusive powers granted to Tatarstan, declared in Article Two, are the establishment of its own taxes (paragraph 2), total control of its natural resources (paragraph 5), the structures of the three powers of any state (paragraph 7), the establishment of its own military service (paragraph 10), (Tatarstan declared itself a de facto neutral zone and nuclear free zone, with the aim of preventing the involvement of its youth in outside ethnic wars and conflicts), the power to establish treaties with foreign states (paragraph 11), the creation of a national bank (paragraph 12), and the establishment of its own political economy (paragraph 13).

The Treaty, which in reality ignored the restrictions laid down in the state-wide Federal Treaty of March 1992, created a relation similar to the associated sovereignty that Quebec reclaimed to Canada. In practice, the Treaty became an instrument that helped to resolve problems through negotiation and therefore lessened confrontation. Because of this the Russian Minister for National and Regional Affairs, Shakhrai, stated that the Treaty could become the model for the 88 subjects of the Russian federation. He especially conceived it as a measure that might resolve the Chechen conflict. Moscow went on to complete similar treaties with Bashkotorstan, also a Volgan Republic, in August of 1994, and in November 1994 and March 1995 with the two Caucasus republics of Karbardino-Balkaria and Northern Ossetia.

However, the 1995 victory of the "party of war" in Russia and Russian military logic towards Chechnya provoked a huge increase in the more radical postures of Russian nationalism; this simultaneously influenced the attitudes of Russian leaders, and the process was stopped in its tracks. Zirinovski's party demanded the disappearance of the ethnic republics, the repeal of treaties such as that of Tatarstan that had already been signed, and the re-establishment across Russian territory of the former *gubernias*, (simple administrative demarcations that dated back to Tsarist times). Meanwhile Minister Shakhrai himself sent a memorandum to President Yeltsin in March 1995 recommending the freezing of Treaties that were ready for signing, Udmurtia in the Volga; Yakutiya in Siberia; Dagestan, Ingush and Karachevo-Cherkessia in the northern Caucasus, arguing that the signing of the treaties would lead the Russian nationalist opposition to accuse the government of breaking the unitary structures of the state and that the treaties could be used by some republics to open up international relations. The Ministerial substitution of Shakhrai by Mikhailov, a believer in the consensus approach, managed to re-establish the Tatarstan treaty to its previous status as the model to be followed. Similar treaties were signed in the same year in June with Yakutiya and in July with Buryatiya. However it is difficult to predict the immediate future, which is determined by the lack of political stability of the Russian Federation (Carrere D'Encausse, 1991; Knippenberg, 1991; Ziatdinov, 1994; Chorbajian, 1995).

The unique way within which history has constructed the Balkan Slavs has tragically led to the conflict that has torn them apart currently. Slavs descended on the Balkan peninsula from the 7th century onwards. Up until that time the region had been occupied by the Greeks, Illyrians (the ancestors of the Albanians), Datians (the ancient Romanians), and the Slav tribes of the Croats, Slovenians, Serbs and Bulgarians. The ancient division drawn up by Emperor Constantine in 395 between the eastern and western Roman Empires became the dividing line in the territory. Religiously, the destiny of Croats and Slovenians was tied in with Latin civilisation (which enveloped both Catholicism and Protestantism); politically, their destiny was first associated with the Charlemagne, then with the House of Hapsburg and latterly with the Austro-Hungarian Empire. The Serbs and Bulgarians constructed their kingdoms in the territories of the Byzantine Empire, adopting its orthodox culture and religion, and became part of the Ottoman Empire in the 14th and 15th centuries.

Within the Ottoman Empire Muslims were the only citizens with full rights; initially they set up a military oligarchy that soon became an urban and educated governing elite, that lived off the taxes charged to non-Muslim peasants. However, the Ottomans did not impose an ethnically exclusivist policy, and were incomparably more tolerant than the Christians of the age who created aberrations such as the Holy Inquisition: they respected the life and religious beliefs of "The Men of the Book" (Jews and Christians). Across the region and especially in cities, the Ottomans reached an understanding with the respective religious authorities who, in accordance with the *Millet* system, were recognised as the political representatives of the ethnic communities.

Throughout the Middle Ages the Ottoman Empire confronted the Austrian Empire of the Hapsburgs. Permanent attachments of infantry made up of Serbs were deployed in the front line in border regions, largely on the banks of the Danube. It is notable that such areas are the current Serb enclaves in Croat territory, Voivodina, and also Krajina, Baninja, Slavonia.

Following the defeat of the Turks in 1683 at the entrance to Vienna, Ottoman political power declined. In the 19th century the Turks became the "West's sick man." Mid-century the Serbs became the first Slav people to gain autonomy from the Turks and, supported by the pan-Slavism of Russian Tsars, they defended the *Nacertanje*, a union of all Slavs in the Balkans, Serbs, Montenegrins, Macedonians and Bulgarians, against the Ottomans. This project became that of Great Serbia after the withdraw of Bulgarians.

"Yugoslavism" (the Union of Southern Slavs) was in reality a project of Croat intellectuals of the Austrian Empire, driven by the ideas of the French Revolution and the German *Aufklärung*. The project promoted both political unity and Serbo-Croat linguistic re-unification.

In the mid-nineteenth century, Bosnia-Herzegovina was the least developed region in the Balkans. The area was 50% Muslim (mainly concentrated in urban areas), 40% Orthodox and 10% Catholic. Contrary to Serbs and Croats, Bosnians had no national project nor reference in the era of nationalisms. The Austro-Hungarian Empire, which occupied Bosnia-Herzegovina from 1877 onwards, did so with a Croat army, and initiated a process of *Croatization* of the western zone, with numerous Catholic churches being built in Sarajevo. From this moment on the associated identity of being Catholic and Croat, and Serb and Orthodox, emerged. The new project of Greater Croatia also emerged, as opposed to Yugoslavism as the Greater Serbian project. The two projects were tied in with two respective "Big Brothers"- the former to Austria-Germany and the latter to Russia.

Following the defeat of the Austrian and Ottoman empires in the First World War the Kingdom of Serbs, Croats and Slovenians was set up, with the Slav territories of the former empires. However, intellectual Yugoslavism was quickly stifled by the Greater-Serbian centralism of King Alexander, who simultaneously provoked the emergence of the Croat Fascism of the Ante Pavelic's *Ustachis*. The Nazi occupation of the Balkans allowed the formation of the Greater Croatian satellite led by Pavelic, which took in a large part of Bosnia-Herzegovina, and whose detention centres performed acts of genocide against the Serbs similar in terror to the Nazi concentration camps of central Europe.

The Communist Resistance in the territory was the only movement in Eastern Europe to defeat fascism without outside assistance. Accordingly, it subsequently challenged Stalin as much as it did the Western powers. The movement was led by Tito, a Croat and fervent follower of *Yugoslavism*. As a result, the Yugoslavian Constitution of 1946 was based on the equality of rights of the Federal Republics.

Tito's death in 1981 and the subsequent crisis of state socialism over the whole decade, pulled *Yugoslavism* back into a process similar to the fall of *Sovietism* of the former USSR. Effectively, the Balkans were returned to the same situation they had faced at the beginning of the century. The political discourse of the emerging forces hid their true direction. The Serbian communists directed by Milosevic, far from being the heirs to Titoist *Yugoslavism*, were the champions of a new *Nacertanje*. Serb repression was especially apparent in the Autonomous Regions of the former Federal Republic, towards the Hungarian citizens of Vojvodina and, to an even greater extent, the Albanian majority in Kosovo. With regard to the Croat followers of General Tudjman, their proclamations in the defence of democracy and their Western credentials hid the reality of their Greater Croatian project.

The direct confrontations between Croatia and Serbia have taken place in the former enclaves of the Turk-Austrian military border, Krajina and Slavonia. Slovenia, on the contrary, has managed to preserve the peace and independence it gained in 1991. In Bosnia, the Bosnian Croats aligned with the Muslims to compensate their numerical inferiority, which had not prevented them to impose a feud in their own territory, above all in Herzegovina. Reminiscent of the situation in the Middle Ages, the armed conflict between the Bosnian Serbs and the Bosnian Muslims became a war between the rural and urban, the vast rural areas sparsely populated by Serb orthodox farmers, who closed in on and bombed the cities of educated Muslims. Ethnic cleansing became an interminable terror, because it had torn apart two historic processes: first the secular coexistence of ethno-religious groups in the Ottoman period, and, second the contemporary mobility of citizens promoted by Titoist development.

It would be unjust and incorrect however to put the disasters down to "exacerbated nationalisms" on their own. When the Iron Curtain collapsed a new dividing line emerged in Europe, a line that separated western-Latin civilisation from eastern Orthodox civilisation. The guardians of the former are Germany and the Vatican, and of the latter, Russia. It was this re-establishment of the Constantine's dividing line that split the former Yugoslavia in two, left Muslim Bosnia in the middle of the clash, and left behind the trail of terror and destruction we all know.

Before the war in 1992 Bosnia-Herzegovina had a population of 4.3 million of whom 42% were Bosnian Muslims, 32% Bosnian Serbs (Orthodox) and 17% Bosnian Croats (Catholics). The groups had become polarised since the first multi-party elections in November 1990. Muslims voted for Izetbegovic's Democratic Action Party (SDA), the Serbs for Karadzic's Serb Democratic Party (SDS) and the Croats for the Croat Democratic Community (HDZ), the party of Croat leader Tudjman. The Serb and Croat parties were open about their plans for integration in Greater Serbia and Greater Croatia, respectively.

When, in early 1992, the West recognised the independence of Croatia and Slovenia, it left Bosnia defenceless in the face of the territorial appetite of its more powerful neighbours. Izetbegovic held a referendum in March of that year for Bosnian independence. This was the spark for the bloodiest war to have broken out in Europe since the Second World War, within which the three groups (especially the Serbs and Croats) imposed their territorial base, which they conquered with arms and under the principle of national homogenisation, or "ethnic cleansing." The Serbs biggest advance, due to the support of the Milosevic's Yugoslav Federation, forced Muslims and Croats to unite in a Federation patronised by the West, but which did not avoid serious clashes in the Herzegovina region between the two supposed allies.

Joint pressure on the competing powers within the Yugoslav federation, forced by UN economic sanctions and the Contact Group of Western Powers, brought about, following a number of frustrated attempts, the opening of a process of negotiations between the three adversaries. In February 1995, the Serb authorities recognised the integrity of the Republic of Bosnia-Herzegovina; in the following months the territorial limits of the Croat-Muslim Federation and the Serb Republic of Srpska were defined. The limits, that attributed 51% of territory to the Federation and the remaining 49% to the Bosnian Serbs, consecrated in practice the territorial gains achieved by armed force, and therefore benefited the Serbs.

The three leaders Milosevic, Tudjman and Izetbegovic came together in November 1995 at a meeting arranged by President Clinton in Dayton, backed by the United States, the European Union and the Russian Federation. The resulting Treaties agreed the maintenance of Bosnia-Herzegovina as a unitary state, and to the creation of IFOR, a NATO international force composed of 60,000 personnel (including 20,000 from the US) to guarantee adherence to the accords.

The Dayton accords drew up a new type of state that was reminiscent, in terms of its complexity, to the former Yugoslavia, even though it was certainly more fragile, with the future maintenance of its integrity requiring foreign pressure. Bosnia-Herzegovina was to be made up of the Bosnian Federation and the Republic of Srpska. The presidency of the state would be collective and rotational, and would be made up of three members, a Muslim and a Croat elected by the Federation, and a Serb elected by Srpska. The candidate of the three that received the most votes would be named head of state for two years. There would be three parliaments, a central parliament in Bosnia-Herzegovina made up of 42 members (two thirds elected by the Federation and one third by the Serbs), a parliament of the Federation consisting of 140 members, and a parliament for the Srpska Republic also with 140 members. Srpska would have an elected president, but the Croat-Muslim Federation would not. The Federation would be divided into 10 cantons, each one with its own elected parliament; but the Serb Republic was not to be cantonalised.

Multiple elections were held on the 15th September 1996. The three ethnic parties- the Muslim SDA, the Serb SDS and the Croat HDZ-, were hegemonic in their respective areas, led, respectively, by Izetbegovic, the Bosnian Krajisnic for the presidency of Bosnia, Plavsic for the presidency of Srpska -

owing to the impossibility of Karadzic standing, as he was wanted as a war criminal-, and by the Croat Zubac. Although IFOR guaranteed freedom of passage from one ethnic zone to another, in practice very few Muslims or Serbs ventured across boundaries.

The Bosnian-Serb candidates (and some Bosnian-Croats) did not go without making proclamations in favour of Greater Serbia and Greater Croatia. Izetbegovic's party was contested by the Muslim Siladzic. The Serb Krajisnic was contested by Ivanic's party. The results confirmed the existence of a staunchly vertical Bosnia-Herzegovina, with the votes of the three ethnic parties not falling below 70% in their respective territories (although Ivanic's Serb opposition obtained the best results). Izetbegovic was named head of state with 729,000 votes, followed by the Serb Krajisnik with 690,000 votes and the Croat Zubac with 324,000. Croatia and Serbia, that had established full diplomatic relations in August, have exchanged ambassadors, but this sign of normality might mean worrying aspects for the new state of Bosnia-Herzegovina (Castellan, 1991; Guezennec, 1991; Samary, 1993; Palau, 1993; Tertsch, 1994; Vodopivec, 1994).

Pan-Nationalisms

Pan-Nationalisms defend a *Kulturnation*, a form of civilisation with boundaries vaster than any defined nation state, and whose reference is a community rather than a national society. Contrary to empires they are vehicles of ethnic homogenisation and often of linguistic and religious unification; consequently they do not tolerate cultural plurality. They can be distinguished from nation states in the sense that they do not pursue the construction of a territorial nation-state; the conglomerates of peoples they wish to unify in a *Kulturnation* may or may not be organised into a supra-national entity. Accordingly, Pan-Nationalisms have no confidence in the sovereignty of nation-states.

However, Pan-Nationalisms can become vehicles for traditional or modern empires, or of actual nation states. Examples include Pan-Germanism, which German National-Socialism finally managed to utilise; Pan-Slavism, a vehicle of the Russian Tsars; Pan-Americanism, the instrument of domination of the United States over the American continent; and Pan-Turkism, which was at the service of Kemälist Turkey. When the *Kulturnation* is more ideological than identity based, Pan-Nationalism remains weak; also, as in the case with Pan-Africanism, because the projected community does not exist outside the minds of its ideologists, or because it becomes the verbal proclamation of a political regime, the case with Hispanity espoused by Franco's Spain. Pan-Islamism is the most powerful contemporary Pan-Nationalism, which emerged out of the identity-based reaction of Muslim civilisation; it has now overcome the Pan-Arabism it competed with.

Pan-Germanism emerged over the course of German resistance to Napoleonic domination, promoted by the works of Fichte, Arndt and Jahn. The initially democratic movement advocate of the German *Kulturnation* later moved off in a reactionary direction defending the superiority of the German race. The greatest exponents of this were the Frenchman Gobineau, in the middle of the century, and subsequently H.S. Chamberlain, a German writer of English origin. Having influenced the earlier Germanist phase of the long reign of Austrian Emperor Francis Joseph, the association *Alldeutscher Verband* came to defend under its postulates the creation of a "Greater Germany", which would unify the Germanic peoples and expand towards the east. Nazism took on this ideology.

The objectives of Pan-Slavism were the preservation of the cultural inheritance of the Slav peoples and the promotion of their political unification. Such aims were declared towards the end of the Modern Age by Slav intellectuals in the Austrian Empire, such as the Croat Krizanic and the Czech philologist Dobrovsky. At the beginning of the 19th century Pan-Slavism was the inspiration for the liberal-revolutionary project of creating a democratic confederation of Slav peoples, expressed at the Slav Congress in Prague in 1848. But the Balkan Slav peoples, struggling against Ottoman domination, put their hopes in the protection of Tsarist empire. From this moment on Pan-Slavism was assumed by Russian ideologists that assimilated it with orthodox culture, arguing for a union of Slav peoples under the aegis of the Russian Empire, from which the Polish and Czechs would be excluded for being Catholic and too Western. Tsar Alexander II used this as an instrument of foreign policy between 1855 to 1881; it later became a right-wing opposition movement.

Pan-Turkism emerged at the time of the "Young Turk" movement. In the early 20th century, whilst the Young Turks were advocating the transformation of the worn out and defeated Ottoman empire into a modern nation-state, (a movement whose heir was the Turkey of Kemal Atatürk) Pan-Turkism defended the creation of a cultural ambit that would unify the Turkish-speaking Eurasian peoples.[III] This influenced Turkish policies of genocide against the Armenian people at the beginning of the century, and continues to inspire the current failure of the Turkish Government to recognise the existence of the Kurdish people.

Two conflictive expressions exist of Pan-Americanism, emerged over the course of the American independence wars. The dominant one served the plans of United States, the second proclaimed the union of American nations on the basis of equality. The first, which was formulated by Clay in 1820, culminated in the Monroe doctrine of 1823, which turned Pan-Americanism into the ideological instrument of US hegemony in the American continent. This tendency promoted the doctrine declared by Theodore Roosevelt in 1904, justifying US intervention in the continent. The second expression inspired the Bolivarian conference in Panama in 1826, but had little practical result (although its spirit was subsequently manifested in the Cuban revolution of Martí, and later in the revolutions of Castro in Cuba and the Sandinistas in Nicaragua). Interest in commercial relations between the United Sates and Latin America promoted the 1810 establishment in New York of the International Office of American Republics, which later became the All-American Union following the Buenos Aires Congress of 1910. In 1947, it became the Organisation of American States (OAS), a regional organisation linked to the UN but in reality an instrument of the United Sates; the OAS excluded Castro's Cuba.

During the period between the Cuban war of independence and the Second Republic in Spain the myth of *Hispanity* was consolidated in Latin America, more concretely so in Argentina. *Hispanity* is an ideological instrument of the traditional Creole oligarchies, in competition with pro-US Pan-Americanism and indigenist ideologies. The myth, which reached Spain in the 1930s, turns the concepts of Hispanic ideal, Hispanic race and Christian gentlemanliness into synonyms; it also presents the Spanish Empire's history of conquest and colonisation as a solution for republican disintegration. García de Vizcarra, a Spanish priest from Buenos Aires, proclaimed that the historic mission of the Spanish had been to undertake a Holy War or Crusade against Judaism, Islamism, Protestantism, and American idolatry. By this point Russia was regarded as the enemy; Vizcarra argued that total victory would be achieved "when a man engendered with Spanish race is proclaimed emperor"; at this moment, according to Vizcarra, "the spirit of Spain will stand up in a forthright and gentlemanly manner over the Judaism that tramples on it". Maeztu, ambassador to Argentina between 1927 and 1929, launched the theory of "*Anti-patria*" in 1931; this later became the "Anti-Spain" theory. Maeztu believed that the "*anti-patria* owes its existence to the *patria* (homeland), just as the Anti-Christ does to Christ." The *Patria* was *Hispanity*, traditional Catholicism in permanent action, an ideal which must awake the enthusiasm of ruling minorities in Spain and Hispanic America (Morodo, 1985). Hispanic Day, commemorated on the 12th of October, continues to be a national holiday in Spain.

Pan-Africanism, an ideology that emerged as a reaction to the white colonisation of the continent, has suffered from the extreme ethno-cultural and political heterogeneity of the African peoples. It initially emerged both through the resistance of the Ethiopian churches against white Christianity and in the works of Black ideologists originating from America, who advocated African emancipation. The Christian churches of Ethiopia opposed the image preached by white missionaries of the "white" Son of God. Its missionaries, more commonly "black" than african as many came from the United States, influenced the post-First World War South African and Belgium Congo emancipation movements.

Bryden, a Nigerian who had come from the West Indies, first used the slogan "Africa for the Africans" in 1891. The principle drive of Pan-Africanism came from Du Bois who before the First World War had set up the National Association for the Advancement of Coloured People in the United States. Following the War the organisation decided to extend itself to Africa. Du Bois along with Diagne, a Senegalese parliamentarian in France, set up the First Pan-African Congress in Paris in 1919, which was followed in the 1920s by further Congresses in London, Lisbon and New York. The fifth Congress, held in London in 1945, was attended by African leaders such as Nkrumah, of the Gold Coast, and Kenyatta from Kenya. African emancipation replaced the racial problems of black Americans as the principal theme of those attending the congress.

The strength of the independence movements caused the dissolution of the Congress. Marcus Garvey, a black American of Jamaican origin, accused the Congress of being too conciliatory. His black racism, that argued that Moses was black and that blacks were the chosen people, was widely adopted in Africa, until the 1925 failure of his negotiation with Liberia aiming at founding a Black empire.

Educated blacks in the British and French metropolis continued the movement. In 1920, and within British reformist and parliamentarian traditions, British blacks in Nigeria created the National Congress of British West-Africa, an idea inspired by Gandhi. The Senegalese led the movement in the French colonies. Leopold Senghor, a socialist and Minister in France (he actively participated in the Constitution of the Fifth French Republic), gave the movement intellectual credibility. A French-language poet, he developed, along with the West Indian poet Aimé Césaire, the concept of "blackness", the supposed personality of black peoples that designates mission and confers responsibilities. However, both of these are initiatives of Westernised intellectuals, which had no influence over the masses.

The Organisation of African Unity was formed in 1963, following the conceding of independence by France and the United Kingdom to numerous African states. But the OAU cannot be seen as the expression of Pan-Africanism. The OAU is an inter-Governmental organisation that argue, faced with the fragility and artificial nature of many of the new states, for the defence of the intangibility of the new state borders (which in reality were almost always the former colonial frontiers) (Bertaux, 1984).

Pan-Islamism, emerged as both a defensive reaction and a political-religious re-elaboration of a civilisation that was extremely powerful for centuries, more so than its Christian counterpart, is now, following the collapse of state socialism, the only "strong" ideology that has successfully resists Western-liberal domination of the world. Al Afghani, the first Pan-Islamic thinker, emerged in 1880 at a time of European imperialist advancement. Afghani's criticism of the Western ideas of industrial modernity and progress as barbarity and amputation of the total human being are similar to those that would be elaborated by the Frankfurt School half a century later. Afghani and his followers defended the preservation of Islam as a religious faith and as a historical-cultural conscience of Islamic peoples, who should become brothers against the West. In 1928, the Muslim Brothers emerged in Egypt, and later extended themselves to the near East, striving for a return to the tradition of Islamic societies.

Arab nationalism, or Pan-Arabism, emerged from the dismantling of the Ottoman Empire after the Great War, and consisted fundamentally in the laicisation and "Arabisation" of the idea of *Umma,* or community of believers in the Koran: the *Umma* should become the community of peoples of Arab language and culture. At the beginning, Arabism was supported by the Allied Powers, who wished to use it to finish off the Turkish "sick man" (this is where adventurers like Lawrence of Arabia fitted in). Lebanese Christians, influenced by the Germanic nationalism of Herder, developed this ideology. Sati el Husri's idea that the Arab nation should encompass Arabs united by language, culture and history, was radicalised in the 1940s by Michel Aflaq, the theorist of the Ba'ath. The Ba'ath calls for revolutionary action directed by an elite party that must use its revolutionary ideology against all forms of Western domination, including Zionism, capitalism and colonialism.

These are the aims of the emblematic Nasserite Arab nationalism of the 1950s. Not long before this, in 1945, the Arab League was created, becoming the institutional representation of Pan-Arabism. This deep movement that shook Arab countries from Baghdad to Rabat between 1952 and 1969, was fuelled by revolutions, liberation struggles and challenges to the West, but also by the defeats at the hands of Israel. Unitary Arab nationalism crystallised over the course of these events, in its dual aspect of transforming countries previously submitted to colonialism, especially in Maghrib, and promoting the construction of a single Arab nation under the aegis of Nasser.

However the legendary idea of the creation of an Arab State-Empire in the territory of the classical Islamic Empire of the Omeyas was a spectacular failure. This was not only due, as followers claim, to the Machiavellian actions of the West that would have broken a previously unified *Umma,* but also due to the same reasons that had fragmented the ancient Empires history. It was only sporadically that historic Islam managed to maintain the unity of countries whose realities on the ground pre-dated it, since it encompassed territories of ancient civilisations that were by no means homogeneous themselves, such as Mesopotamia, Egypt, Palestine, Maghrib and others. Because of this, in practice, new nation states were created from the fragments of the former Ottoman Empire (namely Syria, Irak and Arabia).

Despite Nasser's defeat in the Six Day War in 1967, the intellectual Arab left grew in strength until the mid-1970s. Later, however, Arab leadership was transferred to the traditional Saudi Arabian monarchy, whilst the core of the unitary *intelligentsia* reached a crisis of consciousness, especially after the Israeli invasion of Beirut in 1982.

However, it was during these years that took place the re-emergence of a new Islamism that was not exclusively Arab, promoted from 1979 onwards by the Iranian Revolution and impregnated by the subversive and heterodox values of Shiites. This Islamic tendency argues that the Caliphs, who succeeded Mohamed, became illegitimate by perverting the purity of Islam, which had been preserved by Ali, the fourth Caliph and son-in-law of the prophet, assassinated by the Omeyan usurpers. Shiites are Ali's followers, whilst the followers of the Omeyas, the majority, are the Sunnites. The Imams succeeded Ali; according to Shiite authority, the twelfth Imam, hidden by god from human sight, will return to earth at the opportune moment to restore God's Kingdom.

Therefore, whilst the Sunnites have historically been defenders of the established order, which they conceive as an emanation of the will of Ala, the Shiites consider all authority as essentially perverse; it must only be obeyed if there is no other option. Moreover, their insistence on the martyrdom of the legitimate Caliph and on the future return of a hidden Imam (a dual thematic reminiscent of Christianity) has made Shiite the religion of the poor and disinherited, who aspire to liberate themselves from an oppressive power; it is also the messianic dream of those who have nothing to lose but a lot to gain from radical social change.

Historically, Shiites have lost all their battles against Sunnite armies. The only country in which they triumphed was Iran, where they came in the 16th century through the Safevies dynasty of Turkish origin. They brought with them the Shiite erudites, the Ulemas (or Ayatollahs), whose opinions on Islamic law - that makes no distinction between religion, politics and society - achieved universal respect. From 1892 the Iranian Ulemas headed all the mass movements that mobilised to oppose any alliance of the Iranian Kingdom with a foreign power, and especially the Pahlevi dynasty, whose last Shah transformed the traditional monarchy into a modern pro-Western dictatorship.

The victorious Iranian revolution of December 1978 headed by Ayatollah Khomeini (a leading spiritual figure of the Institute of Theology in Quom until his forced exile), consecrated the triumph of the Ulemas. Khomeini was inspired by the philosophies of Chariati, also an Iranian theologian, who was assassinated in London in 1977. Chariati, who had made contact with the Algerian NLF in France, assimilating the Frantz Fanon concept of cultural alienation, developed a political version of Shiite beliefs in the future appearance of Imam: the wait would not be passive, but would consist in the revolutionary installation of a just society animated by the fear of God. Chariati resolved also the differences between Shiites and Sunnites, by defending the stance that Shiism should not impose itself by force.

In the 1980s this explosive cocktail of crypto-socialist ideology and subversive, pure and triumphal Shiism became the dominant reference point for dispossessed Muslims. The idea of Islamic revolution had no problem in replacing the conception of a unified Arab nation, a laic idea in crisis for over a decade. Its followers not only confronted the West and Zionism, but also the political and economic elites of the new Arab states, which were sufficiently distanced from the dispossessed masses for the claim of the illegitimacy of authority preached by Shiites to appear self-evident for the latter. Whilst at first its influence was felt in traditional Shiite strongholds such as south Lebanon, it soon extended to monolithically Sunnite countries like Morocco and Algeria.

Islamic fundamentalism is therefore strong in anti-Israeli organisations such as the Lebanese Hizbullah (1984), and is also embedded in groups such as the Palestine Islamic Resistance Movement Hamas (formed in 1988, the year in which the "Intifada" commenced, but also the year in which the PLO recognised the state of Israel), and the Algerian Islamic Salvation Front, ISF(1989). To such groups the former laicised revolutionary organisations have become elites which have either conquered power for personal gain like the NLF in Algeria, or have aspired to do so by selling-out the Palestinian cause to Zionism, like the PLO, considering them to be just as Western as the French colons or Zionist Jews which they opposed.

On the other hand, the defeats inflicted on Arab states in the past decade, such as Saddam Hussein's Ba'athist regime in the Iraq war of 1991, preceded by the selective American bombing of Libya, have been seen by these movements as a demonstration of the intrinsic evil of the West, reinforcing the hostile anti-Western polarisation of their identity markers (Lewis, 1991; Djait, 1991).

Non-Territorial Ethnic Movements[IV]

There are some human groups that, although they are not associated with a territory, do however socially reproduce features of their personality over the generations. Such groups fall in to two major categories. First they can be collectives that are more religious than ethnic in nature, and whose faith, passed on through family socialisation, is perceived by others and considered by the group itself as an element that sets it apart from other groups. They can also be authentic ethnic groups whose way of life involves territorial mobility and a nomadic existence. In the West the most notable examples in both cases are, respectively, Jews and Gypsies.

In common with territorial-based peripheral mobilisations, such groups cannot be referred to as ethnic (or ethno-national) until the group produces both an identity based reconstruction and an ideological re-elaboration of its historic personality. It has been the era of nation-states, especially destructive for the two aforementioned groups, which has provoked by reaction these processes in both groups. The impossibility of reclaiming territorial institutionalisation led the Jews to demand the creation of a "homeland" in a specified territory. This was the base of Zionism, which for religious reasons identified from the end of the 19th century this territory with Israel, an objective that in the 20th century has achieved the spectacular success known to all. Following the Second World War, the Gypsy movement was much weaker. Although the idea of a national "Romanistan" exists, in essence the movement's programme calls for an end to discrimination, equal access to education and dignified living conditions.

There are 13 million Jews in the world today, 5.5 million in the United States, 4 million in the state of Israel and 1.5 million in the states of the former-Soviet Union. Political Zionism emerged as a response to European anti-Semitism, which was promoted by ideologies like Pan-Germanism (the existence of anti-Semitism had been revealed by the "Dreyfuss-affair" in France towards the end of the 19th century). In 1896, Theodore Herzl, a Hungarian author, wrote "The Jewish State, search of a solution to the Jewish question." He argued that the integration of Jews in nation-states had not got rid of anti-Semitism; in reality quite the opposite had occurred. Accordingly, it was necessary to group the Jewish people into a territory, and create a Jewish state, which would be Palestine. Herzl admitted to confer to the Holy Places of Christianity a statute of extra-territorial nature, but he made no mention, from his closed Western-type perspective, of either the Arabs that inhabited the land or the Islamic religion. Zionism was officially born in 1897 at the World Constituent Congress of Bale. The Jewish community itself protested at Herzl's ideas, but Nazi extermination camps soon eradicated such reservations. This led to Arab resistance (Chatelet, 1986).

The distinguishing characteristics of the Gypsies (who number between 8-10 million in the world, with between 6-7 million in Europe) are their nomadic life and living on the margins of settled areas. Although Gypsy collectives have suffered permanent persecution at the hands of the "settled community" (Payos in Spain, Giorgios in Anglo-Saxon countries), the lack of a bond to unify the Romany community, such as a specified religious faith and a common literary corpus, has, until recently, impeded the emergence of a Pan-Gypsy movement. As with the Jews, extermination suffered at the hands of the Nazis (calculations estimate that half a million Gypsies were murdered by them) laid the foundations for what would be the Gypsy Congress, which was followed by subsequent World Romany Congresses. The First Congress, held in 1971, discussed the possibility of organising a form of 'Gypsy Zionism', being divided over the viability of a Romany nation-state. However, the romantic notion of "Romanistan" was rubber-stamped at the Congress, and a blue and green flag with a red wheel was created. The Second Congress, held in Geneva in 1978 (attended by representatives of 26 countries) established the Indian origin of Gypsies. The Third Congress in 1981 in Gotinga, discussed Nazi persecution in the presence of Jewish observers like Wiesenthal. The Fourth and last Congress in Poland in 1990, laid the foundations for the normalisation of literary Romany language, and mandated the production of an encyclopaedia of Gypsy themes and the establishment of links with other nomad groups.

According to the French Centre for Gypsy Research, 1.8 million Gypsies lived in the European Union in 1989, of whom only 30% to 40% of children regularly attended school. There are 650,000 Gypsies in Spain, 280,000 in France, 160,000 in Greece, 100,000 in Germany, 90,000 in the UK and smaller collectives in other states. (Such numbers are small compared to Eastern Europe, where two million Gypsies live in Romania alone). The EU Council of Ministers has approved a number of measures to safeguard the cultures of nomadic populations, including the construction of campsites for caravans and facilities for the schooling of children and adults (Clark, 1995).

However, Gypsy collectives continue to confront serious problems, which go beyond the current increase in racism and neo-Nazi ideologies in the EU member states, but rather are generated by the prejudices of societies that do not feel they are racist. Spain is a good example of this.

In Spain the "*Calé*" Gypsies are the overwhelming majority, but minority groups such as the "Hungarian" and "*Tsiganes*" (Portuguese) do exist.

Historically speaking the group's way of life has been predominantly but not exclusively nomadic. The majority of gypsies were settled or semi-settled, displacing themselves from a fixed centre to travel. The Gypsies became nomadic following the Industrial Revolution which deprived them of their traditional occupations of travelling sales, horse trading, hairdressing, puppet theatre, humour and seasonal agricultural labour.

Following the Civil War, Franco sent Gypsies out into camps situated on the sub-urban peripheries of Madrid, Barcelona and other big cities. The territories were public property but not built upon, and they lived with the marginalized white unemployed, which generated inter-ethnic conflicts and racism. In the 1950s, large-scale white immigration commenced towards the cities. In the 1960s and 1970s, the decades of development, the Gypsies' territories were those that were declared viable for building on, and *"manu militari"* evacuations of the Gypsies ensued. Gypsies were taken away in Civil Guard vans and left on open land, without materials to build or protect themselves with and subject to the prohibition of settling in other places. In the words of T. San Román (1995), the evacuations were clear demonstration of "barbarity, dehumanisation and stupidity." Shantytowns were built and unsurprisingly became dangerous, given the living conditions of their inhabitants, who could not obtain work or housing due to demands for complete personal documentation and stable jobs, being continually subjected to constant police harassment.

The arrival of democracy in Spain did not improve the situation. The redevelopment of cities in the 1980s and the use of green spaces was once again to the detriment of gypsy shanty towns. This time it was not Franco who pressurised them, but democratically elected councils. When unemployment increased Gypsies were among the first to end up on the dole. This situation, in addition to the increase in "democratic" racism and the competition from new illegal immigrants from Latin America and the Maghrib, pushed some Gypsies in to anti-payos (or anti-settled community) criminality or drug trafficking, or into apologist stances towards it. The proliferation of pseudo-scientific explanations about the increase in crime, and the understandable antipathy towards drugs, were concentrated on the Gypsy small-men, the last link in the chain, and not the big traffickers who were all white. This increase in anti-Gypsy racism has led to the innocent paying for the crimes of others, and has generated a popular anti-Gypsy pressure that has led to campaigns for the expulsion of Gypsy children from crèches.

A-National Ethnic Movements

In settler colonies, the colonialists have tended to exclude indigenous peoples from modernisation, and this has led to indigenous peoples becoming reduced to the status of a-national groups when their low numbers, or the profound deculturation they suffered, prevented the emergence of anti-colonial emancipation processes. Whilst this process is not exclusive to the American continent, it is there where it has most deeply been felt. *Indigenism* was a white paternalist ideology at the beginning of the century, but current native American "indianist" movements, directed by Amerindians themselves, campaign, beyond the current state borders, for respect of their ideas and political-cultural self-government.

These movements present distinctive differences. In some former colonies, now "white" states whose periphery demands independence from the centre, as with Quebec in Canada, they constitute an uncomfortable reminder of the origin of the peripheral groups themselves, who are as much colonisers of their land as the centre which they campaign against. This leads to indigenous' hostility in supporting the periphery's demands whilst their own problems remain unresolved. In the specific case of Chiapas in Mexico, the indigenous movement defends a state-wide project on an alternative Mexico based on equality, self-government of indigenous peoples and the end of neo-liberalism.

American Indigenism (or Indianism) expresses an inverted nationalism as it is the movement of human collectives that have been excluded from the processes of national construction in an entire continent; 'a-national' collectives whose objectives and forms of organisation ignore the logic and the borders of American nation-states. Their mobilisations find their *raison d' être,* exposing it, in the historical fact that the discovery of America brought genocide to the Amerindians, the biggest genocide known to humanity in the current millennium.

The thesis of the superiority of Christian civilisation sustained at the beginning of the 16th century by imperial theorists like Ginés de Sepúlveda, argued that civilised peoples should take over savage peoples. (This is why the propaganda of the colonisers painted natives in such a bad light: their supposed characteristics, worshiping pagan idols, making human sacrifices, eating human flesh, knowing no shame and practicing sodomy, allowed them to affirm that they lacked human nature). The thesis was countered (it has to be said in its honour) by the Spanish *Ius-snaturalist* school in which Bartomé de Las Casas and Francisco de Vitoria stood out. Their arguments led Pope Paul III to proclaim in 1537 that the Indians were real human beings.

So, the only remaining pretext to justify their conquest and domination was the Christian mission of conversion of the infidels. On announcing the evangelical message the "conquistadors" offered the Amerindians a chance for redemption; if they resisted the preached word, war was justified, and the Indians could be compelled by force. The natives, who practiced pagan religions, were all candidates for hell in the eyes of the conquistadors, who regarded the officiants of their cults as emissaries of the Prince of Darkness, and who assimilated their ascetic practices to infernal rites. As with the persecutions of the Holy Inquisition in Europe against heresy, witches and warlocks, the accused could only be purified through death by burning.

The apocalyptic catastrophe that beset the Amerindian peoples during the 16th century is illustrated by the following demographic data: records show that of the 75 million indigenous peoples thought to have been living on the Mexican Altiplano (high plateau) and Central and South America in 1492, only 9 million remained by 1570 (the indigenous population of the West Indies had completely disappeared by 1530). A number of factors explain this holocaust, besides the immunodeficiency of natives who contracted the diseases that had been imported by the conquistadors: the persecution and torture of peoples at the beginning of colonisation with the aim of taking from them the precious metals that would finance the European wars of the Catholic Monarchs and Charles V; the system of forced labour installed in mid-16th century through the juridical forms of land-division, known as *mitas* and *encomiendas* (that meant land and inhabitants granted to a conquistador); and the enforced collapse of a system of beliefs. Accordingly, neo-feudal dominions were created and governed by white elites over native populations, discriminated against through the Modern Age through postulates like that of the "pureness of blood".

The national revolutions undertaken at the beginning of the 19th century by the "Creoles" (descendants of the conquistadors) against the "peninsular" Spanish did nothing to remedy the situation of the indigenous populations, who remained on the margins of the processes of construction of the new states. Some of the liberal measures adopted[V] made matters worse by pauperising them even more through favouring the formation of huge "*haciendas*" (big ranches). Each modernisation process produced new spoliations: the rubber boom, which lasted from 1870 to 1920, brought massacres and indigenous slavery of the peoples of the Higher Amazon basin in Brazil, Colombia, Ecuador, Peru and Bolivia, at the hands of "rubber barons" like Fitzcarraldo.

It is currently calculated that there are 18.5 million indigenous people in South America, of whom 17.3 million live in the Andes region and 800,000 in the tropical jungles. In Venezuela there are 150,000 indigenous peoples, who belong to thirty linguistic groups, concentrated in the basins of the tributaries of the Amazon and Orinoco rivers and the banks of Lake Maracaibo. Invasions of their lands by colonisers, ranchers and miners are common. There are 300,000 indigenous people in Columbia, making up 60 "nations". Coffee growers have expropriated their lands in the Cauca region. In Ecuador 3 million indigenous peoples belonging to the Quechua and Aymara ethnic conglomerates inhabit the "Highlands"; there are also 20,000 Shuar and the Amazon groups. Since the last century white landowners have occupied the best lands.

Half of the 18 million people in Peru are Quechua and Aymara indigenous peoples of the Highlands, constituting the most important indigenous collective in South America. There are also 100,000 Amazon Indian peoples spread over 60 nations, who have not only been deprived of their lands, but have also had to suffer, especially in the Ayacucho region, the effects of the war fought between the Maoist "Sendero Luminoso" (Shining Path) Guerrillas and the Peruvian Army in 1980. In Bolivia more than half the population belongs to the Quechua (3 million) and Aymara (1 million) indigenous groups, which must be added to the 150,000 indigenous peoples in the lowlands, divided into 30 nations. Three out of four Bolivian indigenous people are illiterate, four out of five are ill and those that inhabit the banks of the Titicaca Lake have suffered attacks from the army.

One million Araucanians (mapuches) live in Chile, ranging from South Santiago to the Archipelago, in addition to 15,000 Aymara and other indigenous nations in the Tierra del Fuego; the Araucanians have become the victims of political harassment. 350,000 indigenous people live in the north of Argentina, on the border with Bolivia and Paraguay, made up of diverse Quechua groups and the Guaranis of the Misiones (Missions): the vast majority do not possess the deeds of their land. In Paraguay the official number of indigenous peoples is just 80,000 (the Guaranis of the West and the nations of Chaco), but the Guarani language is still spoken by 80% of the Paraguayan population; 75% of the Chaco Indians have no land. In Brazil the 225,000 Indians are divided into 225 indigenous nations. Whilst they are concentrated in the tropical Amazon jungles, they are also present in the savannah and costal areas, and offer a broad ethno-cultural spectrum. Of the 316 officially recognised indigenous areas 70% of them have not been subjected to territorial demarcation; so that, on the commencement of the large expansion that from 1968 accompanied the Brazilian "economic miracle", they were left subject to constant violations of rights (including collective massacres) carried out by multi-national companies and "development" projects.

"Indigenism" as a political ideology emerged at the beginning of the 20th century, when rapid South American industrialisation attracted a large number of indigenous peoples to the cities. Intellectuals such as González Prado and Mariátegui advocated, above all in Peru, an integrating nationalism with indigenous peoples that, together with a humanitarian and positivist ideological inspiration, would include the social and cultural ideas of the indigenous peoples.

The Amerindian liberation movement developed itself in the 1960s, 1970s and 1980s. It was created by indigenous people for indigenous people in defence of their lands and their way of life, and called itself "Indianism" to differentiate it from white paternalism that accompanied old Indigenism. The process of indigenous associations fostered by it (the National federation of Shuar Communities of Ecuador, 1964; the Indigenous Regional Federal Council of the Columbian Cauca, CRIC, 1971; and the Rupac Katari movement of the Aymara organisations in Bolivia, 1970) did not remain within the borders of each nation state, promoting the emergence of trans-border organisms such as the South American Indian Parliament, based in Paraguay (1974) and the Indian Council of South America based in the Peruvian Cuzco (1980) etc.

Mario Juruna, the first Indian to win a seat in the Brazilian Federal Parliament, defines the objectives of his movement as "[permitting] Indian culture to rest on community traditions and land, which is sacred [...] The Indians can and wish to choose their own path, and this path is not that of the civilisation made by whites. Indians today [...] want political power" (Gray, 1991).

The native "First Nations" of Quebec and of Canada, have opposed Quebec's demands because of the fear that their realisation would signal the deferment of their own demands. The history of native peoples in Quebec and Canada has been that of suffering and devastation. Since 1867, a law on Indians authorised the Ottawa government to rule them and "the land reserved for them"; the system of central Canadian Treaties stripped the natives of their lands and confided them to reserves. This legal framework fragmented the indigenous world in to four groupings: those Indians registered as groups having proved their tribal origin (the current First Nations); the Inuit (Eskimos); nomadic or unregistered Indians; and the *Half-breed.* If recognising their rights was equivalent to allowing them to organise socio-politically, the 1982 Constitution and its amendments did little to radically modify the previous situation.

In Quebec the 60,000 registered indigenous people are represented in 10 Amerindian nations and 14 Inuit peoples. The provincial governments had ignored them until the 1970s, feeling that they were exclusively under the jurisdiction of the Federal Government. When contact was finally established, conflict dominated relations as the natives' concept of political autonomy did not concord well with national sovereignty for Quebec. The conventions signed with the indigenous peoples (the James Bay convention of 1975 and the North East Quebec convention of 1978) given in exchange for purely symbolic compensation, were only aimed at re-affirming the rights of Quebec over forests and natural resources, with the goal of reaping the benefits of its exploitation along with some trans-national companies. In the same way, the adoption by the Government of Quebec in 1983 of fifteen regulatory principles in their relations with indigenous nations, which insisted on territorial integrity and the supremacy of the laws of Quebec, presented a rigidity that could not fail to irritate the groups. In fact, there have only been negotiations with any of the Ten Nations when they were necessary for the pacific solution of a specific conflict.

Chiapas is a state in the south east of Mexico covering of 75,000 sq. km, with 3.2 million inhabitants and enormous wealth (it produces 55% of the electricity and 40% of the crude oil of Mexico) that has not benefited its inhabitants in any way. Six families own three million hectares of land (half of the state's area); the remaining three million hectares are shared between 200,000 labourers. 26% of the population are indigenous and heirs of the Maya culture, from the ethnic groups of *Tzeltal* (36%) *Tzotzil* (32%), *Chol, Tojolobal, Zoque, Maya Lacandon, Mame Chcohó* and *Cakchiquele.* Only 68% speak Spanish and 46% of indigenous peoples are illiterate. In Chiapas 77% of infants suffer from malnutrition, and 44% of adults earn less than the Mexican minimum wage (10,000 Pesos); these percentages increase within indigenous peoples. Political power in Chiapas is not in civil hands but in the hands of the PRI military (PRI is the governing party in Mexico for the last half century) who have been responsible for mass indigenous expulsions to the jungle. Since 1974 the Indigenous Congress that brings together peasant organisations has demanded rights to land and a dignified life.

The *Zapatista* National Liberation Army (EZLN) formed from groups that radicalised after a long period of using pacific and legal means, came to light on the 1st of January 1994 when they took the main square of the Capital of Chiapas. Sub-Commander Marcos, spokesperson for the Clandestine Revolutionary Indigenous Committee, declared war on the federal Mexican army, "the basic pillar of the dictatorship monopolised by the party in power," and revealed the nature of the movement "that is not just for Chiapas, but national [...] We are Mexicans and this unifies us, along with the demand for freedom and democracy" (Borjas, 1996).

The sympathy for the movement in Mexico and beyond forced the government to declare a ceasefire twelve days later, which the EZLN agreed to. The San Cristóbal de las Casas cathedral was chosen for bilateral negotiations with the Government's commissioner, with the prestigious Bishop Samuel Ruiz acting as intermediary. When dialogue broke down in August 1994, a National Democratic Convention was held in the Lacandona jungle, which decided to extend itself to all Mexican states and participate in the elections against the PRI. In October, Chiapas indigenous organisations proposed the establishment of multi-ethnic indigenous regions across Mexico. In January 1995, the EZLN proposed the formation of an opposition front, the National Liberation Movement, whose leadership was offered to Cuauhtemoc Cárdenas (founder in 1988 of the Democratic Revolutionary Party). But in the same month the Convention dissolved, which the Mexican Government took advantage of by re-invading the Lacandona jungle with around 40,000 soldiers. In August 1995, the EZLN called on the Mexican people to form a National Consultancy for Peace; and in August 1996 pressed for civil society to create the *Zapatista* National Liberation Front, an organisation that would be "civil, pacific and Mexican", but did not adopt the form of a party (Borjas, 1996). The appearance of a new more ideological armed group, the Popular Revolutionary Army (EPR), has recently lessened the military pressure on Chiapas and the EZLN.

Notes

[I]   However, when the West complains when faced with such phenomena, it forgets that almost all nation-states are based on ethnic bastions and were formed historically through the (now forgotten) bloody domination of some groups over others. What the new Balkan nations have done in reality is compress into the space of a few years processes that developed in Western Europe over centuries.

[II]   Kakhimov (1994) describes differences between the Volga and Northern Caucasus regions as follows: "The peoples of the northern Caucasus are not just an ethnic mosaic, rather they differ dramatically from other Russian groups in their conception of the world, value system, social structures and especially in the persistence of their clannish identities. Religious factors are also important in the Chechen conflict; Islam, dominant in the northern Caucasus, reinforces ethnic divisions between traditional orthodox Russians. Islam is also a dominant religion in Tatarstan, although it has not been a significant factor in political events".

[III]   The Turkish family in the Altaic language group is considered to include Osmanli (from which current Turkish is derived), Azeri (spoken in Azerbaijan) Uigur (spoken in the Chinese Xinjiang) Kirguiz and Kazakh from former-Soviet central

Asia, Tartar from Kazan, and Bashkir and Karakalpak in the current Russian Federation.

IV   This does not refer to ethnic immigrant collectives.

V   In 1824 a decree raised the possibility of selling the Aillus lands, communal territory that the indigenous peoples of the Andes regions had managed to maintain during the *Ancien Regime*.

PART II
NATIONALISMS AND VIOLENCE

10 Nationalisms and Political Violence: State Violence and Response Violence

State Violence

Politics presupposes a conflict of interests generated by an unequal distribution of resources, resolved through relations of power which emerge out of the alliance between coercion and legitimacy. In this way, violence stalks politics, ready to break through the fissures resulting from an inadequate welding of the coercion used by power and its social legitimation. The less legitimate the power is, the more violent it will be.

In the political sphere the distinction between violence as a denial of good and violence as a source of harm is largely absent (which is not the case outside of this sphere). "Structural violence" as defined by Galtung, understood as every type of social control which frustrates an aspiration, imposes an opinion or upsets a style of life (i.e., every negative difference between the possibilities for achievement and real achievement), refers in reality to the nature of politics, a conjunction of the different aspects, economic, symbolic and force, of domination, screened by legitimacy (Braud, 1993). In modern societies, coercion and legitimacy are fused in the state; this is the meaning of Hobbes' profound myth of Leviathan.

Violence emerges when legitimation fails, either because domination, lacking legitimacy and therefore without consent, is deeply resented by the individual or the collective, or because it is a source of harm. This latter form of political violence consists, according to Nieburg's (1969) definition, in "acts of disorganisation or harm, which tend, in terms of victims and consequences, to modify others behaviour within an interaction that has effects on the social and political system."

The essential expression of legitimation in modern societies is the rule of law, which is identified with a certain number of democratic norms: respect for representative democracy based on formal equality, plurality of parties and majority government, and the preservation of a sphere of human rights that is not subject to state arbitrariness. Political violence flourishes in those states which have done away with the aforementioned norms, i.e., authoritarian and totalitarian states.[1]

There are two other sources of state violence regardless of whether or not it is democratic: those which derive from social exclusion and those which arise from nation-building. Violence carried out by the specialised apparatus of the state have their own geography, a double geography, social and territorial. The violence of state apparatus cannot be seen or felt at the point in which both centres meet, the state-national political centre and the pinnacle of social hierarchy. On the other hand, the further the groups are from the social pinnacle, or the more peripheral they may be from the ethno-social perspective, the more they feel under attack. For this reason, the groups most subject to the specialised violence of the state are those in which low socio-economic status and an acute territorial peripheral nature are linked.

Behavioural research on political participation is responsible for the downfall of more than a few normative myths; it has confirmed a positive correlation between high social and economic status and a high degree of political participation, and a converse relation between political abstention and social exclusion. Studies carried out by Milbraith (1965) proved that the politically active population that did not exceed 10% of the total population, of which only 1-3 % would be "contestants."

Elitist and pluralist perspectives,[II] as well as theories which criticise the "ungovernability" of the welfare state, accept indirect and low levels of mass participation in politics and discrimination of some ideologies: this is why the interests of specific groups are not taken into account, being systematically excluded.

In democratic states the interests and privileges of economic, cultural and bureaucratic elites are preferentially protected by the coercive state apparatus;[III] and they are protected from the excluded, "apathetic," or non participant sectors (shown, for example, by the different perceptions of the police in wealthy residential areas of New England compared with neighbourhoods such as the Bronx and Harlem in New York).

The coercive apparatus, present in every state, consists of the armed forces which defend the state against foreign attack and the repressive apparatus responsible for maintaining internal order - these include the police, justice and prison system (although in those states where there are intense national conflicts, the armies almost always play a key role in keeping internal order, becoming the last defensive bulwark of territorial integrity in the country). The repressive apparatus is composed of specialised, uniformed and territorialized officials that work in police stations, courts and prisons. Legality and legitimation coincide strictly in these apparatus.

The police apparatus is a permanent, operational and military-type service (well developed in states with a continental tradition, although somewhat less in those of an Anglo-Saxon tradition). In advanced states, the police account for five sixths of the total penal time; the other sixth by the institutions of justice and prisons. The state's fear of not being able to control them is behind the multiplicity of police apparatus, particularly in composite states. Currently, the repressive logic of the police in the liberal gendarme state has been replaced by a preventive logic. A broad concept of the information and surveillance aspects of public order, particularly in the centralising phase of the welfare state, has generated the threatening vision of a police society in which all movements would be under the control of the Orwellian Big Brother. This has revived a neo-liberal critique of the dangers which computer files represent for civil liberties (Soulier, 1985).

With the exception of very unstable or revolutionary situations, the coercive apparatus tends to situate itself psychologically and operationally, in relation to the diverse cleavages which divide modern societies (rural-city, religion-laicism, capital-work, centre-periphery), on the winning side. This is particularly the case with regard to national and class cleavages (the position of the forces of law and order in the latter type of conflict is all the more striking since their members usually come from the lower social strata). The "deep conviction" of judges applying the law is guided almost always by this kind of logic.

At present the armies belong to active international alliances in peace, while the police apparatus has become the irreducible core of state sovereignty. Its members are very open to the centralist reactions provoked by the international position of the state, by class conflicts and, particularly, by peripheral conflicts. It is also inclined towards reactive racism regarding groups of immigrants, especially if these groups come from states belonging to different civilizations.

In so far as penitentiary institutions are concerned, prison has become a state within a state, with a specific territory, population and use of powers. It is the world of infra-law, ongoing discretionary interpretation of rules, in which inmates in crowded conditions and lacking fundamental rights are subject to permanent security measures that are backed up by electronic controls (the fulfilment of the panoptic dream of total surveillance).[IV] However, with certain exceptions, most of those in prison are suburban young people of working class background drawn into a life based on drug related crime, or individuals belonging to harassed minorities, such as gypsies or "quinquis" in Spain, or extra legal groups such as illegal immigrants from outside the European Union, as well as members of subversive armed groups that are often peripheral nationalist in nature.

The construction of the national society and community by the nation-state also generates harmful processes for subjects and specific groups, which when they are not accepted lead to political violence.

From a historical perspective, state penetration in the periphery damaged groups whose ways of living were linked to the old regime; on occasion their reaction gave rise to armed legitimist movements. The political integration of the territory can be particularly coercive in recently decolonised states composed of heterogeneous and centrifugal vertical societies, bringing about the unification of the state discrimination or the open persecution of specific ethnic groups, carried out by other groups or promoted in Machiavellian style by the state centre. In the same way, several crises of mass political integration have been resolved in the West through external imperialism or majority harassment of internal minorities.

Original capitalist accumulation, prior to the generalisation of market relations, requires the proletarian transformation of former peasants and craft-workers through eviction from their lands and confiscation of their work tools. The unity of the market leads, likewise, to the emergence of one or several centres, or to a constellation of peripheries, together with internal colonialism; and, where there are ethnic differences, to the cultural division of work. In those states in which access to industrial society has been authoritarian - the Soviet Union under Stalin and some Third World societies - intense violence has been required; the generation of rural surplus for the industrial world of the USSR in the 1930s was produced through the compulsory collectivisation of land and the physical extermination of numerous kulaks (small landowners).

The creation of community through the processes of national socialisation and acculturation has historically brought about situations of cultural lag, leading to discrimination against marginal social groups and territorial linguistic collectives. The latter were condemned to diglosy and the devaluation of family pedagogical capital depending how far or close the ethno-linguistic groups might be regarding the culture of the national centre; that resulted in different degrees of negative stereotyping of their ethnic characteristics.

War between states involves the exchange of violence. Locke stated that relations among states do not belong to the sphere of the social contract, but to the sphere of the state of nature, meaning by that that the nexus between violence and legitimation disappears in them (the legal system established by the international community since the modern era has been a pale and inadequate substitute for this binomial, as is evident from the limited effectiveness of the great international organisations of the 20th century, such as the Society of Nations and the United Nations, in situations where peace keeping is required). Universal ethics, this outstanding product of the human spirit, goes back in times of war to the tribal concept of morality perceived as a group attribute, not being applied to the enemy. (Military training of armies throughout the world aiming to turn the actions of the soldier into mechanically obedient reflex actions in response to commands and to eliminate personal initiative, which could divert or lessen combat potential, is directed towards the creation of the qualities that soldiers should demonstrate in the battlefield as machines for the destruction of the enemy).

According to Schmitt (1972), it is up to the state to decide who is the friend and who is the enemy; the state is the exclusive holder of *ius belli*, the option of fighting against a designated enemy, because it is the only one which can exercise this supreme act of violence: the right to decide between the life and death of a human being[V] (as long as another group does not assume this position, mirror-imaging the essence of the state).

Socio-Political Response Violence

According to a performative definition, violence only refers to behaviour designated as unacceptable by dominant groups within a particular society. However, Braud (1993) distinguishes between three subgroups: state violence, protest violence and inter-social violence. This allows the author to deal with violence from three different perspectives: violence as a construction of identity and as a means of political affirmation; the distinction between political and instrumental violence; and the systems of reinforcement of social control that leads to the laying aside of violence in contemporary societies. (This scheme will be used in the description of response violence, although modifications to certain aspects will be made.)

The most basic form of response violence is anger, this destroying "acting out" that is the outpouring of aggression produced by frustration. This response has been studied from different points of view (Lorenz, 1970; Gurr, 1974; Bandura, 1973). The ethological school of Lorenz views it as a defensive reaction by families or threatened groups defending their vital space. The theory of relative frustration (Gurr, 1974) argues that all interference in the satisfaction of a pleasure considered to be accessible or legitimate provokes anger. When a target is identified (i.e., the real or imaginary agent of frustration) and harm is inflicted, the aggressive impulse is lessened because the feeling of frustration is partially satisfied. Nevertheless, the damage inflicted may be physical - vital space, personal integrity or property - or symbolic. Additionally, it may be personal in nature, consistent with damages to the rights of individuals, or collective, i.e. attacks against the identity markers of a group (Fromm, 1975). Relative frustration depends on the distance between the living conditions the members of a group believe they should have the right to, and their real existence as perceived by them.

The processes of frustration can develop when aspirations increase but the possibilities of achievement remain the same, when aspirations remain the same but the opportunities of fulfilling them decrease (Down and Hughes, 1982), or when a group that has undergone an improvement faces, in the curve-shaped process described by Davies (1970), an impediment or breakdown in the process. Relative frustration, be it economic, political or cultural, only turns into collective violence when the group is numerous in size and occupies a strategic position. Brinton relates the Puritan English and French revolutions with prolonged periods of prosperity and civil liberties which were suddenly brought to an end. (Nevertheless, frustration does not always lead to anger induced violence -the aggression experienced by individuals can be turned against themselves or lead to apathy and resigned resentment.)

The second perspective studies cultural mechanisms of learning and appraisal of aggressive behaviours which give rise to the creation of "subcultures of violence" (Bandura, 1973); these types of mechanisms make these subcultures relatively autonomous from the frustration-aggression cycle. A society, or a social subgroup, may produce very aggressive people, even where there are low levels of frustration, if it values violent or repressive attitudes and rewards them. (However, the subcultures of violence can also emerge in societies that are permanently subjected to repression, "communities of fear" that make a ritual out of certain aggressive behaviours in order to exorcise the fear, serving as a release valve).

In the course of "volcanic" violence the rational calculation of the costs is put to one side since it does not set clear or coherent goals for itself (these goals are set after the onset of violence, and not always by the protagonists). Events such as riots instigated by young urban and suburban people in Germany, the United Kingdom, France etc., are frequently chaotic and driven by an orgiastic urge to destroy everything, to provoke the police etc. (Braud, 1973). The paradoxical characteristics of the "fusion group" described by Sartre (1960) - freedom and terror - were clearly present in the violent outbreaks in ethnic ghettos in the United States, and even more so in some anti-colonial riots or in strikes to topple hated dictators from power. The first phase of the integration of the "fusion group" brings about an abrupt resurrection of freedom that is often expressed as violence and rioting. This not only requires members of a group experiencing an awareness of the alienation and powerlessness in which they live; according to Sartre, a particular historical conjuncture is usually required which is expressed through the threat of death or the disappearance of the collective.

However, volcanic violence can search for a replacement victim to substitute the causal agent of the furore (a process of deviation sometimes consciously encouraged by the state). The mechanism for selection of the sacrificial victim, described by Girard (1972), requires that the victim be different from the group but not alien to it. The victim may be located at the top of the social hierarchy (the prince, the greats), or it may be an ethnic or religious minority that is hated and despised. These processes are always accompanied by a projected image of the imaginary enemy, which consists in the collective form of a psychological process through which the individual projects outside in the world as evil deeds the aggressive tensions arising from their inner recesses. Therefore, everything that is violent and dangerous for the individual or the group is attributed to the potential victim. Democracies offer up cheap scapegoats through the bloodless sacrifice of persons in command who are forced to resign as "fuses", such resignations contributing to the exorcising of the frustration factors which gave rise to the violence.

Instrumental violence, on the contrary, is carried out without passion or uncontrolled aggression in order to achieve rational goals. It adheres to a logic of calculation and efficiency subject to laws of rational choice, which implies an adjustment of the means to the desired goal (Braud, 1983). Although instrumental violence, above all, is characteristic of the behaviour of the state, it also appears in response violence to the state. (Sometimes, vertical societies of the Third World, or the federal parts of a composite state, politically organised and with their own apparatus of repression, use rational violence as an instrument of secession against their former states. In these cases violence resembles the use of force by opponents in a war waged in order to destroy the power of the enemy and impose their own will).

In horizontal societies of the West, legal organisations such as trade unions, pressure groups, political parties, etc., may engage in sporadic low intensity violence of an instrumental nature, allowing them to demonstrate their representative character in the street and their ability to mobilise; "soft" violence which constitutes one of the means of political participation, in the sense that it attempts to influence governmental decisions (Braud, 1983). This kind of violence (present in demonstrations against the Vietnam war in the 1970s, in French farmers' protests against EU decisions and in some anti-nuclear marches) tends to be in response to specific government actions or decisions; it appears when the negotiation process is blocked, being the groups engaged in this violence usually the radical wings of mainstream movements (Downs and Hughes, 1982). These attempts to draw attention to their demands are only effective if the process of action is controlled, and if the goals can be negotiated within the framework of the political system. The achievement of agreement does, in fact, bring about the disappearance of this violence.

Inter-social violence emerges in the course of interaction between social groups in conflict due to the appropriation of limited resources. These types of conflicts originate from the four cleavages previously mentioned - the rural-city divide, between social classes, between secular and religious groups or within these latter groups, and between ethnic, linguistic or cultural groups. However, it is rare to find conflicts in which political power is not one of the limited resource wanted. For this reason, the majority of these violent processes go beyond the boundaries of civil society, being associated in one way or another with the construction of the nation-state or with opposition to it. Due to the laws of hostile ethnocentrism, the closer the internal relations of the group members, the more distorted, negative and stereotyped will be the perception of the external groups.[VI]

Political violence as carried out by ethno-national movements does not usually originate from sudden violent outbursts, especially in the west. This violence is of an instrumental nature incorporated into the rational-programmatical dimension of the movement; but it is not a form of political participation, meaning by that a means to modify a widely accepted political framework. With the exception of isolated cases (such as south Tyrol), it hardly ever consists in an attempt to draw attention to the need to resolve concrete problems; for this reason, the use of this violence, far from being sporadic and of low intensity, is permanent, and its thresholds begin to be gradually broadened.

The mimesis, or mirror-game nature, of these nationalisms with regard to the nation-state differs from non-violent nationalisms in its radical stance and scope. While the latter only mirror-image the national community and, on occasion, the society, violent nationalisms that contest the legitimacy of the use of force by the state mirror-image the state in its role as a monopolising centre of violence. In its mature phase, the double mirror-imaging of the state and the nation leads them to build up a dual structure formed by an organised armed group as a counter-state and by a civil legitimating community of this group as a counter-nation. (This is an area of particular interest to political scientists, although it has not been systematically explored until now.)

It is important to differentiate the political violence of peripheral nationalisms from social pre-modern forms of violence such as banditry, as well as from contemporary political phenomena such as the social-territorial, revolutionary and ideological violences.

Pre-capitalist banditry of peasant origin emerges, according to Hobsbawn (1972), in any of the following three forms:

1. The generous bandit who avenges injustice.
2. The primitive rebel, the most well known of whom would be the mythical Robin Hood; and
3. The guerrilla unit, that is to say, a group of free men who have been evicted from their land. Bandits are not revolutionaries; their programme consists in maintaining or re-establishing the traditional way of life, or "the way things should be," eliminating the "excess" of coercion and repression with regard to tradition. In capitalist societies, modern banditry, a consequence of the prison system, not only does it not attack the basis of the state, but it is used by state power.

Urban political violence (as waged by all Western violent peripheral nationalisms) is characterised by the totally clandestine activity of the armed groups, a lack of control in rural or urban areas, an absence of active institutionalised support from sectors of the population, an activity aimed at achieving civil objectives, and the priority of symbolic results over military ones.

It is the lack of control over specific areas of territory and over clearly defined sectors of the population which differentiates political urban violence from political-social violence in which a territory is occupied, as was the case in the Mexican revolution, and as has also been the case since the 1970s in some areas under the control of rural guerrillas in Latin America (Peru and Columbia) as well as territorialized confessional groups, as in Lebanon. The specific processes of degradation of violence appearing in the aforementioned cases, a consequence of the estrangement of the primitive social-political objectives (the phases of which have been growing autonomy, privatisation, and finally the commercialisation of violence)[VII] have, therefore, hardly impacted on peripheral violent nationalisms.

Revolutionary action (social and anti-colonial) is a combination of state-wide political violence and of an extensive and profound social action arising from the struggle of classes and social groups (Wiewiorka, 1988), or from the rejection of the metropolis. When social revolution is triumphant it brings about a rapid and violent change of the values, institutions, social structure and leaders of the state in question (Huntington, 1970), the social class in power being substituted. When anti-colonial revolution is triumphant, the substitution of the colonial power of the metropolis by the former colonized power is a factor to be added to the aforementioned. What differentiates it from peripheral nationalist violence is the nature of the conflict, generated by class struggle or by anti-colonial riots, its state-wide dimension and its consequences, in the case of triumphant mobilisations.

Ideological violence tends to be revolutionary, but it lacks the mass mobilisation caused by social action inherent in revolution, making its goals unfeasible and isolating its activists. It is this type of violence which has been described specifically as "terrorism." It appeared in the 1970s, some time later than Latin American guerrillas, in parallel with the emergence of the new social movements and new political agents described by Ingelhart (1991). In this same period, Lacqueur (1979) distinguishes three sub-classes of political violence: nationalist (Ulster, Near East, Canada and Spain), Latin American of a *sui generis* type, which evolved after Che Guevara's failure in Bolivia, and urban terrorism, typical of the United States, western Europe and Japan. Ideological violence refers to this sub-group.

Authors such as Targ relate revolutionary action to industrial society and "revolutionary terrorism" to post-industrial society. The latter represents, to some extent, the end of hopes for the revolutionary capacity of the proletariat and their transfer to new social movements, those which it aims to radicalise and globalise with a view to transform them into the new revolutionary subject. According to Targ, revolutionary terrorism is not a synonym for a revolutionary social movement, since its importance is inversely proportional to the revolutionary potential of the workers. Terrorists acts take place more frequently in pre-or post-industrial societies than in industrial ones, because the latter facilitate the development of reformist or revolutionary mass movements, while the other two types of societies do not. Terrorism appears when the very social movements of industrial society enter into decline. As Targ argues, the social process of post-industrialism interferes with class consciousness and undermines the formation of radical or revolutionary movements of social change. For this reason deviant social action manifests itself as terrorism, and terrorist acts become permanent elements of the social landscape.[VIII]

In central Europe, particularly in Italy and Germany, the first generation of these ideological groups, "Red Flag" in Italy, the "Red Army Fraction" in Germany, constituted a nucleus of armed fighters that represented the violent expression of the Leninist vanguard party. In Italy they assumed a more anti-industrial and anti-capitalist stance, while in Germany they were more anti-imperialist and sceptical of the German proletariat. The second generation of these groups, which emerged at the end of 1970s and in the 1980s ("Prima Linea" and Communist Fighter Formations in Italy, June 2nd Movement and Revolutionary Cells in Germany), saw themselves as the revolutionary sector of the radical wing of new social movements and *lumpen-proletariat* groups (manual workers in the tertiary sector, unemployed, sub-urban youth, etc.) Della Porta (1996) describes the course which these groups have followed toward isolation and organisational impermeability with regard to the social groups out of whose environment they emerged. This isolation has linked their ideological production to the development of cryptic messages for internal consumption and led to the inflection of their actions, which were initially a vehicle for external propaganda of the goals of social movements, but later directed towards the internal goal of survival. These groups, focused on a private war against the state, without any social support, have either disappeared or have been reduced to very small collectives composed of no more than a dozen people.

New nationalisms that chose political violence emerged in the same years as the ideological violent groups of the first generation, and learned from the same theoretical sources (anti-imperialism, Leninist vanguardism, anti-capitalism, influences of the new social movements in their most radicalised form). However, the fact that they were linked to intense centre-periphery national conflicts assured them of a very different organisational and ideological context, guaranteeing their survival and exempting them from some elements of decline which may eventually surface (mercenaries at the disposal of third state, or the transformation of their activists into "minor civil servants of death"). [IX]

Can the political violence practised by nationalist groups be defined as terrorism? An affirmative response is always the case in the heart of public opinion in nation-states. On the other hand, for the legitimating community of the armed groups, which is often much smaller in number than the former, it is about struggle for national liberation. In the core of the state, the negative interpretation of the term "terrorism," identified with what is intolerable and inexcusable, entirely fulfils its performative function; there is no room for compromise or neutral ground between the majority that condemns and the minority that approves of violence.

From an academic perspective, political scientists hesitate when it comes to making this judgement. Crenshaw (1983) provides the following neutral definition of terrorism: "systematic use of unorthodox political violence, carried out by small conspiratorial groups with the aim of influencing and manipulating political attitudes rather than physically defeating the enemy; for this reason their goals are more psychological and symbolic than material". From this perspective, terrorism is another form of political participation. Given that violent peripheral nationalisms hardly ever set out to defeat militarily the enemy, rather they attempt to force them into political negotiations, they could be included in this definition, which is more descriptive than normative.

Another tendency, inspired by Touraine, establishes a closer relation between terrorism and the decline of social movements than with political participation. Terrorism is defined as an anti-movement arising from the inversion of the characteristics of the social movement: in the anti-movement the principle of identity no longer refers to a true social force, but it makes reference to an antagonistic essence; the principle of opposition, which defines the social adversary, turns into an image of war; and the principle of totality, that situates contemporary society in the field of historicity, begins to project itself into a beyond as a communitarian utopia. Wiewiorka (1988) concludes, as a consequence, that in so far as peripheral nationalist violence continues to have links with the social movements that brought about its emergence, it cannot be defined as terrorism. [X]

Notes

[1]   The historic origin of authoritarianism is found in the response of political elites to the demands of the masses for greater political participation. Once the first phase of state integration is over, instead of dealing with them extending suffrage and means

of representation, elites react through the coercive demobilisation of the masses. Totalitarianisms are "ideocracies", paroxystic expressions of the autonomy of the ideological; the group that holds the "true" ideology transforms state power into the instrument by which it subjects civil society to a corpus of ideas, the law of racial selection, the laws of historic materialism, etc., being itself their single interpreter. The concentration of power in a single point and the total atomisation of society allow totalitarianism to spread its implacable logic (Hanna Arendt, 1972).

II   Elitist theories have a unitary vision of the elite and an accumulative concept of its powers. According to pluralist theorists, decisions are made by plural and consensual elites through negotiation and compromise (Dahl, 1963).

III   However, the fluidity of the relationship between the "social centre" and the apparatus of coercion does not mean identity of interests, particularly in the sphere involving the police force and prison guards. Elites use these institutions, made up of members of social classes which are always lower than their own, although they fear them.

IV   Foucault (1975) states that the arguments used from 1820 to 1845 to criticise prisons, "this monster of bourgeois reason," are still relevant today: prison leads to relapse, favours the organisation of delinquents, generates misery in the family of the detainee, and the living conditions of the prisoners that are freed, subject to a range of restrictions and police surveillance, favour their continuous presence in a criminal environment. So what is the use of the proven failure of the prison system? Foucault argues that it manages the economy of illegalities, draws the boundaries of tolerance and excludes some, while taking advantage of others. The delinquency as closed illegality can be used by power, - and this happens frequently -, to produce groups of informers and "agents provocateurs" as a reserve army of power in order to perfect the system of surveillance of delinquents, making the gradual identification of the rest of society possible. Foucault indicates that there is a profitable delinquency in circles where arms, alcohol and drugs trafficking takes place, used in its entirety to obtain illegal benefits for political ends.

V   Aron (1976), paraphrasing Clausewitz (1955), claims that the idea that dominates military thinking is the destruction of the armed forces of the enemy and that, in that sense, war is an act of violence aimed at making the adversary bend to the will of the aggressor. However, he adds that in a war among states the principle of annihilation seldom triumphs, because war ends when the adversary respects the will of the triumphant state, surviving as a state even though it has been defeated. The principle of civil war, nevertheless, is different. It arises, according to Schmitt (1972), when in one state the conflicts among the different sides finally occupy the entire field of political antagonism, being civil conflict determined by the regrouping among friends and enemies. For the army which engages in rebellion, what has to be destroyed is the very possibility of the opponents organising themselves as an enemy. So, there is no intention of allowing the opponent to survive as an antagonist, which means that the principle of annihilation is present here in all its purity.

VI   Stereotypes, according to Maisonneuve (1973), do not exist as such unless they are widely extended within a group or a population. They are simple, the images and representations they create are poor and, almost always, contain some physical features that resemble caricatures; they are above all of a socio-functional nature.

VII   Waldman (1996) describes the phenomena of the degradation of violence that may emerge in situations of conflict among social groups in which the insurgents control specific territories. In the first phase, the group's desire for self-perpetuation and the need to obtain financial resources, a motive which mixes political and private elements, makes them forget their initial political goals, replacing social consensus with coercion. In the second phase violence becomes privatised, placing itself at the service of motives such as personal gain or revenge; the internal tensions bring about the emergence of leaders who become mercenaries acting at the disposal of the highest bidder. The population is no longer considered to be the base of social support, but seen as something to be plundered, and the territories, inhabited by "local subjects", are re-feudalised. Certain groups, such as the confessional ones in Lebanon, become, through administrative and political party control, states within the state. In the third phase, violence is commercialised, becoming an element of market relations and penetrating every aspect of social life. The paradigmatic case is that of Colombia, where there are a large number of violent groups: paramilitary groups against rural leftists, death squads against the urban gangs, drug trafficker's henchmen, guerrilla organisations, radical youth groups, self-defence urban collectives, militias paid by property-owners... All of this has led to the proliferation of mercenary "death offices" in Bogotá· and Medellín.

VIII Charnay (1981) describes the phenomenon of armed violence in a coincidental way. He argues that terrorism, even if it gives expression to its commitment to promote a revolutionary process, arises from several phenomena related to the loss of hope. In Europe, the failure of May 1968 had impacted on the hopes for widespread, post totalitarian, non-bloody and perhaps non-violent revolution. Outside of political parties and trade unions, a break-up of the doctrine of political philosophy had occurred, which had taken refuge in more heated and restricted settings such as ethno-nationalism, ecology movements, - or terrorism -, in order to take the reins of history once again.

IX According to Charnay (1981), "to commit oneself to a clandestine life means breaking with professional, friendly and family relations. Terrorists live a fragmented life divided into two types of activities: one set that is visible, daily and ordinary, while the other set is hidden, related to the struggle. It is, in the end, the latter which gives rise at the same time to fear, attraction and dedication". Wiewiorka (1988) adds that "the terrorist is exceptionally a hero or a lord of darkness; he usually leads the miserable existence of an insignificant civil servant of death. The experience of clandestine activity, isolating him in a group, restricting him to circles which determine his most personal acts, gives rise to specific mentalities and develops certain aspects of the personality, while atrophying others". The "lived in experience" of members of nationalist armed groups does not coincide with that described by Wiewiorka and Charnay, since rarely do they remove themselves entirely from their social environment.

X According to Wiewiorka (1988), extreme left-wing terrorism would be the opposite of revolutionary action. Whilst the latter combines social and political action, terrorism avoids mass actions in order to place itself only at the political level. However, terrorism also differs from political violence, because it has no relation to popular demands. Wiewiorka suggests that organisations such as ETA cannot be classified as terrorist: "its armed struggle, he argues, is linked to the radical side of popular protests, to claims that are aggravated by the economic and cultural crisis, to a social anger, not being entirely unconnected to the sensibilities it claims to represent... Although it may strike a discord, it has to be said that ETA's violence, as a whole, is not sociologically terrorist". Ibarra (1987) considers that both the derogatory term "terrorism" and the exalting term of revolutionary violence should be eliminated; it is about an armed political struggle that establishes political objectives of a symbolic nature through the use of weapons. According to this writer, the correct term is the abbreviated version "armed struggle."

11 Nationalisms and Violence: Case Studies

Macedonia

The Ottoman Empire turned Macedonia and other parts of the Balkan region into an ethnic mosaic. Macedonia's multi-ethnic nature and the competing national projects of Serbia, Bulgaria and Greece led to the territory being heavily disputed; in the late 19th century the IMRO, Europe's first ethno-terrorist group, emerged in Macedonia. Two inter-Balkan wars were fought in the territory, pre-cursors to the First World War, which led to Macedonia being divided between Serbia and Greece. Following the First World War the allies created the Yugoslav Kingdom, an entity into which Macedonia was integrated; at the end of the Second World War the territory became part of Federal Yugoslavia, which had been unified by Tito's Communist party. In the 1990s, the former Macedonian Federal Republic achieved status as an independent state and did not suffer the horrific ethnic cleansing that took place in the rest of the former Yugoslavia. However, this process did not occur without causing apprehension within the historic imaginary of Greece.

Historically Macedonia covered an area of 60,000 square kilometres, considerably more than the 25,000 square kilometres of the current ex-Yugoslav Macedonian Republic. The geographical term Macedonia was first used in the 19th century to describe an area which bordered Serbia and Kosovo to its north, Bulgaria to the east, and had a coastline on the Aegean Sea, an area that is currently part of Greece. The centres of historic Macedonia included Thessalonica (or Salonica), Monastir and the current capital, Skopje. Homeland of the legendary King Philippi II and his son Alexander III the Great, it was part of the Byzantine Empire until successive takeovers by the Bulgarian Empire of Tsar Samuel in the 10th century, and the Serb Dusan in the 12th century. In the 15th century the Ottomans occupied the territory, along with the rest of the Balkans. Following the defeat of the Ottomans at the gates of Vienna in 1683, Christians fled Macedonia over fears of Turkish reprisals. The two groups that repopulated the area were both Muslim, namely the Kosovan Albanians and the Islamic Serbs, both of whom had come from the north. The subsequent predominance of Muslims did not prevent the vast ethnic fragmentation in the territory, which was inhabited by Serbs, Bulgarians, Greeks and Albanians, with Valaks, Aromani (of Rumanian origin), Sephardim Jews and Gypsy minorities, as well as Turk collectives, either originated from the Anatolia peninsula or having been part of the Ottoman administrative apparatus. The ethnic complexity of the territory led to the emergence of the phrase the "Macedonian Fruit Salad".

In the 19th century, Macedonia, historically lacking a political entity of its own and the territory occupied by the Ottoman Empire for longest, (until the end of the First World War), became the flashpoint for the nationalist projects of the three Christian states that had achieved liberation from the Turks, namely the Serbs, Bulgarians and Greeks. The Greeks had rebelled against the Ottoman Empire between 1821-1833, and established an independent state. Their Project (or *Megale Idea*) of Greater Greece naturally had to include the birthplace of Alexander the Great. This brought them in to direct conflict with the Serbs, whose objective of establishing a *Nacertanje* involved a Pan-Slav alliance (a union with Bosnia-Herzegovina, Montenegro, Bulgaria and Macedonia) against the Turks.

However, the Bulgarians own national project, which emerged in 1870, also encompassed Macedonia, with its roots in the renaissance of the Bulgarian Orthodox Church: over a thousand years before, the power of Bulgarian Church in both culture and religion had allowed it to implant Slav-Bulgarian as the Orthodox liturgical language, and to spread the Cyrillic alphabet all over Eastern Europe, including to Russia. An additional factor was that Macedonia, although a poor country, was of great strategic importance, as it controlled the traditional routes that brought the Danube to the Aegean Sea, mainly through its major port of Thessalonica.

Towards the end of the 19th century, the region was inhabited by 1.15 million Muslims, 620,000 Orthodox Greeks (subject to the patriarchy of the Greek Church), and a further 620,000 Bulgarians (subject to the Bulgarian "exarchate"). The Bulgarian groups (helped from abroad by Russia) influenced most Macedonian nationalism, whilst the Greek state was absorbed over the Crete question and the Serbs, who were promoting the creation of Serb schools in Macedonia, lacked the necessary ecclesiastical backing.

The Internal Macedonian Revolutionary Organisation (IMRO) was established in 1894. Headed by Joce Delcer (1872-1903), the nationalist group's objective was to establish a "Macedonia for Macedonians," an autonomous Macedonia in a Balkan Federation. The IMRO, which fiercely controlled its own militants, organised an underground military network, revolutionary taxes for finance and an executive police force. The Greeks reacted by creating their own armed organisation, Philiki Hetaira, whose members attacked targets in the Ottoman administration and Exarchate churches, but were much less efficient than the IMRO, which between 1898 and 1902 carried out 132 attacks, causing the death of 4,373 ethnic Turks.

Despite Sofia setting up its own "External Macedonian Organisation" from 1885 onwards, an organisation that pursued the territory's annexation into the Bulgarian Kingdom, Bulgaria was still used as a refuge by the IMRO's *Comitadji,* and financed partially the Macedonian armed group. In 1903, the IMRO launched a full offensive geared at liberating Monastir from the Ottomans, with the intention of advancing from there to the rest of

Macedonia. The offensive initially succeeded in creating the Republic of Krusevo, but two months later it had been defeated and Delcer was killed. The Turks unleashed appalling reprisals against the Macedonian population.

The European powers, preoccupied by both Turkish and IMRO atrocities, demanded that the Sultan establish Bulgarian, Serb and Greek ethnic zones in Macedonia, controlled by an international gendarme. However, in their own respective zones the groups embarked on the physical elimination of the other two (pre-empting the current "ethnic cleansing"), and the IMRO's armed activities did not cease. In addition, the Turkish officials of Macedonia, including Captain Kemäl Ataturk, opposed the international gendarme. From 1905, these officials formed an alliance with the exiled "Young Turks" movement in Paris; in 1908, threatened with the march of the Macedonian II Army into Istanbul, the Sultan re-established the constitution.

However, the Macedonian problem did not go away, and from 1908 Russia promoted the establishment of a Serb-Bulgarian "Balkan Bloc" with the aim of confronting the Ottomans. In October 1912 a war broke out that, for the European powers, was effectively a rehearsal for World War One. Within the space of a few weeks the Christians took almost all Ottoman territories in the Balkans. However, the territorial appetites of the different Balkan states clashed in Macedonia, leading to the June 1913 outbreak of an inter-Christian war between the Bulgarians and a Serb-Greek alliance. Following Bulgaria's defeat in July, Serbia and Greece carved up Macedonian territory between them, splitting it into northern and southern zones and essentially bestowing it with the territorial configuration it has today.

Following the First World War a new Kingdom was established under Serbian hegemony. The Kingdom, consisting of Serbs, Croats and Slovenians, left Macedonians out of the equation, granting the nation no autonomy whatsoever. Such moves explain the sporadic collaboration of the IMRO with Ante Pavelic's *Ustachis*. However, the carving up of Macedonia between Bulgaria and Italy decided by the Nazis in the Second World War, led Macedonians to side with Tito's anti-fascist resistance. Following a conflict with Bulgarian communists that was resolved by Stalin in Yugoslavia's favour, Macedonian communists integrated themselves into the Yugoslav Communist Party. After the victory of Tito's Resistance movement Macedonia became one of the Six Federal Republics of Yugoslavia (Castellan, 1991).

The newly independent Macedonian Republic of the 1990s did not suffer the bloodshed of the other territories that emerged from the dismemberment of the Yugoslav Federal Republic. However, it met with resistance, not only from the Serbs, but also from Greece's opposition to an independent state internationally using the same name as a Greek region that, being the birthplace of Alexander the Great, has a special place in the Greek national imaginary. The international pressure placed on Greece to accept a compromise formula in which the name "Ex-Yugoslav Republic of Macedonia" would be employed (which Macedonian President Kiro Gligorov accepted also with reluctance) was due to fear that instability within the country could lead to fresh Balkan conflicts of inter-state proportions, that could simultaneously flare up across Eastern Europe. That is why, at a stage when the conflict in Bosnian appeared to be drawing towards a solution, the assassination attempt on Gligorov towards the end of 1995 sent a shudder through all European governments.

Palestine

Jewish Zionism led to the emergence of a Palestinian national consciousness, especially after 1917 when the British, who favoured a divided Palestine, offered the territory to the Jews as a national homeland. The triumph of the Israeli Haganah after the 1947 British withdrawal, and the subsequent mass expulsions of Arabs from their lands, led to the emergence of two tendencies within exiled Palestinians. The first of these was Pan-Arabic and was controlled by the first Palestine Liberation Office, of Nasserite character, while the second was the pragmatic and Palestinian tendency of Al Fatah. Following the Six Day War in 1967 the PLO moved closer to becoming a Palestinian nationalist organisation; there was an emergence of armed groups who exported political violence to the West. After its expulsion from Jordan in the 'Black September' of 1970, the PLO made Beirut the centre of its operations, which led to the Israeli invasion of the Lebanon in 1982. The civil resistance movement known as the Intifada emerged in the West Bank and Gaza Strip territories in 1987, the moment when the PLO substituted its goal of transforming Israel into a secular state, for the creation of a Palestinian state alongside its Jewish counterpart. This led to the establishment of the Islamic group Hamas, which rejected the project and weakened the PLO. The 1991 Peace Conference, held under the auspices of the United States, culminated in the 1993 Oslo Accords, in which the Israeli state (then governed by the Labour party) authorised the establishment of limited autonomy in the West Bank and Gaza Strip, with the National Palestine Authority projecting its future transformation into a Palestinian state. The electoral victory of the right-wing Likud party cast a serious shadow of doubt over the global viability of this project.

> The Ottoman Empire governed Palestine, and the rest of the Middle East, from the 16th Century to the end of the First World War. From the beginning of the 19th century onwards, despite the fact that 80% of the Palestinian, who shared the Sunnite Muslim religion with the Ottomans, were small peasants (*fellahin*), the Ottomans privileged the urban based Palestinian elite who lived in Jerusalem, Akko, Nablus, Gaza, Hebron, Jaffa, made up of absentee land lords, politicians, the clergy and intellectuals (*effendi*). As was the norm in the Ottoman Empire the non-Muslim peoples were organised into *Millets*, the most important of which were the *Millets* of the orthodox Christian communities and the Jewish community (*yishur*).

From the beginning of the century the emergence of Zionist parties, who claimed the biblical lands for the Jews, drove the emergence of a Palestinian national consciousness, which had not existed up until that point; a consciousness that increased after the British took control of Jerusalem and other territories during the First World War. British interest in the territory was strategic, namely the protection of the north eastern flank of the Suez Canal, which is why they preferred an ethnically divided Palestine, feeling it was less of a risk. It was in this context that the 1917 declaration of the then British Foreign Secretary, Balfour, was pronounced, guaranteeing Baron Rothschild the creation of a "Jewish Homeland" in Palestine, alongside other "non-Jewish communities".

Form this moment on the Arab-Israeli conflict was essentially set in place. From 1920 the Zionist Haganah organisation encouraged Jewish immigration, which was stimulated by the deterioration of Jewish conditions in Europe during the interwar years. Meanwhile, Palestinians mobilised in opposition both to the Balfour declaration and to the torture of Palestinian detainees. 1932 saw the creation, non controlled by the *effendi,* of the first Palestinian party Istiqlal (independence), and of the Palestinian Communist Party which towards the end to the 1920s followed "Arabisation" policies. Following a number of general strikes the Great Palestinian Revolt broke out in 1938, and developed into a guerrilla war. In response the Irgun, the Jewish military organisation that split from the Haganah in 1935, increased their attacks. In 1935, the British disarmed the Palestinians, but did not disarm the Zionists.

The conflict became irreconcilable. The Jews (who in 1938 possessed just 6% of land around Tel-Aviv) pushed for a mass removal of the Palestinian population to produce ethnic homogenisation, the British dithered over the proposal, and the Palestinians attacked both. Following the Second World War Britain decided that its presence in Palestine was too costly. The Jews (in 1947 one third of the population, as opposed to two-thirds of Arabs and Christians) armed themselves; the Haganah's Dalet Plan advocated the expulsion of Palestinians from their homes. In April 1948, Menahem Begin's Irgun massacred the 254 inhabitants of the town of Deir Yassin, crushing their bodies in mills.

The end of the British mandate in May 1948 marked the beginning of the war, in which the Jewish Haganah was crushingly victorious. 700,000 Palestinians, 60% of the Muslim and Christian inhabitants of historic Palestine, suffered expulsion, and crammed into overcrowded refugee camps in the Lebanon, Jordan and Gaza. Those Palestinians who were granted authorisation to stay in northern Israel lived under military occupation until 1965, separated from their Jewish colonisers.

Palestinian political culture changed. Two tendencies emerged in the Palestinian diasporas. The first was the Nasserite Pan-Arabism of the Arab Nationalist Movement led from the early 1950s by the Christian Georges Habash. The Movement was anti-imperialist, anti-Zionist and opposed to the conservative Arab countries; it also advocated a united Arab state, arguing that the creation of a nation state within the territorial boundaries of Palestine was unviable. The second tendency, which emerged in the late 1950s, was the pragmatic nationalism of Yasser Arafat's Al-Fatah that proposed the liquidation of Zionism in Palestine. In order to obtain the support of the Arab states, Al-Fatah considered it necessary to remain impartial in inter-Arab conflicts and differing ideologies. The Arab Summit of 1964, held by Nasser in Cairo, mandated the creation of a Palestine Liberation Office in Jerusalem, ideologically Pan-Arabic, that was accordingly directed by Chukairy, a diplomat from the Arab League; its creation was reluctantly approved by Arafat and Habash. In 1967, Habash promoted the formation of the Popular Front for the Liberation of Palestine (PFLP), which, in common with Al Fatah, embarked on an armed struggle against Zionism.

Israel's abrupt and successful victory over Egypt, Jordan and Syria in the June 1967 Six Day War, involved the conquest of the Sinai and Gaza Strip from Egypt, the West Bank from Jordan and the Golan Heights from Syria. 400,000 Palestinians were once again expelled from their homes, meaning that over twenty years 75% of the population had suffered the same fate. The Pan-Arabic dream sank without trace in Palestine. In Jordan Al Fatah increased its prestige through their 1968 defence of the Karameh refugee camp; in 1969 Arafat took control of the PLO and made the organisation less Pan-Arabic and more Palestinian nationalist, stating its objective as the setting up of a "democratic secular state" to replace the Zionist regime. In 1969, a faction of the FPLP, led by Nayev Hawatmeh, founded the Democratic Palestinian Liberation Front, which soon developed pro-Soviet leanings.

The prestige of the PLO within Palestinian groups allowed it to create its own organisations within refugee camps, which led it into direct confrontation with Jordan's Hachemit Dynasty (half of Jordan's population was Palestinian). King Hussein's forces attacked Palestinian positions in the "Black September" of 1970, which forced the PLO to move camp from Jordan to the Lebanon, a country that became the organisation's centre of operations for over a decade (with grave repercussions, especially with Pan-Arabic leftist groups attacking worldwide the Western powers that they felt were complicit with Israel).

The October 1973 war demonstrated that Israeli military invincibility was a myth, which only exasperated the distance between extremists and those open to compromise. Whilst Arafat gave his famous "rifle and olive branch" speech at the UN in 1974, the FPLP and other leftist groups formed the Rejection Front in Baghdad. In 1975, Beirut was the scene for the re-emergence of the conflict between the Palestinians who wished to support Lebanese progressives against the Christian Phalangists, and Arafat's "pragmatic" wing, who did not. However, the Phalangist blockade of the Tel-al-Zaatar refugee camp implicated Arafat's grouping in Lebanese affairs until 1982.

The Peace Treaty signed in 1979 by the Israeli Prime Minister Begin and Anwar-el-Sadat, which handed back the Sinai peninsular to Egypt, led to an increase in the repression of Palestinians in the occupied territories, which accordingly intensified their resistance. In August 1982, Begin and Defence Minister Ariel Sharon launched an invasion of the Lebanon to destroy

Palestinian bases in the country. The PLO abandoned Beirut for Tunis in August 1982, after which Falangists, coordinated by the Israeli army, massacred thousands of Palestinians, levelling the Sabra and Chatila camps.

The Lebanese resistance, who forced the Israelis to set up a security zone on their northern border, spread to the occupied territories in the late 1980s. One of the territories, the Gaza Strip, is one of the poorest and most densely populated regions on Earth. In 1987, some 650,000 Palestinians lived in Gaza, 65% of whom were under 15; half of the labour force in this Palestinian "Soweto" worked in poor conditions in the Jewish zone of Tel Aviv, whilst the few thousand Jews that did live in Gaza owned a third of its land. The possibility of emigration to the Arab gulf states had acted as a pressure-release valve to the region until the 1980s, but when the fall in crude oil prices led to this option disappearing, the national and social frustrations of the region's youth were addressed only with repression. In December 1987 following the death, at the hands of the Israelis, of four Palestinians in the Jabalia refugee camp, the Intifada (or "liberation from Israeli occupation") commenced in Gaza. The civil resistance movement soon spread to the West Bank and even crossed over the "green line" that marked Israel's 1948 border. The *Intifada* followed a policy of national construction and created Popular Committees that were outlawed by the Israeli government; Palestinians left their seats in local government and traders fixed their own hours and prices.

The initially hard-line PLO moderated its programme, and in 1988 accepted UN Resolutions that recognised the state of Israel with its 1967 borders. The PLO replaced its objective of substituting the Zionist regime for a secular state with the goal of establishing a Palestinian state next to Israeli state in the occupied territories. Public negotiations between the PLO and the United States began in Tunis in late 1988, and ran up until shortly before the Gulf War.

However, it was at this moment that the political orientation of the territories changed drastically. Until this time the activity of Islamic groups, the Palestinian branches of the Egyptian Muslim Brothers (who broke away from Islamic Jihad in 1975), had been limited to the social and religious spheres, and had been tolerated by the Israelis who saw them as competition to the secular and revolutionary PLO. But the Intifada provoked the creation of the Islamic Resistance Movement Hamas, which was established by Ahmad Yassin, the head of the Muslim Brothers, in December 1987. The calls by Hamas for a revolt against Israel and its accusations that the PLO had sold out, echoed by the leftist Damascus-based Rejection Front, were favourably received by the disinherited youth of Gaza (Welty,1993).

Arafat's error of judgement in supporting Iraq's Saddam Hussein deprived him of the financial support of conservative Arab states; a combination of the PLO's weakness, the power of the Intifada and the height of Palestinian Islamism motivated the United States to pressurise Israel into accepting the opening of public peace talks in late 1991. The talks, which first took place in Madrid and then moved to Washington, hinged around the phrase "*peace for land*". Over the course of the negotiations a number of Palestinian provincial leaders came into public life, such as Professor Hanan Ashrawi.

The electoral victory of the Israeli Labour Party in 1992, coupled with the direct intervention of the PLO in the secret Oslo negotiations, led to the Washington Accord of September 1993. Within the accord the Israeli state, which recognised the PLO, authorised the establishment of Palestinian self-government that was restricted to the Gaza Strip and the city of Jerico, but would later be extended to the whole of the West Bank. Although the Accord affected less than one third of the five million Palestinians who lived in Israel and the exiled communities, and was riddled with numerous ambiguities,[1] it did arouse initial enthusiasm, since it meant the end of open war.

However, a number of factors led to initial enthusiasm drying up, specially the continuous postponing of the timetable. The coming to power of the Palestinian National Authority did nor occur until May 1994; a slowdown occurred in the withdrawal of Israeli forces; Elections in Gaza and the West Bank for a Legislative Council and a President for the Palestinian Authority (boycotted by the Islamites and the Rejection Front) were finally held on the 20th of January 1996 after a very long delay. Other factors of disaffection were the failure of the West to carry through pledges to finance the economic revitalisation of the territories; the authoritarianism (albeit perhaps forced by circumstances) of Arafat as Head of the Palestinian National Authority, and of police forces made up largely of former PLO guerrillas; the continuation of appalling attacks by both sides, such as the Hebron and Tel Aviv bombings (perhaps the most important factor); and the internal conflicts that tore the two sides apart. Such conflicts led to the assassination of the Prime Minister Yitzhak Rabin by Jewish extremists on the 4th of November 1995. The Labourite Shimon Peres replaced him.

Five parties (or movements) had taken part in the Palestinian legislative elections: Arafat's Al Fatah; the Palestinian Peoples Party (the former communist party) led by Abed Rabi; the Democratic Coalition, headed by Abdel Shafi, the former head of the Palestinian negotiation delegation; and the General Women's Union. Both Arafat and Jalil ran for the Presidency, with Arafat and the Al Fatah party winning handsomely.

The Israeli army's withdrawal from the West Bank town of Hebron, whose historic district was inhabited by 450 Jewish colonisers surrounded by 150,000 Palestinians, was put off "until a later date"; but the results of the Palestinian elections strengthened the position of the advocates of peace in Israel as the country prepared for elections to the Knesset (national parliament) in May 1996. However, Palestinian groups opposed to the peace accord embarked on the terrorist bombings of buses and central business districts in Israel which left 60 victims and infuriated the Jewish population. In the last two weeks of April a new Lebanese War was unleashed by the Israelis against the armed Hezbollah group, through the bombing of Lebanese villages. This pushed the Arab parties in the Israeli state to withdraw their electoral support for the Labour Party's Shimon Peres. The right wing Likud party campaigned on the slogan "Netanyahu will gain a secure peace", a programme which in practice replaced the principle of "land for peace", which the Oslo accord had been based on, with the principle of "peace for peace." Zionist Jews turned in support Netanyahu.

On the 29th May Peres suffered defeat in the Prime ministerial elections at the hands of Netanyahu by the tiny margin of 30,000 votes, just 1%. In the Knesset the combined votes of Likud and the rightist religious parties (the Judaist Shass and Mafdal parties and the Israel Be Aya party of Russian immigrants) exceeded those of the Labour Party and other left wing parties (the Mertez and communist parties). The Religious-Laicism cleavage, which is particularly notable in Jerusalem, divided Israel into two.

The peace process stagnated as a result of this, and dialogue between the Palestinians and Israelis was frozen, leaving the implementation of the Oslo accord deadlocked: an accord which had foreseen not only the Israeli withdrawal from some territories, but also a three phased process of releasing 4,000 Palestinian prisoners. During September 1996 there were mobilisations of Syrian troops along the Israeli border, and Mubarak, the Egyptian president, publicly voiced his disgust at Israeli policy. Following foreign pressure, especially from the United States, Netanyahu and Arafat did physically meet for the first time, but the encounter brought no results.

The Israeli policy aiming to make Jerusalem all-Jew achieved the numerical dominance of Jews over Arabs (165,000 Jewish inhabitants to 150,000 Palestinians), even in East Jerusalem. In late December, in an open gesture of provocation, Netanyahu's government decided to unite the Jewish Wailing Wall with the Via Dolorosa, by reopening the Asmoneos tunnel which passes underneath a Muslim neighbourhood and under the Mosques of Al-Aks and Omar, one of the oldest and most revered monuments of the Islamic World.

Faced with this aggression against one of their most important symbols of identity the Palestinians rebelled, being supported by Arafat, the president of the Palestinian Authority. The Israeli army entered West Bank towns such as Ramallah, Belen and Nablus; and armed Palestinian police defended the population. Seventy people were killed in the clashes, including a number of Palestinian police and fifteen Israeli soldiers, and a thousand people were injured. 35,000 Palestinian Police, who had been accused of violations including the killings of Palestinian dissidents in detention centres, were hugely popular following the clashes. The unrest spread to Gaza.

The Israeli army oscillated between supporting reluctantly Netanyahu's policies and a project known as operation Crown of Thorns, which stepped into three phases: first, the reinforcement of troops deployed in the West bank; second, the closing off of cities protected by the Palestinian police; and, finally, the establishment of mass detention camps.

In early October President Clinton invited Netanyahu and Arafat to meet in the United States to restart the peace process. Mubarak turned down an invitation to the meeting. At the time of writing talks between the Israeli and Palestinian Authorities have not yielded visible results. Israel has not closed the Asmoneos Tunnel, has continued its policy of establishing Jewish settlements in the West Bank and has intentionally applied pressure by closing its borders to prevent 10,000 Gaza Palestinians from going to work in Israeli territory. The West however, and particularly the European Union, have put pressure on the Israeli government.

Kurdistan

The boundaries drawn up after the First World War left the Kurds divided across four states. Most Kurds ended up within the borders of either Iraq or Kemälist Turkey, but there were also pockets of Kurdish population in Syria and Iran. The Kurds are an Islamic mountain people who share a Persian language, but they are not totally linguistically and religiously homogenous; having been divided and deprived of any recognition of their identity markers, the various ethno-national Kurdish movements have not transcended the borders of their respective states. In Iraq Barzani's conservative DPK and the more leftist PUK of Talabani have opposed both the traditional monarchy and the Ba'athist regime of Saddam Hussein that succeeded it. The relative support given to the Iraqi Kurds by Iran during the Iran-Iraq war, led to systematic destruction of Kurdish towns after the 1988 armistice. In Turkey the Marxist-Leninist PKK, which was headed by Ocalan, began a resistance campaign in 1978 which led to the militarization of Kurdish provinces, the criminalisation of elected Kurdish representatives and acts of genocide; the Turkish forces have had no qualms over crossing into Iraqi territory in search of Kurdish guerrilla bases. Following Saddam Hussein's defeat in the 1991 Gulf War, UN support facilitated the establishment of a Kurdish autonomous region in Iraq, in an exclusion zone to the north of the 36th parallel. The exclusion zone has prevented neither the incursions of Turkish troops into the territory nor inter-faction Kurdish conflicts. The conflicts between Kurdish groups intensified in autumn 1996, becoming instruments in a war of delegation between the Middle East states.

Kurdistan has a surface area of 410,000 square kilometres which covers the southeast of Turkey, the northwest of Iran, and the northeast of Iraq and Syria. Of 22 million Kurds, 11 million live in Turkey (25% of the Turkish population), 6 million in Iran (16%), 4 million in Iraq (28%) and 1 million in Syria (9%). Two of the four current states that cover Kurdistan are of Arab descent, with the others having descended from the Persian and Ottoman empires; all four have subjected the Kurdish people to acts of genocide, victim of countries that themselves have been victims of the West.

Kurdish is a northeast Persian language, not an Arab language. However, whilst the Turk and Syrian Kurds have used a Latin alphabet from the 20th century, the Iranian and Iraqi Kurds use an adapted Arab alphabet. This factor, combined with the existence of three dialects (Gorani in western Turkish Kurdistan; Kurmanji in the east; and Eastern Kurd in Syria and Iraq), makes inter-Kurd communication difficult. The Kurds along with their neighbouring peoples are Sunni Muslims. However Kurdish beliefs stem from the Shafiit family, which is not the case for other peoples in the region: the Turks and Arabs are Sunni Muslims from the Hanafit School, and the Persians and Azeris are Shiites. Kurdish religious minorities include those from the Halevi School in Turkey, although the biggest group is the Yezidi, a non-Muslim religion that incorporates beliefs from ancient Iran, and has suffered persecution at the hands of both the Ottomans and the Muslim Kurds.

Having conquered Kurdistan in the 16th century, the Ottoman Empire established a clientelist network of tribal *Aghas* (small feudal lords) who competed to gain favour from the Ottomans. Contrary to one particular and "oriental" vision of Kurdish society, which painted the Kurds as a savage warrior people, agriculture-based Kurdistan was in fact a bastion of Islamic education until the 19th century. In the late 19th century Kurdish intellectuals embarked on an incipient nationalist campaign against their corrupt Ottoman rulers and the Kurd Sheiks and *Aghas* that collaborated with them.

Following the collapse of the Ottoman Empire the majority of Kurdistan fell under Turkish control, with South Kurdistan, the former Ottoman province of Mosul, being taken by the British. There were a number of rebellions in both parts of Kurdistan, and in 1920, following the Treaty of Sèvres, the allies decided to establish a Kurdish state in Turkey, which was to be united with South Kurdistan a year later. However, Kemälist Turkey (that had used the Kurd *Aghas* against the Armenians) prevented the project in 1922. The British subsequently incorporated Mosul into Iraq.

Such circumstances were detrimental to Kurdish language and culture. The Kurdish language was outlawed in Turkey, Iran and Syria, and the establishment of universities in Kurdish territories was prevented, meaning Kurdish education could not progress above an elementary level. Whilst Kurdish tribes in marginalized areas remained intact, the establishment of large landed estates made rural farmers poorer and proletarian. All Kurdish movement fell into a state of paralysis after a number of revolts were put down.[II]

Following the Second World War Iranian Kurdistan became a neutral zone between the Soviet and British blocs. In 1945, the Kurds proclaimed the territory the Kurdish Democratic Republic of Mahabad, which enjoyed Soviet support. However, in 1946 Iranian, Iraqi and British troops entered the territory and the Republic was dissolved. The Kurdish movement in Iran has been weak ever since.

The Kurdistan Democratic Party (DPK) was formed in Iraqi Kurdistan in 1945 with Barzani as leader. Following talks with the Communist Party and other groups on the Iraqi left, the party called for a Kurdish Democratic Republic within the Democratic Republic of Iraq. However, the abolition of the Iraqi monarchy in 1958, which had been engineered by the Ba'ath Arab nationalist party, far from satisfying Kurdish national aspirations, in fact forced the PPK to embark on a defensive war in 1961. In 1970, the Ba'athist regime and Barzani did agree to the establishment of a Kurdish autonomous zone, and held talks over a period of four years; however, differences over the location of the zone's borders, the Ba'ath parties refusal to accept a co-federal regime in Iraq and the DPK's switch of alliances (the party became closer to Iran and the United States following the Iraqi government's close relations with the USSR), led to the failure of the talks, and the outbreak of a fresh war in 1974. A new Kurdish party, Talabani's Patriotic Union of Kurdistan (PUK), emerged over this period and opposed the conservatism of Barzani, but this did not impede the PUK and DPK uniting their *peshmergas* (fighters) in a guerrilla force. The Iraqi government responded by destroying Kurdish villages.

The Iranian revolution was followed by the Iran-Iraq war that lasted for eight years between 1980 and 1988. Iran supported the Kurds against its Iraqi enemy, although it did support the DPK to a greater extent than the leftist PUK. When the PUK moved closer to the Iraqi government in 1983, the Iranians facilitated the convergence of the two groups, which in 1987 re-emerged as the Kurdistan Front. However, the August 1988 armistice left Saddam Hussein with a free hand, and over the following two years he destroyed almost all Kurdish towns with chemical weapons.

At this time left-wing Turkish political parties, that had shared the Kemälist line of simply denying the existence of the Kurdish people, embarked on a search to a solution to the Kurdish problem. In 1978, Abdullah Ocalan, known as "Apo", formed the Kurdistan Workers Party (PKK), a Marxist-Leninist (but not pro-Soviet) grouping that recruited its members among Kurd workers in both Kurdistan and Turkey. The PKK's Maoist discourse argued that Kurdistan was a four state colony of classic colonial capitalism, with a collaborating feudal class of *Aghas* and Sheiks. Accordingly, it was deemed necessary to fight not just against the Turkish state, but also against the traditional sectors of Kurdish society that supported it.

The military coup in Turkey in 1980 meant the end of the line for most PKK leaders. Those that survived, including Ocalan, set up base in the Lebanon, where they established contact with the Palestinian resistance. In 1984, the PKK set up the Armed Forces for the Liberation of Kurdistan (HRK) that launched attacks against Turkey from bases in Iraqi Kurdistan. Having obtained the consent of Saddam Hussein, Turkish government forces crossed over the Iraqi border to destroy HRK bases; in 1987 Turkey turned Turkish Kurdistan into a militarised super-province. Such harsh repression did nothing but increase the PKK's influence over the Kurdish masses. In 1988, a number of Kurd Deputies from the Turkish Social Democratic party split and set up the People's Socialist Party (HEP), which the Turkish government has accused of being the political wing of the PKK (Ibrahim, 1995).

In April 1991, following the Iraq's defeat in the Gulf War, the Kurdistan Front factions agreed the terms for the establishment of a provisional Kurdish government and instigated a general insurrection over the whole of the Iraqi Kurdistan that took control of the entire territory; but Saddam Hussein, who had succeeded in clinging on to power, launched a counter-offensive that led to a huge number of Kurdish refugees fleeing the area, and obliged UN Security Council intervention to protect Kurds from Iraqi repression. The United States, United Kingdom and France forced the establishment of an air exclusion zone to the north of Parallel 36, which covered 10,000 square kilometres and the majority, but not all, of Iraqi Kurdistan (a similar zone was established to the south of parallel 32 to protect the Shiite population). In May 1992, elections were held under UN supervision to establish a Kurdish parliament, with the DPK and PUK each winning half of the seats. The DPK and PUK presented a common front in the Iraqi National Council and became the centre of Iraqi opposition to Saddam Hussein.

Since Turkish government was sworn enemy of Saddam Hussein, its policy towards the Kurds appeared to have changed by 1991. However, isolated gestures such as the lifting of the prohibition on speaking Kurdish lacked continuity, and a fresh Kurdish rebellion broke out in May 1992. Turkey established relations with Iraqi Kurdistan with the objective of frustrating PKK attacks from the territory. Talabani subsequently succeeded in obtaining a ceasefire from Ocalan, although the Turkish Kemälist military blocked any agreement and forced the government to imprison elected Kurdish representatives on charges of separatism and collaboration with terrorism. On the outbreak of a new insurrection in 1992 the PKK, its prestige intact, progressed from its previous Marxist-Leninist base becoming a 'people's' party, ceasing its attacks against the traditional Kurdish elites.

In early 1995, Turkey, which had already launched an offensive in its own part of Kurdistan, took full advantage of a full-scale conflict between the DPK and the PUK, and launched a fresh incursion into Iraqi Kurdistan. The operation, which proceeded despite opposition from the EU and Arab world, involved 35,000 soldiers who penetrated some 25 miles into Iraqi territory in a tactical agreement with Saddam Hussein. The dual objectives of the incursion, namely ridding Iraqi Kurdistan of PKK guerrillas and establishing a stable "security zone", were not met.

The alliance between the two Iraqi Kurdish factions of Barzani and Talabani proved transitory, and in December 1994 a civil war broke out between the two factions for control of the zone, leading to 2,000 being killed. A US brokered ceasefire broke down in July, and Iranian troops, circumstantial allies with Talabani, assisted the PUK in reinforcing its control over the zone.

Despite the Iraqi regime's assassination of his own family, DPK leader Barzani nevertheless called for the Ba'athist regime's assistance in late August. In early September Hussein, who feared the CIA's support of the Iraqi National Council (aimed at provoking his downfall) and who wanted to secure control of the oil pipeline that transported Kurdish petroleum across Turkey to the Mediterranean, sent in 15,000 troops from his own Republican Guard north of Parallel 36 to assist Barzani's DPK in taking control of the Kurdish city of Arbil, which had been held until then by Talabani. Hussein argued that the Kurds themselves had called him there in order to repel an Iranian invasion. Yet, the United States, that maintained 23,000 troops and 20 warships in the Persian Gulf, launched air strikes against Iraq with the stated aim of punishing Hussein and under the pretext of defending the Kurds: but, in a gesture of cold cynicism, the US bombings were targeted in reality at the anti-aircraft defences south of Parallel 32 (defences that actually threatened Kuwait), and not Iraqi positions in Kurdistan.

On September the 8th, after Arbil had been secured, the PDK took control of the Kurdish city of Sulaymaniya, a city of over a million people that is situated south of parallel 36. The operation led to the total control of Iraqi Kurdistan by the PDK and provoked a mass exodus of Kurdish PUK followers towards Iran (Iran and the US, who are irreconcilable enemies, found themselves on the same side in this particular conflict).

Turkey then massed troops in the Diyarbakir region of Turkish Kurdistan. The objective was to launch an incursion into Iraqi Kurdistan and establish a five to ten mile deep security zone along the length of the border (similar to the Israeli zone in south Lebanon), to protect the territory from attacks from PKK guerrilla bases (which were protected by Barzani's DPK.) The project was rejected by the Arab Gulf states, and Iraq believed it could use Barzani to justify its opposition to the Turkish expansionist plans. However, the DPK leader, who had no faith in the assurances of Saddam Hussein, met Turkish leaders in Ankara and attempted to re-establish relations with the United States. Days later the Talabani's UPK took back the city of Sulaymaniya.

The Punjab Sikhs

The Indian sub-continent provides a good example of a civilisation that has suffered discrimination at the hands of the West. Its destiny, however, has been very different to that of the Arab world, due to the sub-continent's early and exclusive colonisation by the British, which prevented the fragmentation common in such cases. In the Indian civilisation-state, which is made up of numerous Indo-European linguistic groups and other non-Indo-European groups, e.g. Dravidic and Tibetan-Burmese, integrated by the multi-ethnic National Congress Party, religion has been the only divisive factor in the state (namely the clash between the confessional minorities, especially Muslims, and the Hindu majority). A number of ethno-national movements have emerged in post-colonial India, the most violent of which shook the Punjab, a region of Sikh religion (which emerged in the 16th century as a synthesis of Hinduism and Islamism). The Government's refusal to grant the moderate religious and socio-economic demands of the Akali Dai, a Sikh denominational party, provoked a more radical approach and led to the occupation of the Golden Temple in Amritsar by the movement's most extremist wing. In an effort to win-over the Hindu majority, Indira Gandhi sent in troops to take the Temple in 1984, massacring its occupants. Following the assault Gandhi herself was assassinated by Sikhs among her own bodyguards. A pogrom against all Sikhs in India ensued, and a full-blown conflict commenced in the Punjab that continues to this day.

The ancient Punjab region encompassed the plains of northeast India that lie between the Himalayas, the Indus and the Ganges high plateau. The region is irrigated by five rivers, making it agriculturally rich. Whilst Muslims inhabit the East Punjab, situated in Pakistan, West Punjab is home to 70% of India's Sikhs, who live alongside the Hindi minority (twelve million of the Punjab's 20 million inhabitants are Sikhs, some 60% of the total). The philosopher Nanak founded the Sikh religion in the late 15th century, regarding it as a link between Islam and Hinduism, although Hinduism has been more dominant. In 1700, Govind Singh, the last of ten Gurus, established a military theocracy that successfully resisted the power of the Mongols. The ensuing military tradition led to Sikhs being the last group to be dominated by the British, preventing British completion of their occupation of the Indian sub-continent until 1849. The tradition also explained the high proportion of Sikhs in the colonial-era army, 30% of the total force, an anomaly that has not changed in post-colonial India.

The Sikh religion believes in the *Khalsa* or fraternity of the "pure men", who made up the military elite that governed the Punjab; the faith has a sacred text, the *Adi-Granth*, which is recited in places of worship, the most well known of which is the Golden Temple in the Punjabi city of Amritsar. Followers do not shave of their beards, and tie up their hair and cover it with a turban. Nowadays the Punjab Sikhs tend to be landowners or affluent farmers, whilst the Hindi population tends to be largely urban based.

At the time of partition, the Sikh party Akali Dal collaborated with the Indian Congress Party against the pro-Pakistani Muslims, a collaboration which led to thousands of Sikh casualties. Akali Dai in the 1970s, under the leadership of Sant Singh Longowal, elaborated a non-secessionist programme known as the Anandpur Sahib Resolution, which was ratified in 1978 when Akali Dai governed the Punjab as a part of a federal government coalition led by the rightist Janata party, a rival of the Indian Congress Party. The resolution, that was largely political-economic in nature, called for Chandigarh to become the exclusive capital of the Punjab, a greater degree of control over water resources (a source of tension with the neighbouring states of Haryana and Rajasthan), and the redefinition of the state to include Punjabi speaking areas. Only small sections of the resolution contemplated subjects related with the Sikh religion, whose followers rejected its categorisation as a variant of Hinduism.

When the Congress Party was returned to power in 1980, Indira Gandhi exploited the issue to mobilise the Hindu majority, and refused to implement any part of the resolution. Gandhi relied on the support of the Punjabi Hindis, who being urban based were unconcerned over the problems of irrigation. Akali Dal launched a campaign of mobilisations known as the *morcha*. However, Singh Longowal's moderates were displaced by the radicals of religious leader Sant Singh Bhindranwale; in July 1982 200 of his followers occupied the Golden Temple of Amritsar. A Holy War ensued which left 300 dead over three years.

In October 1983, Indira Gandhi suspended the Punjabi Government, and on the 2nd of June 1984, with the unanimous consensus of the Indian population, launched an assault on the Golden Temple with 10,000 soldiers, killing Bhindranwale and hundreds of his followers. The Sikhs clamoured with indignation at the sacrilegious death of these *'jata'* (religious martyrs), and rebellions ensued in the Punjab along with mutinies by Sikh troops across the north of India. Four months later, on the 31st of October, Indira Gandhi was killed by two Sikh members of her own personal guard. In the wave of hysteria that followed, Indian mobs killed 1,500 Sikhs across India.

In December 1984, Indira Ghandi's son Rajiv, having made pledges to combat Sikh separatism, secured the biggest electoral victory in Indian history. However, in 1985, the same year Sikh radicals downed an Air India jet off the Irish coast with the loss of 329 lives, Rajiv Ghandi surprisingly recognised the non-secessionist character of the Anandpur Sahib Resolution and opened up dialogue with moderates from the Akali Dai party. Ghandi, however, reneged on these commitments; as the conflict swung once more to the extremes the moderate Singh Longowal was assassinated, being proclaimed "an Indian patriot" by Rajiv Ghandi. In 1992, the Indian government mobilised the army against Sikh radicals, and destroyed them in a few months. However, as is often the case, military success has not translated into political resolution of the conflict, with the opposite having in fact occurred (Bianco, 1994; Gupta, 1995).

The Tamils in Sri-Lanka

The Buddhist Cingalese majority has maintained hegemony over Sri Lanka since independence in 1948, and has discriminated against the Hindu Tamil minority, which includes Tamils indigenous to the isle and Tamils from Dravidian southern India who were forced to migrate there by the British in the 19th century. The institutionalisation of populist Singhalese Buddhism in the 1972 Constitution proved the catalyst for the emergence of radical Tamil nationalism. This bolstered the post-1975 formation of a number of radical armed groups, of which the Tamil Tigers is the most well known. The ethnic polarisation that followed, the establishment of Tamil bases on the northern Jaffna peninsula, the anti-terrorist measures adopted by the government of Sri Lanka, and the mass exodus of Tamil refugees to the Indian province of Tamil Nadu, provoked Indian Government intervention through the Indo-Lanka Accord of 1987. The Accord forced the establishment of territorial Provincial Councils in Sri Lanka with the objective of addressing Tamil demands. However, the establishment of a Tamil state in these provinces led to its dissolution by the government of Sri Lanka, and the withdrawal of Indian troops, who had been besieged by both sides. The war continues today and is at its bloodiest stage yet.

Sri Lanka, formerly Ceylon, is an island situated off the southeast coast of India with a multi-ethnic population of 17 million, 75% of whom are Cingalese and 18% Tamils (13% indigenous and 5% brought over by the British). The Tamils, Dravidic speaking Hindus, inhabit the northern provinces (the Jaffna peninsula) and the eastern part of the island, where they make up 92% and 45% of the population respectively. The Singhalese, who are Buddhists and Sanskrit-speakers, live in the other seven provinces, popularly known as the south. There are also Moor and Malay minorities and Muslim and Christian groups.

The coastal regions of Ceylon were occupied by successive European colonisers, namely the Portuguese, Dutch and, from 1796, the English. A Singhalese Kingdom ruled the inner part of the island until losing its independence in 1815. The British set up a mass-cultivation colonial economy on the island that was based first around coffee, then from 1870 around tea, and latterly around coconut and rubber; which required the importation of a semi-slave workforce from the southern part of the Indian sub-continent (from the region that is now the Tamil Nadu province, with a population of 50 million and with Madras as its capital).

In February 1948, Sri Lanka followed India in gaining independence. The constitution of Sri Lanka, which had been modelled on the British, constituted a centralised state, and established a system of checks and balances which maintained Singhalese and minority Deputies at a ratio of three to two in parliament. A constitutional clause safeguarded minority ethnic rights. Senanayake's United National Party, a Singhalese anglophile conservative party, won the elections; pressure from Singhalese peasants in the mountains led to a 1949 law that excluded non-indigenous Tamils from citizenship and the vote. In 1952, Tamils in the UNP abandoned the party to create a federalist party.

Bandaranaike's leftist United Popular Front won the 1956 elections; but the coalition was dominated by the United Front of Monks who made Buddhist nationalism the new dominant ideology in Sri Lanka and pushed through legislation that made Singhalese the island's only official language. Bandaranaike himself worked with indigenous federal Tamils in an attempt to find a solution to the situation, which led to his assassination in 1959 by a Buddhist monk. His widow took power in 1960, but under policies aimed at resurrecting "the great Singhalese culture." Although the UNP won the 1965 elections through a coalition that included federalist Tamils, their federal demands were once again ignored. A combination of Bandaranaike's widow being returned to power, and pressure from the armed Cingalese National Liberation Front, led to the approval of a republican constitution in 1972, legislation that brutally imposed the dominance of the Buddhist Singhalese majority.

As a consequence, Tamils began to demand an independent state from 1972; a new Tamil leadership emerged, firmly based on the Jaffna peninsula, and more populist and anti-Western than its federalist predecessor. The new Tamil nationalism, inspired by the examples of Palestine and Vietnam, embarked on armed struggle with the objective of securing control of the territory it inhabited. A number of separate groups emerged, which launched attacks on police barracks and Tamil banks that collaborated with the Singhalese. The strongest group was the Eelam Liberation Tigers of Tamil

(LTTE) formed in 1975 under the leadership of Prabakarhan, which achieved dominance in the 1980s.

The 1977 elections were polarised between the UNP, which unified Singhalese nationalism, and the United Tamil Liberation Front, which took all twelve seats in the northern province and four of the twelve in the east, and hence became the island's major opposition party. Government attempts at negotiation with Tamil moderates, which tried to separate them from the radicals, were unsuccessful; in 1978 the constitution was reformed to establish an elected president with the aim of suffocating the insurrection in the north. In 1979, a Prevention of Terrorism act was passed and the army was subsequently sent into the Tamil territories, whilst Tamils living in the south suffered increasing social violence. This led to the ethnic conflict becoming a full-blown civil war. Following the death of 13 soldiers in a Tamil Tiger attack in July 1983, the Singhalese masses turned on the Tamils in the south, massacring a thousand of them and provoking an exodus of a further 90,000 towards the Indian province of Tamil Nadu. Overflowing refugee camps on the outskirts of Madras became hotbeds for Tamil paramilitary groups to recruit their fighters (whose numbers quickly rose to 6,000); following an attempt at uniting the five most important groups in 1984, the Tigers achieved dominance through the efficient method of killing rival leaders. The Tigers had little difficulty obtaining dominance over the Jaffna peninsula, whose 750,000 people lived without water, electricity or telephones, fenced in by the army and bombed by the air force, and where unemployment stood at 40%.

The situation led to solidarity from southern India, where a Dravidic Nationalist Movement emerged, which in turn led to Indian government involvement in the conflict. India acted as mediator to bring about negotiations between the two sides, which took place first in Thimpo, then in Bhutan and finally in the Sri Lankan capital Colombo. An indirect result of the meetings was the Indo-Lanka Agreement of 1987, in which the Sri Lankan government agreed to establish Provincial Councils that would transfer certain powers to all provinces of the island (and not just the Tamil provinces). It was felt that such a measure would end the conflict. India agreed to send in a peacekeeping force of 45,000 troops that would supervise the conclusion of the agreement and disarm the Tamil groups.

The agreement, which with all its limitations did at least institutionalise for the first time the multi-ethnic nature of the island, was at first not contested by either the Singhalese opposition or the Tamil Tigers, who maintained a ceasefire for some time, although they did not stand in the elections that came out of the agreement. The elections were held in the Singhalese south in April 1988 and under Indian supervision in the Tamil north and east that November, where the Eelam People's Revolutionary Front won handsomely.

However, the truce lasted less than 18 months. The Provincial Councils criticised legal limitations imposed on them by 13th amendment to the constitution, a clause that returned powers over rural planning, policing, security, education and public administration to central government. In addition the Indian peacekeeping forces quickly became forces of repression that fought against both Tamils and Singhalese nationalists. In 1989, the Sri Lankan government requested the withdrawal of the Indian troops (who had suffered 1,500 fatalities) and began fresh negotiations with the Tamil Tigers; meanwhile, the Provincial Councils of the north and east proclaimed an independent Tamil state, being subsequently dissolved.

In March 1990, two months after the Indian withdrawal, negotiations were broken off, and full-scale war resumed in the north. The Tamil Tigers, who were blamed for the 1991 assassination of the Indian President Rajiv Ghandi, killed Sri Lankan president Premadas in 1993. Kumaratunga, the widow of a Popular Front leader who had been killed by the Tigers in 1988, and daughter of the assassinated Bandaranaike, won the November 1994 elections against another widow, whose husband UNP leader Disanayake had been killed the previous month (Bastian, 1994).

In contrast to the UNP, the Popular Front programme considered the re-initiation of talks with the Tamils, which had been broken off in 1990. The objective was to re-launch and improve the 1980s programme and turn Sri Lanka into a union of autonomous provinces with a high level of self-government. However, the ceasefire agreed to by the Tamil Tigers in January 1995, was broken four months later when disarmament was demanded as a pre-condition to initiating negotiations. An offensive launched by the Sri Lanka army in October 1995 culminated in December 1996 with the taking of the Tamil headquarters in the Jaffna peninsula. However, this has not put an end to the extreme manifestations of a conflict that has already claimed 50,000 lives in the last twenty years.

On the 30th of January 1996 a Tamil Tiger attack on a Colombo bank left 100 dead. The Tamils, having been dislodged from their Jaffna base, had set up camp in the jungle, and in July attacked the Mullaitiva base, killing 1,500 Sri Lanka army personnel. The land, air and naval offensive launched by the government against the Tiger's base in Killochi has not ended their resistance.

Corsica

Corsica is a mountainous island that was occupied by Genoa until French occupation in the 18th century. The island's structure revolved around a Mediterranean system of rival clans who acted as intermediaries between the invaders and the population. By the 19th century the Clans had evolved into French parties, and by acting as clientelist agents to the metropolis contributed to the under-development of Corsica. During the Second World War right-wing Corsican nationalists collaborated with their Italian occupiers. In the sixties modernisation plans placed the development of tourism in non-Corsican hands, and assigned vineyard cultivation to the *pieds noirs* -French colonialists who had returned from Algeria. Reaction to these moves led to the emergence of a new nationalist tendency that was little influenced by the arguments of the French left, and led to the 1976 formation of the paramilitary Corsican National Liberation Front (FLNC) which engaged in a struggle that was based more on "armed propaganda" than on killings. In 1982 the French Socialist party, with the objective of securing nationalist complicity against the right-wing clans, drew up a *Statut particulier* (Autonomous Statute) that permitted limited autonomy. The failure of the French parliament to recognise the existence of the "Corsican People" has polarised Corsican society into two extremes, conservative Clans and radical nationalists; although in recent years the armed groups have fragmented, become more clan based, and attacked each other. The ceasefire announced by the original FLNC in January 1996 collapsed in September of the same year following the lack of response by the French government; the ceasefire was reinstated when the French Socialist party took power.

> Corsica is a beautiful Mediterranean island that is geographically closer to Italy than France, and whose eastern coast is home the only island's flat land; the island has been occupied throughout its history by a number of foreign powers: the Vatican, the Muslims, the Pisa Republic from the 11th century and the Genoa republic until the 18th. Corsicans speak an Italic yet non-Italian language, and the island's indigenous population have either lived from sheep farming in the mountainous regions or from the subsistence agriculture economy that was developed on the island's middle ring of land. The island's invaders built up towns on the coast.
>
> Competition for scarce resources, especially water, led to large families evolving as the basic unit in Corsica (along with other Mediterranean societies) as they were the most efficient grouping to ensure success: the grouping of families with a dominant family at the head led to clans. In all of its centres Corsican society was divided into two bitterly opposed *partitu* (clans) whose members were united to each other through a sense of "Honour". All personal conflicts were considered as an attack on the honour of the clan, which could be resolved in vendettas that could last for generations, although they were frequently ritualised to prevent them ending in killings (the Corsicans, however, reserved violence for themselves, as visitors to the island were treated with the utmost hospitality). The Clan chiefs became mediators between the political power of the colonisers and the Corsican population.
>
> The French took control of the island from 1768 having taken advantage of a rebellion against Genovese occupation organised by the renowned Corsican Paoli. Despite initial resistance from the clans, their leaders, who were ennobled by their new colonisers, accepted the situation and assimilated the political divisions that had emerged from the French Revolution. The Jacobin Bonapartes successfully took on Girondin Paolists and made their son, Napoleon Bonaparte, a household name. This however did not stop

Napoleon from putting down numerous rebellions in his native Corsica through the imposition of States of Emergency.

Throughout the 19th century the Clans took on the role of providing for state resources such as jobs and subsidies, evolving into political parties similar to those on the continent. In reality, however, they continued to be apolitical, distancing themselves from ideology or class conflicts, and switching their position in accordance to rival clan stances. In the late 19th century Corsica was the poorest region of the French state, and its population had dropped from 260,000 to 160,000 mainly through emigration. However, the clientelist relationship between the clans and the French centre, excellent from the reign of the Corsican Napoleon III, alleviated the consequences of the crisis. The Clans were given jobs in the administration and customs posts of the metropolis, and became "*pieds-noirs*" colonialists in Algeria and black Africa; the French also facilitated Clan members access to the highest ranks in the army, police and even the French government, a tradition which continues today.

In the inter-war years groupings that were excluded from Corsica's clans emerged as the island's first nationalist tendency; the movement was institutionalised in 1927 as A Muvra, Partitu Corsu Autonomista (Corsican Autonomist Party). However, the grouping's extreme right wing ideology led it into partnership with Italian occupation during the Second World War; the party was totally discredited when the island was liberated. Corsica was the only French territory to liberate itself through its *maquis* (a Corsican word), and a sentimental relationship emerged between General De Gaulle and the Corsicans, who felt "ultra-French". One Clan's mass-membership in the Communist Party in 1946 secured it a victory in the island of proportions that were only matched by the scale of his subsequent downfall, when the Clan system abandoned communism having begun to perceive it as a threat to their interests.

The clans failed to produce desperately needed economic reconstruction in the 1950s, a necessity that appeared to have be met by the regional policy of the 5th Republic which produced the Regional Action Plan of 1957 (PAR). The PAR prioritised agricultural modernisation and the development of tourism in Corsica (as with the Balearic model) and consequently set up two joint economic societies, the SOMIVAC (*Societé de Mise en Valeur Agricole de la Corse*), which dealt with agriculture, and the SETCO (*Societé d'Equippement Touristique de la Corse*), which developed tourism. The two largest clans, the Gavinists and the Landrysts, led by Rocca-Serra and Giaccobi respectively, joined the societies. The measures did not, however, solve the island's problems; it quickly became apparent that the real benefactors of the SOMIVAC, (which was not actually run by Corsicans) were 17,000 *pied-noirs* repatriated from Algeria to whom had been delivered the vineyards of the eastern plains, whilst the major benefactors of the SETCO were in fact the large multi-national companies that created sophisticated tourist resorts, and hence marginalized Corsicans.

Gaullist *grandeur* policy, which in 1959 led to the establishment of an atomic experimentation centre on the Corsican coast (as in the Pacific Isles) and the closure of the island's only rail service on economic grounds, added to the disappointments of SOMIVAC and SETCO, and Corsican anger boiled over. A new alliance emerged of Communists, former regionalists and a new wave of journalists, that in 1959 set up the 29th of November Movement. Whilst their programme advocated the islands integration in the modern French economy, the movement did lay the foundations for its immediate nationalist successor. Two tendencies were apparent within the new nationalism. The first emerged within Corsican exile groups in Paris, socialist in nature, which proposed the establishment of links with the French left. The second tendency was indigenous to the island, and emerged from resentment to the subordination of Corsicans on the island; whilst this group was more ideologically moderate it was also more combative in its day to day struggle.

The Corsican Union was founded by Charles Santoni, Gisèle Poli and a number of students in Paris in 1960; its concern with economic and social questions and its Mediterranean projection coincided with demands from those in Corsica who campaigned for the establishment of a university on the island. In 1964, the leaders of the second tendency, Pail-Marc Seta and the Simeoni brothers, established the CEDIC (Committee for the Study and Defence of Corsican Interests), an organisation inspired by the ethnic federalism of Guy Héraud which demanded local administrative structures that would safeguard Corsican ethnic identity. Whilst the Marxism of the Corsican Union and the widespread anti-Maghribian racism of the CEDIC were incompatible, their youth, their criticism of French centralisation and a desire for radical change brought the two groups together, eventually leading them to unite as the Corsican Regionalist Front in 1966 (FRC).

The union broke down after several months, with the Corsican Union holding on to the FRC name. FRC analysis in subsequent years, especially in their 1971 book *Main basse sur une île* (destruccion of an island) was based on Occitan Lafont's "internal colonialism" thesis and on the May 1968 French spirit. The analysis gave a precise definition of Corsican problems which included the proletarianization of Corsicans through detrimental emigration, the dispossession of Corsican lands by the SOMIVAC, the destruction of small local trader and industries, the French monopoly over maritime traffic, and the self-colonisation promoted by the clans. The FRC, a small group of intellectuals, lost unity and disappeared in the 1970s, with Santoni affiliating himself with the Socialist Party. Their denouncement of French colonialism would, however, re-emerge in the radical nationalism of the mid-seventies.

In 1977, Simeoni's CEDIC adopted the name it holds today, the Unione di Populu Corsu (UPC). The UPC was strong in the Bastia region and among those who had lost out under external development. In "*Autonomy*", a 1974 UPC pamphlet, the organisation rejected all elements that it considered divisive to the Corsican ethnic group, listing Marxism and class struggle speculator, which resulted in two deaths following clashes with the CRS) won them the sympathy of the Corsican youth.

A central armed group, the Corsican National Liberation Front (FNLC), finally emerged in 1976. From the early 70s there had been a number of attacks launched by shepherds and mountain farmers, peoples whose situation had deteriorated with modernisation. The attacks continued a secular tradition of struggle by the Sheppard communities against the settler farmers of the eastern plains which had precipitated the semi-spontaneous formation of a number of paramilitary groups, including Paolina Justice in 1973 and the Corsican Peasant Liberation Front in 1974. At the height of Aleria in 1976 these two groups merged into the FLNC, a group clearly inspired by the Algerian National Liberation Front (FNL). FLNC ideology was a synthesis of FRC ideology, without its regionalism, and UPC ideology, discarding its ethnic nature (that allowed it to have as activists within the movement pure nationalists, third-worldists, and former communists) its programme demanded the "Corsification" of large properties and tourist trusts, the recognition of the Corsican people and their right to self-determination.

In a very Corsican style the FLNC was aimed mainly at spectacular demonstrations of force dramatised in numerous press conferences. Towards the late 1970s the FLNC destroyed a Boeing jet in Ajaccio airport, bombed military bases on the island, launched so-called "blue nights" when hundreds of bombs were planted against Frenchmen's properties and symbols of the French state. However, the number of fatalities until the 1990s was only 16, scarcely one a year, a figure well below that incurred by ETA or the IRA. This self-contention was written into FLNC theory in its distinction between "armed propaganda" and "armed struggle", with the FLNC regarding the latter as a phase that would arise subsequent to armed propaganda. As in other conflicts, the FLNC began to establish its own legitimating community, and in 1976 Nationalist Committees emerged which became the FLNC's civilian mouthpiece.

The FLNC's campaign led to pro-French violence from the *barbouzes* by the *Front for Action Against Independence and Autonomy (FRANCE* in its French acronym), which from 1977 embarked on bombings and the kidnapping of Corsican activists. By the early eighties Corsica had become a *'pudrière'* (powder keg) for France; on taking power in 1981 the French socialists attempted to try a different approach to resolving the problem. In the 1970s, the historic centralism of the French left had been channelled by the ideas of worker self-management and the "right to difference" of May 68. Of the 110 proposals contained in the Créteil Socialist Manifesto in 1981, the 54th set out details for a "Particular Statute" for Corsica and a Department for the Basque Country. Whilst promises made to the Basques were never kept, the Socialists saw Corsica as a testing ground for greater French decentralisation.

The Corsican Particular Statute of 1982 established a 61 seat assembly that was elected by universal suffrage. The Statute removed the executive from the hands of the Prefect and placed it in the hands of the President of the Assembly, and established two Consultative Councils, one for socio-economic development and the other for culture and education; with three new Offices putting development in Corsican hands in contrast to SOMIVAC and SETCO. However, the assembly's laws were not binding for the central government, which could veto them at whim, and the statute made no mention of the Corsicans as a people. The project, therefore, had the virtue of both defrauding, Corsican nationalists and unsettling the Clans who feared that modernisation induced by the Statute would prejudice their clientelist relationship with the centre.

Simeoni's UPC, critical of violence, decided to participate in the March 1982 elections, and picked up 11 % of the vote and 7 seats. Although the FLNC had observed an intermittent truce since the Socialists had taken power to facilitate a prisoner amnesty, its civilian mouthpiece described the Statute as an "instrument of colonialism" and decided not to participate. The Gaullist Clan, which most strongly opposed the Statute, won the most seats with 20, but the leftist radical Prosper Alfonsi was elected President of the Assembly as a result of alliances. The right succeeded in blocking all the Assembly's resolutions in favour of Corsican language and personality and created a pro-French association which campaigned for a French and Republican Corsica opposing nationalism.

The Nationalist Consulta, criminalized by the government in September 1983, changed its name to the Corsican Movement for Self-Determination (MCA) in October; the Corsican Worker's Union was created from within the MCA and quickly became the island's second most voted union. FLNC attacks on "continentals" intensified, being often targeted on teachers or shopkeepers. In 1983 and 1984 the supposed kidnappers and killers of FLNC militant Orsoni were killed. Some press libellously alleged FLNC links with the mafia (mafia power had increased on the island following the tourist boom), with police chief Broussard applying anti-gang strategies to the island.

The UPC, which perceived that its prestige within Corsican youth was diminishing, withdrew from the assembly in early 1984 precipitating fresh elections in August. The MCA, standing for the first time, gained 5% of the votes and three seats, as did the UPC. The rightist clan parties, aligned to Arrighi's National Front, obtained 46% of the votes, and the Gaullist clan chief Jean-Paul de Rocca-Serra ascended to the presidency of the assembly. The MCA and UPC joined forces as the Unitá Nazionalista against the right and stood in the August 1996 elections, obtaining 9% of the votes and six seats. The right won again, and used the years in power to impose its stance against Corsican identity. The clans managed to persuade the rightist French government to criminalize the MCA in January 1987, which then changed its name to *A Cuncolta*.

Following the Socialist triumph in the 1988 presidential election the new Interior Minister Joxe attempted to win over nationalist support with a reform *Naziunalista Corsa* (ANC) emerged. The ANC proved influential in the teaching and cultural spheres, and set up links with a new armed group named "Resistanza".

The Socialist government's October 1990 proposal of a new reform to the statute based on the Corsican assembly's August 1988 motion led to the formation of new pro-reform factions in both the civil and paramilitary arenas, the Movement for Self-Determination (MPA) and the "Habitual Channel" FNLA (opposed to the "Original" FNLA) respectively. Whilst Article One moderately spoke of "a Corsican people as a component of the

French people", the French assembly nevertheless narrowly passed the law, with 276 votes to 262. The Senate, opposed to the project, amended it in May 1991, and in the end the French Constitutional Council declared Article One anti-constitutional. Modifications to the Statute were therefore minimal.

The failure of the Socialists to act polarised the Corsican assembly elections of March 1992 between the right and the nationalists who campaigned in two groupings: the MPA and the Corsica Nazione made up of the Cuncolta, the UPC and the ANC. The right held 51% of the votes effectively blocking any change; the nationalist vote rose from 9% to 22% (Corsica Nazione with 15% and the MPA with 7%); the Socialists disappeared of the map (Dressler-Holohan, 1981; 1990; Arrighi and Pomponi, 1984; Tafani 1986; Loughlin, 1987; Culiooli, 1990).

This explosive situation led the armed struggle to burst beyond its normal self-imposed limits. Forty people were killed in 1992 alone (although the press includes mafia killings in this figure). From 1994 on A Cuncolta and the FLNC demanded the application of the Statute of Overseas Territories to Corsica, but their protagonist stance provoked the exit of the UPC and ANC from the Corsica Nazione. The nationalist armed struggle seems to have receded into history, mirror-imaging the archaic violence of vendettas between clans. Among the reciprocal accusations of betrayal and corruption the paramilitary groups have begun to attack each other, with six FLNC and Cuncolta activists, including their Secretary General, and five MPA activists killed in two years.

The conflict appeared to change in 1996 when the French government declared Corsica an exempt zone with the aim of attracting investment; on the 12th of January the main armed group, the Original FLNC, declared a three-month ceasefire that was officially prolonged on April and July. The ceasefire declaration was announced at a press conference in the Tralonca Mountains in northern Corsica and was guarded, in a demonstration of force and in a spectacle in the best of Corsican traditions, by 600 balaclava-clad activists armed with machine guns and grenade launchers. (So, and spite the small population of Corsica at 250,000 inhabitants, the number of FLNC volunteers seems similar to the IRA or ETA). In addition, this was undertaken just a few hours before the arrival to the island of the French Interior Minister Debré, whose strident silence on the matter suggested the existence of secret talks with the nationalist group (which demanded a political solution whose first point would be recognition of the Corsican people).

However, there were still bombings throughout the year of Gendarme stations and French public buildings carried out by the Ribellu Front, a new paramilitary group that had emerged within the environment of the Original FLNC. On a July visit to Corsica the Prime Minister Alan Juppé declared that no nationalist demands would be met; as a consequence the FNLC truce collapsed in September. In early October a FNLC bomb destroyed the halls of the Bordeaux mayor's office, a post held by Juppé, who the FNLC held responsible for the failure of dialogue.

Consequently the Ribellu Front called on Corsicans to rise up against the "oppressor French state" (language which reveals the will to pass from a phase of "armed propaganda" to one of "armed struggle"), and announced its intentions to kill "informers, elected representatives and collaborators."

October's symbolic sentencing to four months imprisonment of François Santoni, the General Secretary of A Cuncolta and alleged interlocutor of the FLNC in negotiations with the French government, was followed by attacks in Nimes and on the Law courts in Bastia. The FLNC threatened a "response of unexpected proportions" to "any attempt to physically eliminate" Corsican militants. Ten months after the Tralonca press conference, the government initiated legal proceedings against the participants, whose number plates had been noted by Gendarmes. Suspicions were raised within French and Corsican public opinion over the reasons for the long delay in the process. The ascension of the French socialists to power has opened up a new chapter of hope in Corsica.

Southern Basques

In the Basque Country, nationalism, a result of the transformation of Carlist legitimism into populism brought about by Sabino Arana at the end of the 19th century, arose as a reaction to the double Spanish state-national project of the Basque oligarchy, dependent on the Spanish market, and on the Biscayan working-class movement. At the beginning of the 20th century its half clerical and half democratic ideology divided into two tendencies: one was autonomist and right-wing, and the other, independentist and Sabinian. During the Second Republic, both based themselves on the double banner of "statutism" and Christian-democracy. During the civil war, the Basque government was led by the nationalist José Antonio Aguirre; the Nazis bombed Gernika in 1937 to break the Basque resistance. After the end of the war the "atlantics" and pro-church ideology of Basque nationalism entered into crisis; out of this crisis EKIN emerged and, in 1959, went on to found ETA (*Euskadi ta Askatasuna*), Basque Homeland and Freedom. Initially Christian-Democrat and European federalist, ETA developed under the influence of Cuba, Algeria and Vietnam into an anti-imperialist and Marxist movement. Torn apart by pro-Spanish splits, the core of the militarists within ETA began to assume control after killing Admiral Carrero Blanco in 1973. ETA politico-military (pm) and ETA military (m) confronted each other over whether or not to use the institutions of Spanish democracy in the Spanish; the former were in favour, while the latter were against. ETA-pm finally dissolved in 1985; ETA-m set in motion the dynamic which would generate its own legitimating community. KAS, or the leading bloc, was established in 1976; *Herri Batasuna* (HB), a People's coalition of electoral nature, was founded in 1978. After the Spanish Army's coup of 23rd February, 1981, ETA-m rooted its strategy in the slogan "to resist is to win" with the aim of forcing the Spanish government to negotiate the KAS Alternative, whose primary objective was the self-determination of the Basque people. The Spanish government responded by passing a series of draconian laws which were to accompany its never recognised "dirty war."

The Basque Nationalist Party (PNV), which had been politically hegemonic since the institutionalisation of the Basque Autonomous Community, adopted the same stance as Spanish socialism in 1985 with regard to ETA, which led on to the Ajuria Enea Pact in 1988. Neither the negotiations in Algeria in 1989 between representatives of ETA and the Spanish Ministry of the Interior nor the dialogue between HB and the PNV in 1992 led anywhere. The deployment of the Autonomous Police, the *Ertzaintza*, from 1981 onward generated an increase in street violence carried out by young Basque radicals. This development has been counteracted to some extent by the emergence of pacifist movements and pro-dialogue pressure groups from both sides, that is, those in favour of the Ajuria Enea Pact and those anti-system nationalists, although the latter disapproved of the pro-dialogue movement which arose from within its own ranks. The socialist government, in crisis from 1994 onward due to press revelations about its role in GAL and, above all, because of the numerous cases of official corruption which became public, was replaced in 1996 by the Popular Party (PP). The new government did not bring about any appreciable change in policy toward the Basque controversy; it is, at this point impossible to determine which line it will follow: direct confrontation or a negotiated solution to the conflict.

Basque nationalism emerged at the end of the 19th century, when Sabino Arana set up the Basque Nationalist Party (PNV), which is the oldest peripheral party in Europe. After the death of Arana in 1903, the aim of the party was no longer formally that of independence, but "foral" integration (which represented a return to the pre October 1839 situation when, after the end of the first Carlist war, the law declared the supremacy of the Spanish Constitution over the Basque "fueros"). Nevertheless, this goal was in harmony with the aim of Basque independence. In this way, a strictly legal battle, which had as its objective the repeal of a Spanish law, became one with the struggle for independence, integrating independentists and non-independentists and facilitating conciliation between the "regionalist" policy of the leadership and the independentism of the grassroots.

During the twenties, Eli Gallastegui came to symbolise this latter tendency. His thinking, removed from the thesis of "foral" reintegration, was a revival and radicalisation of early Sabinianism, in reaction to the increasing proximity between the official line and the Spanish right-wing; but the political and emotional developments to which this line of thought responded belonged to the global anti-imperialist movement of that era.

At the start of the Second Republic, the reunited PNV put forward a synthesis predominating "statutism." Its social policy consisted in denouncing the excesses of capitalism, defence of industrial and agricultural private property and proposing, on behalf of the working-class, a programme of social reforms based on the social doctrine of the Church; although during these years the PNV demonstrated, with regard to the Church hierarchy, a heterodoxy that was as clear as it was involuntary.

The singularity of the Basque government at the time of the Spanish civil war, in the course of which the Nazis bombed Gernika, was related to the hegemonic position the PNV occupied within it. The government opposed the execution of the rebels and was an ardent defender of the Church. Its social programme was, without doubt, the most moderate in the Republic. During the early days of Francoism the PNV strengthened its position in exile within European Christian democracy, coming to occupy a predominant position within it.

The anti-Franco street violence in Spain linked the post-civil war years with the decades of the fifties and the early sixties. When, in the seventies, a consensus was formed across the Spanish state which opened the door to democratic change, its "atlantism" and Europeanism facilitated the incorporation of the PNV into this consensus. On the other hand, the features of new Basque nationalism (its independentism, anti-statutism and, particularly, its admiration for Third World liberation struggles) polarised by the EKIN-ETA phenomenon, made it incompatible with the state consensus which had been evolving during those years.

Mutual ill-will arose between the PNV and EKIN-ETA, which increased in 1961 after an attempt was made to de-rail a train, the PNV strongly condemning the attack. For its part, the hostility that ETA felt toward the PNV transferred itself to the Basque government in exile, which it denounced as an institution operating at the behest of the Spanish interests, proposing its replacement by a National Resistance Front.

Nevertheless, the emergence of the new nationalism impacted on the PNV which, from 1964 onwards, began to show signs of growing activity, consistent with the illegal celebration of the *Aberri-Eguna* (Basque National Day) and mass rallies. In January 1964, the PNV made public a new document in which it defended both the right to self-determination, thereby breaking with its traditional Sabinian approach, and the integration of native Basques and immigrants.

The almost permanent state of emergency which the Basque Country had experienced since 1968 until the death of Franco in 1975, consolidated the formation of a new anti-repressive nationalist community. Given that the community suffered repression not only because of the fact of being Basque, but also because it inhabited a territory under emergency legislation, there was an opportunity for immigrants to become integrated into the new community through participation in the resistance struggle; at the same time, the Spanish police became the negative image within this new community. A new relation of identity was established between the symbols of Basque difference, the flag, national anthem and the Basque language, and democracy; and the new nationalism, of an underground nature, experiencing its aspiration for freedom through numerous debates and transgressions and strengthening the role of utopia within its nucleus. At the same time, members of ETA's Military Front, began, mainly in exile, to transform themselves into a state-group, becoming an ideal-group of the new Basque anti-repressive community through a process of transformation of negative identities projected onto them by Francoism, in other words, making the fact of being working-class, coming from Basque-speaking areas and speaking Spanish poorly, into positive identities.

The ETA of the early seventies, which arose out of the Burgos trial, assumed the radical and utopian features of the newly consolidated nationalism, and became mesmerised by the mirage projected by the Burgos events, which would have a profound impact on an entire generation of young Basques, according to which the working-class movement had already merged with the national resistance struggle. Ideologically, it was the result of two tendencies; it modelled itself on armed anti-colonialist resistance movements which were renowned in those years (Cuba, Algeria, Vietnam and, later, Brazil, Uruguay and Palestine), and took as its own the hard-line Marxist-Leninist language which had spread throughout Western political groups as a result of events in France in May 1968.

ETA's Military Front predominated over ETA's Workers Front after the attack on Admiral Carrero Blanco in December 1973. However, the approaching death of Franco brought about a new split among veteran members of the organisation; on the one hand there were the "military," many of whom had been exiled for years in the Iparralde (French Basque Country) and, on the other the new wave of radicals, the "military-politicals." Against the opinion of the latter, the "military" decided to separate armed activity and political activity, and called for the formation of a civil front which would operate within the framework of "bourgeois" legality.

After the death of Franco, at the end of 1975, Spanish anti-Franco opposition opted for reform through a pact with the powers of the old regime, making it impossible to meet the utopian (in the noble sense) and national demands which had been generated in the heart of new Basque radicalism. The reform did not entail a break with the police-judicial or military apparatus of Franco. The Spanish army, even though it had difficulties coming to terms with the legalisation of the Communist Party, accepted it in spring 1977 as a *fait accompli*. From that point onwards, anti-separatist, more so than anti-communist, attitudes became increasingly prevalent within its ranks. Therefore, in a state which had historically been polarised into two extremes, centrism and moderation came to the fore while the right to self-determination disappeared from the opposition's political programme. The left-wing nationalist movement (*abertzale*), a result of the utopias conceived during the Franco era, found itself on the margins of the evolving political set up and, in this way, became the "internal enemy" of the flourishing democratic consensus.

The constitutional text developed after the first democratic elections following the death of Franco, in June 1977, and approved in a referendum at the end of 1978, emerged out of the aforementioned circumstances. The logic of the agreed reform was not only reflected in the essentialist definition of the Spanish nation, which embraced with difficulty a balanced acknowledgement of the right to autonomy of the nationalities and regions of Spain, but also in the reverential allusions to the Spanish army. The defence made by the Basque left-wing in the Spanish parliament of the right to self-determination was rejected by the all Spanish political groupings, who also rejected the PNV's call for the reintegration of the historic rights of the Basque Country and their recognition through negotiations in which Basque representatives and the central government would have equal roles. However, the State of Autonomies set up a devolved power situation to regional governments throughout the state, for which reason the PNV declared its intention to observe the constitution with a view to taking advantage of any opening offered by the future Statute of Autonomy.

Basque national radicalism chose and adapted a new group of identity markers. Deception born within the Basque Country out of frustrated national hopes in the constitutional debates created a space for a response within which alternative movements, proper to post-industrial society, located themselves. At the end of the seventies, this response merged with nationalist radicalism to form an anti-institutional opposition front. However, on the other hand, the increasing institutionalisation of the Basque Country from 1978 onward, and the hegemonic position within this process of moderate nationalism, made the growth of an anti-repressive nationalist community unviable. For this reason, ETA-m initiated an offensive of deadly attacks of an intensity previously unknown; the effect of the violence guaranteed the consolidation, not of the entire nationalist community, but of the socio-political community which provided legitimation for this violence (a community out of which *Herri Batasuna* would emerge in 1978). HB obtained between 15% and 20% of Basque votes in successive elections, as against 30 to 40% received by the PNV (from 1986 onward this latter percentage was shared with *Eusko Alkartasuna*, which emerged from a split within the PNV).

The chasm within Basque nationalism became public in October 1978, shortly before the referendum on the constitution. After the approval of the Statute of Autonomy at the end of 1979 by the majority of Basque voters, the PSE-PSOE (the Socialist Workers' Party of the Basque Country), the second largest party within the region, proposed the transformation of the old Basque nationalist-Spanish nationalist dichotomy into a new dichotomy of Basque democrats (those who supported the Statute) against Basque fascists (anti-statute radical Basque nationalists who supported armed struggle). The ongoing repression, the "illegalisation" of *Herri Batasuna,* and the progressive siege to which it was subjected in the Basque Country, transformed its social sectors into a community of fear.

Following the military coup of 23rd February 1981, ETA, which viewed the Spanish state as a continuation of the military dictatorship and the coup as a self-coup, did not see any reason to halt its campaign. ETA-pm, on the other hand, started to dissolve, a process which concluded in 1985. ETA's strategy of "a war of attrition" provoked a drawn out counter-insurgency war that was both legal and illegal on the part of the Spanish state. The insurgents' slogan, that was "to resist is to win", was met by the Spanish state by attempting to isolate the armed nucleus; the social sectors which gave legitimation to ETA' struggle were, likewise, subjected to marginalisation and stereotyping. However, during that period, and up until 1987, in spite of the fact that the goal of those responsible for the counter-insurgency war was "the recovery of the Basques for Spain" and that the majority of the population disagreed with ETA's violence, the state was unable to win over the hearts and minds of the Basque people.

Since its victory in the October 1982 elections, PSOE decided to carry out the counter-insurgency policy of the previous government. It put into practice the Plan for the Special Zone of the North (Plan ZEN) which made the principle of "suspicion" a governing criteria for all its political activities in the Basque Country. From the end of 1983 onwards a terrorist group began to carry out attacks in Iparralde later known as the Anti-Terrorist Liberation Group (GAL), which was to develop a reputation for kidnapping and murdering Basque refugees and activists. However, this unacknowledged dirty war had little to do with anti-terrorist instrumental rationality; from the very beginning Basque public opinion overwhelmingly attributed GAL's activities to the Spanish state, which consequently "de-legitimised" in the Basque Country, and continues to de-legitimise it today, given the absence of public acknowledgement of the state's role in GAL, the Spanish government's discourse which draws an insurmountable line between 'democrat' and 'violent' sectors. Such a strategy appears to have been conceived as a short-cut to win the good will of the apparatus of the state, the army and the police, not to give support for another coup; in fact GAL dissolved in the period 1986 to 1987, when those temptations disappeared as a result of the entrance of the Spanish state, and therefore the army, in NATO.

These events made ETA's claim that the Basque Country was in a war situation more credible, and generated the traits of a community at war within the social sectors which legitimated it. The Basque National Liberation Movement (MNLV) assumed a counter-state form (ETA), and a counter-community form (HB), which operated according to their own rules in the Spanish state and Basque society, which they mirror-imaged.

However, ETA's plans for bi-lateral negotiations with the Spanish state, based on the armed organisation's belief in its capacity to act as representative for the entire Basque people in dialogue with the Spanish government, were in fact unfounded, as they ignored the central role which moderate Basque nationalism had come to play in Basque political life. Neither did it acknowledge that from 1985 onward traditional Basque nationalism had aligned itself with the Spanish state in order to confront ETA; the chasm between both sides was wider, since the humanism, pacifism and pluralism of moderate Basque nationalism strongly contrasted to the Third World and revolutionary ideologies of the MLNV.

The convergence of the nationalism of the PNV and PSOE, with the aim of combating ETA and its support base in civil society, led to a joint Basque government programme in 1985, a PNV-PSOE coalition government in 1987 and, above all, to the Ajuria Enea Pact for the Normalisation and Pacification of the Basque Country at the beginning of 1988. Its objectives were to obtain sufficient political and social support within the Basque Country to increase police action against ETA and to marginalize those sectors of civil society which legitimated its activities, pressurising, particularly, the MLNV, so that ETA would engage in dialogue on the terms set out by the Spanish government. However, the different degree of support given to Spanish State of Autonomies political model by the signatories of the Ajuria Enea Pact, which was absolute in the case of the PSOE and other state-wide parties, but relative and with reservations in the case of nationalist parties, was reflected in the content of the Pact. For this reason it was not out of the question for representatives of the community to enter into discussion relevant to the juridical-political framework in the Basque Country. Therefore, point 10 of the Pact, although it categorically separates dialogue with ETA on a resolution to the violence through political negotiations, it does recognise that there may be a parliamentary solution to it.

In the two years prior to the negotiations in Algeria between representatives of ETA and the Spanish Ministry of the Interior, the MLNV updated its tactics elaborated in 1975-1978, developing the theory of "accumulation of strength" as a means by which to approach the political negotiations. The MLNV had to gather strength on three fronts by: deepening the contradictions within the Ajuria Enea Pact, generating mass struggle, and intensifying the armed struggle. The negotiations could not question the KAS alternative, which was unquestionable, but they could discuss the speed and the method of institutionalising its content. The interlocutors in the negotiation would be ETA and PSOE (to whom the role of representing the remaining political parties was assigned). Once political agreement had been reached, HB was to prepare the staff for the future Basque National Administration and assume the leadership of the institutional activities. With regard to ETA, its role was to become guarantor for the fulfilment of the agreement, as well as training the officers of a Basque popular army. The failure of the 1989 negotiations in Algeria was due to several reasons: the exclusive perception which ETA held of itself as the only agent with whom the Spanish government should negotiate; the impossible demands of the Spanish government (which led however to the end of ETA-pm in 1982), for the handover of weapons, without any political concessions; and perhaps the bad will among some responsible for the negotiations to find a solution to a violence, the ETA's violence, whose end would have meant also the end of a source of immense personal wealth to them. So, the Ajuria Enea pact continued to be an instrument of siege of radical Basque nationalism.

The deployment of the *Ertzaintza* (Basque Autonomous Police) in the period 1987 to 1991 in parallel with, but not symmetrical to, the withdrawal of state Police forces, strengthened the autonomous power of the PNV. However, this was bringing the Ertzaintza into confrontation not only with that group considered to be its natural enemy, ETA, but also against civil organisations, especially youth, who were taking part in demonstrations in support of ETA's political objectives. ETA's attacks were being carried out on two fronts, against the state and against autonomous institutions, becoming more and more indiscriminate and choosing as targets *ertzainas* and political leaders; the movement's leadership concentrated on a single point while it closed ranks and silenced internal dissidents. On the international level, Western states, especially France, openly supported the all out offensive of the Spanish state against ETA.

From the mid 1990s, the PNV and *Eusko Alkartasuna* (EA) adopted a posture in favour of dialogue with the MLNV. Therefore, dialogue took place between the PNV and HB during the months of June and July of 1992 with the dual aim of breaking the lack of communication between the two parties and developing a common solution that would lead to the end of violence. But once again, the conversations broke down. ETA continued to give priority to military logic, and the PNV did not state clearly enough its opinion on what could be politically negotiated with HB; that is, a peaceful road to sovereignty.

Nevertheless, from the end of the eighties onwards social processes were developing that would facilitate agreement. During this period a tendency, which had always been present within the left-wing *abertzale*, began to evolve; it was open to alternative movements such as ecology, feminism, anti-militarism, etc., and, therefore, also pacifism, and its leaders thought that the term "negotiation" should have a consensual and participatory meaning. The social movement *Elkarri* put forward a new interpretation of the Pact which stressed the negotiation aspects of the Basque conflict included in the text. This interpretation undermined the sacred anti-MLNV union of the Basque and state-wide parties signatories of the Pact. The impossibility of making the Basque Country into an autonomous framework for labour relations within the existing political set-up also favoured the convergence of the two Basque trades unions, ELA and LAB. This was a link which facilitated closing the gaps among the different sectors of the Basque community, and provided a broad social basis of support for the new perspective on a solution to the Basque conflict.

The end of the PSOE government (which was brought about, in part, by revelations in the media over its links to GAL and the consequent trial of high-ranking members of the Ministry of the Interior), and its replacement by the Popular Party (PP) forced the PNV to turn to radical nationalism in its search for an alternative alliance, in spite of the obstacle which hostile polarisation with respect to the *Ertzaintza* represented. But this new strategy required, in order to be even minimally viable, a radical change in the policy of the left-wing *abertzale* which demanded a move away from reliance on street violence, which was becoming increasingly "inter Basque" in nature, and the announcement of a ceasefire (Letamendia, 1994).

Northern Ireland

Ireland was Britain's oldest colony. In the Modern Age "plantations" were the property of Protestants, Catholics were subject to a range of legal exclusion measures; however, what the English had established was a colonial economy for the exclusive benefit of the metropolis, to the equal detriment of Catholics and Protestants. For this reason, the 1798 rebellion, which took place in the wake of the American and French Revolutions, was led by Protestants.

But the struggles in the 19th century in favour of Catholic emancipation and, above all, for ownership of the land, which was in Protestant hands, made Irish nationalism into a Catholic movement. In mid century, the first armed group, the Fenians, emerged simultaneously in Dublin and New York, where thousands of Irish had emigrated, decimated by the Great Famine. Out of this armed tendency, supported by a strong Catholic peasant movement, Sinn Féin, initially a conservative organisation, emerged at the end of the century. The support of British Liberals for "Home Rule," or Irish self-government, led to a reaction on the part of Unionist nationalism of the Protestants of Ulster, the industrial north-east of the island.

During Easter 1916 there was a rising led by Sinn Féin, allied to James Connolly's Marxists; the British sent sixteen of the leaders to the firing squad. This brought about the electoral victory of Sinn Féin, the proclamation of the Republic of Ireland, the formation of the Irish Republican Army (IRA), the military wing of Sinn Féin, and to the start of the War of Independence. This culminated in partition, that is the division of Ireland on the basis of an artificial border, drawn up to separate the Free State of the south from six of the nine counties of Ulster, in the north-east of the island (which became Northern Ireland), dominated by a clear Protestant demographic majority. While the south became a Republic, in the north, Catholics were discriminated against in their civil rights and access to employment. The IRA continued to maintain a strong presence in the north although its campaign lacked widespread Catholic support.

At the end of the sixties, the Northern Ireland Civil Rights Association (NICRA) and the repression to which it was subject by loyalist paramilitary groups, the Unionist police and the British army, deployed in 1969, radicalised the IRA and led to the suspension of Stormont, the regional parliament. The nationalist community diversified; the moderate Social Democratic and Labour Party (SDLP), led by John Hume, became the majority Catholic party. Unionism also became fragmented, with anti-Catholic terrorism emerging that was destined to intensify its activities from

1985 onwards, the year in which London recognised Dublin's right to intervene, to some degree, in those Northern Irish affairs that were of common interest to the whole island.

These developments, added to pressure exercised by the European Union, and above all, by the powerful Irish-American lobby in the United States, gave rise to a joint SDLP-Sinn Féin initiative in 1993. This was followed by a new London-Dublin Agreement, which recognised the double right of self-determination of both Northern Ireland and the island of Ireland in its entirety. Hesitation and the setting of new British pre-conditions brought about the collapse of the IRA ceasefire in February 1996. Dark clouds over the peace talks were finally dissipated by the electoral victory of the British Labour Party in May 1997.

The origins of the current national conflict between Catholics and Protestants in Northern Ireland go back for centuries. The British occupation of the island of Ireland acquired its religious aspect in the times of Henry VIII, who proclaimed himself King of Ireland in 1539; four years later he oversaw the imposition of Protestant Anglicanism in Ireland. At the beginning of the 17th century, Mary Tudor reinforced the policy of "plantations", the settlement of Protestant colonialists on lands belonging to Catholic peasantry. After half a century of risings, the Puritan leader Oliver Cromwell landed and his arrival was followed shortly afterwards by the Act of Colonisation, on the basis of which two thirds of land owned by the native Irish was confiscated. At the end of the century the Catholic Jacobites, under the leadership of Talbot, rebelled throughout Ireland against the Protestant king, William of Orange. James II was defeated by the Orangists on the banks of the River Boyne, in Ulster, on 12th July, 1690. The renowned commemoration of this battle, which has come to symbolise an imagined superiority of Protestants over Catholics, being the latter perceived of as sinister and reactionary, continues to be an important date in the calendar for many sectors of unionism.

For over a century, until 1829, the Catholic majority of the population was excluded from parliament and subjected to special penal laws, while the Catholic Church was outlawed, although tolerated to some degree. The entire island of Ireland was subjected to British commercial policy of a colonial nature, which led to the beginnings of a Protestant Irish nationalism (in which, for example, Jonathan Swift, the writer of *Gulliver's Travels*, participated).

At the end of the 18th century, the American and French revolutions made their influence felt in Ireland, symbolising the growth of European nations in the Tree of Liberty. Secret societies were formed, out of which emerged in 1795 the conspiratorial association of the "United Irishmen" to lead a rebellion three years later. The main nationalist leaders, such as Wolfe Tone, were Protestants, but the concept of freedom meant different things to Protestants and Catholics. In the north of the island, where the process of plantation was most intense, and where Presbyterians of Scottish origin predominated, the ethos of the Rights of Man drew its inspiration from David Hume and Adam Smith, leaders of the Scottish Enlightenment; in the south it meant, instead, the handover of Protestant land to Catholics and the end of the repression suffered by Catholics. Nevertheless, British troops repressed both Catholics and Protestants, and killed 50,000 people out of a population of four million. The Irish bards who perceived of Napoleon as a Liberator (because of which they waited in vain on his support) were to sing the Tree of Liberty in homage to 1798[III].

After the rising, the British enacted the Act of Union which suspended the Irish parliament and made the island into an agricultural annex of Britain. An Irish-speaking counter-revolutionary Catholic leader, Daniel O'Connell, opposed this development and managed to get the Catholic Emancipation Act passed in London. Protestants such as Davies and Catholics such as Mitchel collaborated in the insurrectionist movement, the "Young Irelanders", which emerged during the same period. The strong presence of Protestants in Irish nationalism during the 19th century is explained by the Protestant monopoly of higher education; while the majority of the leaders were Protestant, the vast bulk of their followers were made up, almost exclusively, of dispossessed Catholics. However the religious factor made mixed marriages difficult and had, therefore, led over the centuries to the existence of two separate cultural groups which came to form a caste system. In this way, although Irish nationalism was anti-sectarian in its roots, the momentum which led Catholic peasants to confront Protestant landowners soon led to the movement becoming sectarian in nature.

During the period 1845-1849 the Great Famine caused by successive poor harvests exacerbated the miserable conditions in which the peasants lived, and led to the death of one and a half million people and the emigration of a further million to the United States, where they became a sub-proletarian class. These events explain the claim to Irish descent currently made by more than 30 million Americans. (If we add to this the fact that four of those who signed the Philadelphia Declaration, the precursor of American independence, were of Irish origin, it can easily be understood why there is such a high degree of support within the United States for Irish nationalism). During the 1850s, the Irish Republican Brotherhood emerged simultaneously in Dublin and New York; its shock troops called themselves the "Fenians." This organisation, which, apart from the clergy, had widespread support in Ireland, unleashed a new rebellion in 1867 that was destined for failure, and followed by executions, those of the three "Birmingham martyrs".[IV]

Nevertheless, owing the Great Famine, the number of labourers went into decline and the number of small landowners increased, while the Catholic Church became a national institution. This meant that in the last third of the 19th century a class of free, Catholic and democratic peasants had been formed acting as the basis for Irish nationalism. Paradoxically, it was a Protestant landowner, Charles Stewart Parnell, who became the leader of these Catholic free-holders; it was significant that he fell into disgrace in 1890, having violated Catholic morals through his adulterous relationship with the wife of one of his Lieutenants.

In the last decade of the 19th century, new political formations began to emerge in Dublin and Belfast. Out of the working-class movement arose the

Irish Republican Socialist Party (IRSP), which was Marxist, and among whose members was James Connolly. In 1899 Fenian nationalism inspired the formation of Sinn Féin ("Ourselves Alone", in Irish), which was, in those times, conservative in its outlook. The British Liberals, led by Gladstone, supported the restoration of Irish self-government in the form of Home Rule; however, their plans were opposed by English conservatives and, above all, by the Northern Irish unionists, who had grown strong throughout the nine counties of Ulster. This area was heavily industrialised, the basis of which was iron and steel and ship building in the Belfast docks. In 1910, in view of the foreseeable enactment of Home Rule, the Orange Order and the Unionist Council of Ulster armed themselves and their leaders, under orders from Carson, collected 450,000 signatures in opposition to the proposed changes. This led to the postponement of Home Rule, approved by Westminster in 1914.

At the start of the First World War, and given the political impasse, the nationalist Fenians and the Citizens' Army, under the leadership of the Marxist, James Connolly, unleashed an insurrection in Dublin on 24th April, 1916, that became known as the Easter Rising. However, the country did not give them its support, and after resisting for five days, 2,500 rebels were taken prisoner by the British and sixteen of the leaders were executed. Among those shot were some of the signatories of the Proclamation of the Republic (Connolly, Pearse, Clarke and McDonagh).

The emotional impact of these events on the population led to the electoral triumph of Sinn Féin in 1918; Eamonn DeValera was appointed the first president of an Irish Republic created in response to the appeal made by the 1916 martyrs in their Proclamation. Fenian leaders such as Collins and Brugha set up the Irish Republican Army (IRA) in 1918, which acted at the behest of the Republic. Irish nationalists and British troops engaged in a war, the end of which, within two years, came to be desired equally by the conservative majority of Irish as well as the British government. Finally, at the end of 1921, the British government passed the Government of Ireland Act, which divided Ireland into two parts. The border was drawn up in the war between Irish nationalists, British troops and Protestant unionists, and was biased in favour the interests of this latter group. Dublin governed the twenty-six counties of the south, including three from Ulster, and a population of three million, of which 88% were Catholic and nationalist. Protestant unionists, allied to the British Conservatives, managed to create frontiers within which they held power for almost forty years because they represented a stable and undeniable 60% of the population, as against the 40% Catholic minority.[v]

The Catholic Church and conservative nationalists used their influence to push for acceptance of the Treaty. De Valera, the majority of Sinn Féin and almost all of the IRA rejected it. In June 1922, the majority in favour of the Treaty was triumphant, a result which led to a civil war that lasted a year, until May 1923, and during the course of which leaders such as Collins and Lynch, who fought on different sides, died. De Valera ordered arms to be laid down and later assumed the leadership in a split which led to the formation of

Fianna Fáil, the majority party from 1932 onwards; this party was to end up accepting the *status quo*. The 1937 Constitution put the Catholic Church in a privileged position and removed references to the United Kingdom; it also substituted the title of Free State for Eire. The three million Catholics in the south created, after a bitter civil war that had impacted more upon the elites than on the general population, a small parochial democracy in which the Church occupied an important cultural role. The "white-boy" tradition disappeared in the south (but not in the north, where it would survive in a specific and more intense form). The power of the Church only began to go into decline in the Republic of Ireland during the sixties, as part of the overall process of urbanisation and secularisation; Protestants became culturally assimilated into the general population.

In Northern Ireland, on the contrary, the Unionist majority blocked the Catholic minority's access to power for several decades. In Belfast, a representative of the British Crown shared power with the Stormont Parliament, which consisted of two chambers, the Congress and the Senate. Legally Catholics were not discriminated against, but in reality sectarian prejudice was widespread. Only 12% of members of the Royal Ulster Constabulary were Catholic; Protestants created their own sectarian police, the B Specials. Many Catholics, who represented the proletariat and, above all, the urban sub-proletariat of Ulster, came to see the IRA as the avenger for the injustices which they suffered and as the defender of their civil rights. Education was segregated; Catholics were discriminated against in housing and social welfare as well as electorally, through gerrymandering. Private companies, almost all of which were owned by Protestants, passed over Catholic workers in favour of Protestants, which meant that the rate of unemployment among Catholics was more than double that suffered by Protestants; neighbourhoods such as Ballymurphy in West Belfast became black holes of unemployment and misery.

The IRA launched an offensive in Northern Ireland, during the period 1942-1943, right in the middle of the World War II, and carried out its border campaigns through attacks against border posts during the years 1956-62. Nevertheless, these actions did not generate the popular support which the IRA would have hoped for. From the 1960s onward, a process of social stratification within both communities began to take place. A middle-class conservative unionist class made up of liberal professionals favoured, without compromising their hegemonic situation, better relations with the Republic of Ireland; they expressed their disquiet in relation to the sectarian activities in which the Protestant working-class engaged (the police was almost entirely Protestant). Within the Catholic community, the newly emerging bourgeoisie, which lived side-by-side with its Protestant counterparts in the more affluent areas of Belfast and Derry, began to distance itself, in the same way, from the Catholic sub-proletariat, who lived in segregated areas where there was strongest support for the IRA campaign. It is in this context that the Irish Catholic movement of the period 1969-1972 should be understood, led by the struggle in favour of civil rights for the Catholic community and inspired by the black leader Martin Luther King in the United States.

In 1967 NICRA, a non-political and non-religious movement, which demanded universal suffrage, the abolition of special legislation and the end of gerrymandering, emerged. Its campaign provoked the hostility of the Orange Order and unionism. On 5th October, 1968, on the occasion of a march organised by NICRA and the Housing Defence Committee in Derry, the unionist Home Secretary, William Craig, ordered the brutal repression of the march claiming that it coincided with an Orange Order Parade. At this point People's Democracy appeared on the scene, arising out of the student body in Queen's University; one of its members was Bernadette Devlin, the future M.P., who organised a march from Belfast to Derry. On 4th January, a number of people were murdered in Derry by radical unionists who were actively supported by the police; that same afternoon the police invaded the Catholic Bogside area of Derry. In April 1969, when London decided to send five thousand troops to Northern Ireland, they were initially welcomed by Catholics, relieved by their presence; on 14 August the troops acted as a barrier between rioting Catholics and Protestants in Derry. However, the situation changed radically within a short period of time; when British soldiers entered Belfast in search of IRA activists, six people died and another three hundred were injured in the ensuing riots. Catholics barricaded themselves into their areas and "Free Derry" and "Free Belfast" were created.

The radicalisation generated by the developing political situation found echoes within the ranks of the IRA, which had moved on towards orthodox Marxism. From 1969 onward, the Official IRA made two proposals that were rejected by its radical sectors: participation in the Belfast and Dublin parliaments and the foundation of a National Liberation Front in favour of a Workers' Republic which would come to an agreement with the Communist Party in the south of Ireland. After both proposals were approved in 1970 by the IRA Assembly, a large number of delegates walked out (out of this number, 180 were from Northern Ireland), and formed the "Provisional Army Council," out of which arose the name "Provisional IRA."

The change of government in Britain, from Labour to Conservative, in July 1970 led to the military invasion of Catholic areas with the goal of defeating the IRA. This organisation killed the first soldier in January 1971; from August of the same year, the introduction of internment, which equipped the police and army with special powers, facilitated the imprisonment of Catholic suspects in concentration camps. A month later, in September, the number of troops deployed in Ulster had risen to 12,000; up until the previous month thirty had been killed at the hands of the IRA. On 30th January 1972, more than six thousand people gathered in Derry for a march against internment; British paratroopers stormed the Bogside and killed thirteen Catholics, The massacre was widely publicised and became known as "Bloody Sunday." The repercussions were far-reaching; fifty thousand Catholics took part in a protest march in Newry. On 22nd February the Official IRA carried out its first operation in England; it exploded a car bomb outside the paratrooper barracks in Aldershot, leading to the death of six people; four days later the unionist Home Affairs minister was killed in an attack.

On 24 March, 1972, London dissolved the Stormont parliament controlled by the unionists and declared Direct Rule from London, appointing William Whitelaw Secretary of State for Northern Ireland. Unionists opposed direct rule claiming that it was a concession to the Republicans.

In June 1972 the IRA proposed a truce if the British army reciprocated the gesture. The Conservative government in London, which had announced at the end of the previous year its willingness to accept the reunification of Ireland if the majority in the North was in favour of it, responded positively. The truce was declared on 27th June, but Whitelaw had previously promised that any change would be subject to a plebiscite in Ulster, which meant that it could in fact be vetoed by the majority Protestant population. The IRA called for the exercise of the right to self-determination for the whole of the island of Ireland, and demanded the complete withdrawal of British troops within a maximum period of two and a half years. But on Sunday 9th July, Catholic families who were taking their belongings from their abandoned homes in the Protestant area of Lenadoon found themselves under threat by unionists and under fire by the British army. The Provisional IRA called off its truce, and on 21st July twenty-three bombs exploded in the centre of Belfast, killing 16 people.

The British army invaded Catholic areas on 30 and 31 July, removing barricades and arresting a high number of suspected IRA members. The armed organisation concluded that the visibility of its members had facilitated the wave of arrests; Gerry Adams and Martin McGuinness were among those detained. (This experience explains the hesitance of the IRA to hold further truces, which were held seldom, in 1975 and again, almost twenty years later).

From this point onwards the urban landscape in Northern Ireland acquired the appearance which would characterise it. The old walls of Derry and the tallest buildings in Belfast, the highest floors of which were reserved for the police and army, with the aim of using residents as a shield against possible attacks, became fortresses which served as surveillance and listening posts. Streets were divided with barbed-wire, along which machine-gun post were installed and British soldiers aimed their weapons from tanks and armoured vehicles at passers-by. Urban areas and much of the countryside became, after the forced movement of about fifteen thousand families belonging to whatever group which was there the minority, mono-religious; so, by the eighties and nineties, 90% of the population were living in mono-ethnic communities. The River Boyne divides the two communities in Derry; in working-class areas of Belfast there are high walls (ironically known as Peace Walls) which separate working-class Catholic and Protestant neighbourhoods. In Catholic areas, British army and RUC patrols carry out raids on a regular basis on local houses and arrest their occupants. Even in prison, Catholics and Protestants are segregated.

At the beginning of the seventies, owing to the intensification of the conflict as well as the process of social stratification taking place within both communities, the system of political parties in Northern Ireland took on its present shape. The hegemonic Official Unionist Party (OUP) was led by West until 1979, from then until 1995, by James Molyneaux, and thereafter by

David Trimble. More fundamentalist-type unionism was led by the Presbyterian preacher, Ian Paisley, founder of the Free Presbyterian Church in 1951 and of the Democratic Unionist Party (DUP) in 1971, which became the representative of the most sectarian and intransigent sector within the Protestant community. Paisley attributed all peace initiatives, as well as the very European Union itself, to a Catholic Church plot, overseen by the Pope. The heroes which his British nationalism gloried in were Oliver Cromwell and William of Orange, as well as historic events such as the Siege of Derry and the Battle of the Boyne in the 17th century, commemorated as symbols of Protestant bravery and examples of Protestant superiority with respect to the Catholic counter-reformation.

Several paramilitary organisations emerged or reorganised themselves within the ranks of radical loyalism. The Ulster Defence Association (UDA), founded at the start of 1972 brought together a number of individual vigilantes and sectarian murder gangs, which used the name Ulster Freedom Fighters (UFF) when it carried out attacks; there was also the Ulster Volunteer Force (UVF) and the Red Hand Commandos (RHC). Contrary to IRA activities, which were technically sophisticated and mainly directed against military and police targets, loyalist groups tended to use home-made weapons indiscriminately against members of the Catholic community.

The Official IRA disappeared, and the "Provisionals" assumed the lead of the republican movement definitively. Out of a split of the former Officials the Irish National Liberation Army (INLA) emerged, which believed that the response of the IRA to the British occupation was inadequate. Co-existence within Catholic ghettos created a close relationship between the IRA and Sinn Féin, even more so than that existing in similar conflicts within Western Europe. In the seventies, the IRA abandoned, for security reasons, the old-style battalion formation, and replaced it with clandestine units of a cellular nature which had minimal communication with each other.

On the other hand, out of the new situation a political party was to emerge within the Catholic Community that would become electorally more successful than Sinn Féin: the centre-left Social Democratic and Labour Party (SDLP). The SDLP was reorganised in August 1970 by Gerry Fitt on the basis of the old more moderate Nationalist Party, its predecessor. The SDLP has been led by John Hume since 1979 and is respected by socialist parties throughout Europe; it has had positive relations with successive Dublin governments, it has opposed Direct Rule and supported the devolution of power to Northern Ireland. Hume has made his pro-European stance the corner-stone of his discourse, for which reason, in spite of its Labour leanings, there are numerous similarities between the SDLP and the Basque Nationalist Party (PNV).

In 1973 the London government held a referendum in Northern Ireland to allow the population to decide on whether to retain links with the United Kingdom. The Protestant community voted massively in favour, obtaining a 58% result, while Sinn Féin, the SDLP and the IRA proposed abstention, which left 41% voting against links. Whitelaw felt it was the right time to declare an end to Direct Rule and set up a non-sectarian executive, in which

Catholics and Protestants would participate almost on an equal footing. In June 1973, the moderate unionist Faulkner was nominated president of the executive. The vice-president was Gerry Fitt of the SDLP; the rest of the government was formed by six unionist members, four representatives of the SDLP and one from the non-sectarian Alliance, the leader of which was John Alderdice. The Sunningdale Agreement of December 1973, drawn up by London and Dublin, created for the first time, although on a rhetorical level, a Council of Ireland which would deal with those areas of common interest for both parts of the island. But the Unionist Council, under pressure from Paisley's party, rejected Faulkner and, in the elections of February 1974, the eleven unionist Westminster MPs were solid in their opposition to Sunningdale Agreement. A general strike organised by the Ulster Workers' Council in May put an end to the non-sectarian executive; in July, London returned to Direct Rule, which would last until 1996.

In March 1976 special category status was withdrawn from political prisoners, being subjected to the same conditions as ordinary criminals. Republican prisoners in Long Kesh refused to wear the prison uniform and, instead, covered themselves in blankets or sheets; they also refused to leave their cells, where they lived surrounded by dirt. In October 1980, prisoners in H Block 2 began a hunger strike demanding political status. A second strike, begun in March 1981, was led by Bobby Sands, the Officer Commanding and MP in Westminster, who continued with his fast until he died on 5th May. Between that date and 3rd October, when the strike was called off, nine more republican prisoners died. The IRA campaign received a massive boost from developments in Long Kesh from sympathisers in the Republic of Ireland and in the United States. Sinn Féin, under the leadership of Gerry Adams and Martin McGuinness, reaped the electoral benefits. In the 1984 elections to the European Parliament, Sinn Féin won 13% of the votes in Northern Ireland (representing 35% of the Catholic community), while the SDLP obtained 22% (60% of Catholic votes).

Given the electoral threat which Sinn Féin represented, John Hume looked to the South for support and persuaded the Dublin parties of the need to establish a New Ireland Forum. The report, published by the Forum in May 1984, recognised the legitimacy of the unionist identity and emphasised the need to draw up a new agreed constitution which would achieve a united Ireland (Articles 2 and 3 of the 1937 Constitution declared that the whole of the island of Ireland was the territory of the Republic). Although this proposal was scornfully received by unionists, it did open the door to direct contacts with London without going through the unionists. The Anglo-Irish Agreement (also known as the Hillsborough Accord) resulted in November, 1985, from these contacts; it arose out of a concerted effort by Garrett Fitzgerald, the Irish primer minister at the time, and his British counterpart, Margaret Thatcher. The Agreement set up a new Inter-governmental Conference which would deal "on Northern Ireland and on relationships between the two Irelands with the goal of reaching regular agreements in matters of policing, security, legal affairs, including the administration of justice and cross-border co-operation". British recognition of Irish government authority in Northern

Ireland stirred up mass unionist opposition; Paisley and Molyneaux denounced it as treason and led the campaign "Ulster Says No"; a similar situation was created to that which arose during Sunningdale. "

Nevertheless, the Anglo-Irish Agreement initiated profound changes, the effects of which would only be felt during the nineties. The Agreement, in conjunction with European regional policy actions, promoted economic co-operation between the two parts of Ireland, which had been weak since 1922, thereby favouring a certain degree of political convergence.

A demographic fact threatens to challenge drastically the relations of power within Northern Ireland: the higher birth rate within the Catholic community gives it the right to 45% of the vote by the end of the 1980s. If the increase continues, then by the year 2010, Catholics will be in the majority in Ulster, putting an end to the majority of Protestants. From the mid-1970s, Northern Ireland has been experiencing industrial decline in the key areas of iron and steel and ship-building. The deficit has made the expense of troop deployment and high police levels in the region even costlier for the British government. The current subvention is estimated at one thousand million pounds per year.

Therefore, the SDLP, dispossessed of all political power (making it immune to the risk that its ideological conflicts with more radical sectors of nationalism would become institutional conflicts) came to occupy the role of interlocutor with the Dublin government and the crucial partner of moderate unionists within the OUP. Its programme, backed by the E.U. and by the powerful Irish-American lobby in the United States, among whom was Senator Edward Kennedy and other such influential figures, began to be accepted as a factor of convergence by radical nationalism.

This exasperated and terrified radical loyalism, particularly Paisley's party, and the number and ferocity of attacks against Catholics greatly accordingly increased; by the nineties the victims of loyalist paramilitaries exceeded the number of IRA victims. Moreover, in 1993, the RUC represented 18% of public sector employees, a figure which had risen from 9% in 1974; this meant that the rate of unemployment could rose as a result of the "pacification" process (in fact, 28% of Catholics were unemployed, while the figure among Protestants was 13%). For all these reasons, loyalist paramilitaries began to re-arm after the Anglo-Irish Accord was signed in 1985, and their operations acquired a level of sophistication which they did not previously have. Until 1988, their arms were smuggled in from South Africa, but by the nineties the main source of their supply was Eastern Europe. In November 1993, British police decommissioned a Polish ship caught transporting arms for unionists from the port of Teesport in Britain (Frères du Monde, 1972; Coogan, 1987; O'Dowd, 1990; Lee, 1992; Garvin, 1992).

After the collapse of the 1992 talks between the SDLP, the Alliance and the Unionist parties - an inevitable collapse due to the absence of Sinn Féin - Gerry Adams and John Hume met in April 1993. On 27th September of that year it was announced that both men had arrived at an agreement with respect to the establishment of a joint London-Dublin government in Northern

Ireland, which Sinn Féin said should be a step toward the reunification of Ireland; the terms of this Hume-Adams initiative were never made public. While Paisley's party warned that there would be an increase in loyalist violence, London claimed that it would not oppose any majority decision of the population in the north of Ireland. In the following days the various pieces of the jigsaw fell into place: the IRA expressed its approval of the initiative; John Hume held meetings with Albert Reynolds, the Irish prime minister (who represented the more progressive sectors of Fianna Fáil); and finally Reynolds met with the British prime minister, John Major.

In spite of attacks carried out by paramilitary organisations from both communities during the following month, which appeared to want to undermine the initiative (on 23 October IRA bomb killed nine people in the UDA headquarters on the Shankill Road; on 31 October a loyalist paramilitary group killed seven people in the Catholic town of Greysteel), Reynolds and Major reached a formal agreement in Downing Street on 15th December. The agreement included a joint preamble by both sides and two declarations, British and Irish. In the preamble, the two governments committed themselves to the promotion of dialogue within Northern Ireland and to the establishment of institutions and structures which would facilitate unionists and nationalists working together. For its part, the British government committed itself to respect the wish of the majority of the population in favour of or against the union with the United Kingdom or a united Ireland; it also declared that it had no selfish economic or strategic interest in Northern Ireland, for this reason it was up to the population, through agreement between both sides, to exercise the right of self-determination on the basis of consent if this was its wish. For its part, the Irish government announced that the right to self-determination should be achieved in agreement with the majority of the population of Northern Ireland, respecting the civil and religious liberties of both communities; it also said that if agreement were reached it would support the corresponding changes in the Irish Constitution. Both governments were in agreement that "those parties committed to peaceful methods and the democratic process could join in the dialogue". (It was stipulated that Sinn Féin could be included three months after the IRA had declared a ceasefire.)

The Joint Declaration did not stipulate the nature of government in Northern Ireland nor did it propose structures for co-operation between the two parts of the Ireland; it was restricted to stating certain principles. However, the impact of the declaration on the radical elements within the two communities was quite different. The London government's support for the right to self-determination within Ireland and, particularly the fact that this stand was backed by Dublin, enraged loyalist paramilitaries and Paisley's DUP; but at the same time, the SDLP expressed its satisfaction with the declaration. The fact that Dublin subordinated the right to self-determination to the consent principle (unionists still retain the majority within the area) allowed the UUP, the main party within the north, not to veto the declaration.

However, it was this aspect of the statement which generated suspicion among the ranks of Sinn Féin and the IRA who, being unable to reject the terms of the agreement, particularly because it recognised the right to self-determination, considered that it did not alter to any great degree the aspirations of the republican movement. Gerry Adams declared that the positive aspect of the declaration was its acknowledgement of the right to self-determination, while the negative was its recognition of the unionist right of veto; he proposed that the two governments co-operate to obtain the maximum level of acceptance of the right to self-determination.

During the first half of 1994 the peace process progressed toward consolidation. Gerry Adams' visit to the United States not only gave the Sinn Féin leader an excellent opportunity for a publicity coup, it also facilitated the commitment on the part of the US administration to a large financial investment (of between 80 and 200 million pounds) if peace were to be achieved. On 31 August, 1994, the IRA finally declared an indefinite and complete ceasefire as it believed that the circumstances existed for the achievement of a "just and lasting peace." At the same time Gerry Adams emphasised that the struggle was entering a new phase, and he demanded the demilitarisation of the north of Ireland, the withdrawal of the British army and freedom for political prisoners.

The IRA statement was enthusiastically supported by Albert Reynolds and Bill Clinton; John Major, however, limited his response to one of cautious optimism. Unionists, on the other hand, were suspicious of the initiative, particularly so in the case of Ian Paisley, although the response of James Molyneaux was more muted. Given that it seemed impossible, at least in the short or medium term, that the population of Northern Ireland would vote for straight forward reunification, the Joint Declaration offered within this time frame British-Irish condominium rule over the region. At the same time, it established the economic unification of Ireland (supported by EU structures) and the basis of a new political culture of reconciliation between the Catholic and Protestant communities, thereby facilitating the medium/long-term institutional reunification of the island. After a period of hesitation, loyalist paramilitary organisations (which had always claimed that their violence was in response to the IRA's activities) also announced a ceasefire.

Nevertheless, over the following year and a half the political weakness of Major's Conservative government, which was reliant on Unionist votes in the House of Commons in order to maintain its majority position, finally brought about the collapse of this opportunity to resolve the conflict. One of the tacit agreements was that the British government would include Sinn Féin in the multi-lateral political negotiations three months after the start of the IRA cessation was declared. Major ignored this agreement and added a new pre-condition which was unacceptable to the IRA, the decommissioning of weapons prior to Sinn Féin's participation in the negotiations. The leaders of Sinn Féin, Gerry Adams and Martin McGuinness, dismissed the British proposal as ridiculous and unrealistic, and the conversations reached an impasse.

In July 1995 republican indignation increased further when John Major decided to free a British paratrooper jailed for the murder of a young Catholic woman in Belfast. At the end of November the establishment of an

International Commission on disarmament was announced, and political conversations were due to start in February 1996. The US president, Bill Clinton, visited Northern Ireland during this period and gave backing to the agreement, thereby re-launching the peace process. However, the recommendation of the International Commission, presided over by George Mitchell, a U.S. senator, in favour of negotiations progressing in a parallel process of disarmament by the paramilitary groupings, and not previously, did not find favour with the British, who rejected it. On the same day, in January, as the Commission made its recommendations public, London announced that there would be elections to a Northern Ireland Forum on 30th May with the aim of forming a Northern Ireland Assembly. This was the last straw for the republican movement, which was very mindful of the discrimination suffered by Catholics for over half a century at the hands of a Protestant majority in institutions which the British government was proposing to restore.

On 9th February an IRA statement announced the end of its cessation; a few hours later a bomb exploded in the financial area of London, injuring a hundred people. A number of other bomb attacks were carried out in the British capital in April and May, and a car bomb exploded in Manchester in June, causing two hundred injuries.

Sinn Féin, which decided to take part in the May elections, obtained a strong vote from the nationalist community; all in all the party won a record 116,000 votes (more than 40% of the nationalist vote). The Forum had its first sitting on 10th June, under British sponsorship, without the participation of Sinn Féin or George Mitchell, who had been rejected by unionists as chairperson; it was therefore condemned to failure.

The unionist prepotence became even more exasperating within republican circles. The Orange Order announced its plans to march through Catholic areas in July; the RUC, which initially prevented the Orangemen from passing through Garvaghy Road, in Portadown, finally gave way and authorised the go-ahead of the march, repressing protesters in violent confrontations which led to one death and a number of injuries.

After the death of an unarmed IRA volunteer in London at the end of September at the hands of the British police, a bomb was placed in Ulster for the first time since the IRA had announced the end of its cessation; it exploded in the British Army Headquarters in Lisburn causing thirty injuries.

Tensions grew within the rank and file of the loyalist paramilitary organisations, and some sections of the leadership were in favour of returning to violence. In September, John Hume and David Trimble, respective leaders of the SDLP and UUP, reached an agreement within the Forum on the parallel decommissioning of weapons and the start of inclusive negotiations; the agreement was similar to the proposal put forward by the International Commission on Decommissioning and represented a mid-way position between the British pre-condition for the hand-over of weapons and the IRA's refusal to disarm. However, all the signs were that any solution would be dependent on a Labour win in the upcoming general elections, which was in fact the case in spring of 1997.

Notes

I   The accord did not mention the future creation of a Palestinian state; it did not make clear fundamental territorial issues, such as the future status of East Jerusalem; Jewish colonial settlements in the occupied territories were maintained, and no indication was given to their temporary or permanent custody by Israeli troops; it did not specify when the Israelis would stop guarding the external borders of the territories; and it did not set a timetable for the total liberation of Palestinian political prisoners.

II   There were revolts in Turkey (1928-30) organised by the Khoybun Party, and in Iraq organised by Sheik Barzani, who was the father of the current Iraqi Kurd leader.

III This theme would, through Padraic Pearse, become an important one for the contemporary IRA.

IV   This experience, the first ethno-national armed struggle in Western Europe, had its roots in the "White Boys" of the 18th century, which defended the people from the Church and landowners and demanded the application of the "rights of the English" to the Irish. It was a non-sectarian movement, in which Catholics and Protestants participated and who used bonfires to send messages from parish to parish. However, in the 19th century, this movement diversified; in the south it took on a Fenian perspective, while in Ulster it gave rise to Protestant sectarian organisations such as the Orange League..

V   However, in the north there were a number of areas where Catholics exceed 50% of the population; besides, from the seventies onwards, the higher birth rate among Catholics would begin to alter the ratio.

12 Peripheral Nationalisms and Violence (1): Identity and Ideology

The violence exerted by peripheral nationalist movements almost always differs from that of movements in vertical societies in the Third World (or Latin America) which exert control over parts of the territory belonging to the state against which they are fighting. It also differs from past forms such as legitimist movements in traditional societies which take up arms under social hierarchies, or the violence of current secessionist parts from states that utilise former coercive structures for their own ends. The lack of territorial control and the symbolic nature of their violent activities allow us to study these nationalisms as social movements, even though their tendency to imitate the state escapes the logic of social movements.

Their violent response emerges in a space previously occupied by the centre-oriented elite ethnic groups, or by traditional nationalist groups opposed to the use of force (and on occasion, both of these groups at the same time). Violence implies serious risks for those who practice it, constituting a point of no return. For that reason, its appearance requires a combination of exceptional external factors and specific internal emerging conditions.

External Factors

The exceptional factors which unleash violence are often related to serious events such as changes of system of government, decolonisation, fragmentation of states, persecution of ethnic groups; in these cases, the theory of relative frustration is less applicable. In the West, however, the spreading influence of Third World liberation struggles and some perverse consequences of the process of modernisation, which translated into internal economic colonialism, became, in some cases, unleashing factors, or combined in other cases their effects with, more serious factors.

In Spain, where some peripheries such as Catalonia and the Basque Country had sheltered historically powerful nationalist movements until the Civil War, the Franco regime laid the foundation for the violent response which would emerge in the late 1950s and early 1960s; the sons and daughters of those who lost the Civil War were ready to accept the ideological influence of some Third World liberation wars (like those in Cuba, Algeria and Vietnam).

In Ireland, the revitalisation of the movement took place in the 1960s, as a result of socio-political objectives such as the struggle for the civil rights of Catholics, encouraged by the example of the black community in the United States. The British army's support for the Unionists, by sending troops to Ulster, returned the national character to the movement in the early 1970s.

The end of French imperialism and the radical expressions of the new social movements pushed the minority nationalist movements in Brittany to the left. In Corsica, added to these influences was the response to the hegemony of client clans and the French government's plans for modernisation of the island under non Corsican direction .

The creation of a Jewish "homeland" in Palestine, which brutally expelled a large portion of the Arab population from its land and homes, provoked the emergence of a new armed resistance force, which was organised into various groups in the Diaspora.

The Kurdish ethnic group had been fragmented into four parts by the Western powers when they set up the borders of the Near Eastern states, following the defeat of the Ottoman Empire in World War I. Its various parts began new mobilisations after World War II, when their national aspirations were once again ignored.

Indira Gandhi's rejection in the early 1970s of the demands of the moderate Sikh party Akali Dal, which ruled in the federal state of Punjab, sparked the emergence of a radical wing of Sikhs which occupied the Golden Temple of Amritsar. The Indian army subsequently took back the temple, and that led to the assassination of Indira Gandhi at the hands of Sikh officers under her command, thereby inflaming the conflict even more.

In Sri Lanka, the emergence of Cingalese Buddhist nationalism provoked the emergence of armed Tamil groups such as the Eelam Tamil Tigers, who gained strength on the Jaffna Peninsula, in the northern part of the island.

Internal Emerging Conditions

There are two internal conditions which lead to organised nationalist violence. The first is the existence of a situation of social or political exclusion (or the two together). Such exclusion can affect the ethnic-national community as a whole; this situation is more frequent in the Third World than in the West -except for the existence of some authoritarian regimes-; it can concern sectors as a whole, or specific deeply rooted national demands on the part of those sectors (political, linguistic, religious). The emergence of violence is accompanied by the transformation of groups within these excluded communities into counter-elites, whose new values, accepted as "the group ideal" by sectors of the community, are precisely those considered "counter-values" by the state (or its dominant culture).

The second condition is the existence of a dense network of nationally oriented groups which protect the violent organisations, accept their values and provide them with people and resources, groups subject to linguistic, socio-political or religious exclusion whose codes of honour prohibit denunciation. (The existence of this "social capital" is what significantly differentiates nationalist violence from purely ideological violence.)

These community groups are what guarantees the strength and viability of the armed organisations; when clandestine parties or trade unions organise underground political life, violence is inhibited, even under authoritarian regimes. This explains the difference in the 1960s and 1970s between the Basque Country and Catalonia: while in Catalonia the anti-Franco forces were led by parties, trade unions and church organisations, in the Basque Country the clandestine social movement was in the hands of "cuadrillas" (peer groups) and mountain groups, *mendigoizales*, in those years hothouses for ETA members (Laitin, 1996).

The violence exerted in the initial phase by these groups is never angry (or volcanic); it is always clandestine and instrumental. It may consist of sporadic violence presented as a warning to the authorities about the existence of problems that should (and can) be solved by them. If the demands do not ask for profound changes and are met, as was the case in the Italian Tyrol, this "soft" violence begins to fade and eventually disappears.

Violence can also be extinguished if, besides government repression, there is no network of groups favouring it, since the nationalist social movement is occupied with another project. Such was the case with the Quebec Liberation Front in the 1970s, which after kidnapping and killing a government minister disappeared due to the strength of the *Parti Québécois*, representing the country's techno-structure.

If initially there are various armed groups, they are obliged to undertake a process of unification, in order to survive state repression and the opposition of local elites. When they face competition from long-standing nationalist groups -as is often the case- they must engage in intense activity to differentiate themselves ideologically, synthesising the response to the exceptional factors. This process of ideological unity is difficult, because of the need to bring together elements as dissimilar as long-standing populist nationalist discourses and new subversive and leftist ideologies. Only the concentration of violence in a single armed centre is capable of consolidating this process of totalizing ideologies and identities.

As a result, the first batch of leaders tends to be students, the only group with all three conditions of marginality, generational renovation and the capacity to synthesise old and new theoretical contributions. (This rule has rare exceptions, due to the total exclusion of ethnic elites, as in the case of Iraqi Kurdistan; due to the long history of radical nationalism, as in the case of Northern Ireland; or due to the brutality of the persecution, as in the case of Palestine.)

But the nature and origin of the leadership group soon changes. The students' calls to action are not successful if they are not heeded by the groups involved, either urban or rural, or in many cases a combination of peasants and workers, depending on the social makeup of the nationalist community; these groups always have an intellectual status, and often a social status, lower than that of the students. The new militants, more skilful in the art of warfare, gradually replace the university students as leaders. But despite the increasingly popular nature of the group's leadership, it becomes hierarchic, depending on the members' willingness to take personal risks, and this feeds elitist discourse (in the writings of the first ETA frequent exaltations can be found of the *gudaris* - fighters - or the "prophetic minority shock forces"). Indeed, the clandestine lifestyle differentiates the militants from the social groups from which they came, thereby reinforcing elitist concepts. Increasing government repression also contributes to these concepts, because it makes return to normal life more and more difficult, for emotional reasons as well as security concerns.

Although most attempted violent activities fail in the early stages, the few that succeed destroy the state's image of invincibility and create the illusion of an ever more favourable balance of forces. The clandestine militants thus become the "ideal group" of an ever larger sector of the national ethnic community, conferring upon it a new dignity, and transforming its negative identity of an excluded, devalued, subjected and stereotyped community into a positive identity. Distance, provoked by the clandestine lifestyle and often by imprisonment and exile, transforms admiration into identification, and finally, into sacralisation. The armed group becomes an organisation mimetic of the state, and the larger social groups which protect its members turn into a broader community which legitimates it as a group-state, and in that sense imitates the nation (because it conceives of itself as the nation to which that state should belong).

The Four Phases of Violent Nationalist Mimesis

Thus, we can identify four phases in the appearance of a political violence of a peripheral nationalist nature, led by an armed group. The first phase is the production of social violence with a defensive-aggressive "response". The second phase is the appearance of an armed group, the product of a double process of fusion and totalization. The third and fourth phases develop simultaneously; they are the transformation of the armed group into a group-state which imitates the state, and the formation of a nationalist socio-political community of an anti-repressive nature which legitimates the group-state, accepting it as such. Through fusion, the groups which move into armed struggle elaborate syncretic discourses which incorporate the various social, cultural and political elements used up to then in a new totality, with totalization being the necessary condition for present and future violence. In this phase, through an operation of theoretical structuring, and fused together through the use of violence, concepts as dissimilar as long-standing nationalist ideas and vanguard Leninist and Third World ideas are amalgamated.

The armed group's quasi-state nature explains its intolerance of the existence of other armed groups, and the sometimes bloody struggles to achieve monopoly in the use of violence; since, according to its logic, several states do not fit in a single territory. In any case, gaining this exclusive right is a process that takes years and is almost never completely finished.

The state in which the armed group is transformed is a state at war. Its perception of reality is filtered, therefore, by the polarity of friends and enemies. The longer its activity is prolonged, the figure of the enemy occupies more and more of the space outside the group, encapsulating the sectors of the national ethnic community which renounce the use of violence. Thus, in the long run, hostile polarisation covers the whole range of its identity field and blocks the instrumental dimension of these nationalisms.

The existence of a real centre-periphery conflict of which the armed group is an expression, as well as the symbolic nature of their actions, make it impossible for the state to achieve victory over the armed group by eradicating it. But at the same time, except in unusual circumstances which have not yet taken place in the West, the objective of national liberation is blocked: the group habitually finds itself in a minority situation within the national ethnic community, whose majority treats the group with the hostility with which the group has treated it, thereby reinforcing the situation of exclusion out of which violence is born. This "eternal stand-off", which can last decades, reinforces the frontiers of the armed group's legitimating community, generating a closed circuit of a counter-state and a counter-community within the real nation-state and national ethnic community.

This blocking of objectives can also be due to external circumstances created by other national projects supported by foreign states operating in the national ethnic territory (such was the case with Macedonia, which was the object of expansionist plans by Bulgaria, Serbia and Greece.)

The logic of the process culminates in the formation of a single legitimating community. But specific situations can emerge which force the armed group to extend its radius of action to diverse civil groups, which may or may not rival each other. (For years, the Koordinadora Abertzale

Sozialista, KAS, the first of several concentric circles around   ETA, contained within its ranks several competing parties - HASI, LAIA, EIA. Furthermore, although the political arm of the Corsica National Liberation Front is A Cuncolta, the Union of the Corsican People, which does not approve of the use of violence, it has on occasion been under the protective mantle of the armed group.)

Polarisation of Identity Reaction

The ideology and identity dimension has specific characteristics in violent nationalist movements, and its elements form a more unified *Gestalt* than in non-violent movements.

When violence intervenes in the formation of a new identity, the community's self-recognition intensifies, and the entrance into a new world of values takes on the nature of a "conversion". On occasion, a violent response gives the first boost to a reaction of national identity against state violence, which was so intense that until that moment no response had been made to it.

Violence becomes the cement for intense group solidarity. It is a catalyst for strong solidarity, especially among those belonging to organisational structures, and intensifies the feeling of group-belonging. Its emergence exacerbates the dividing line between friends and foes and simplifies (through a psychological and social process that is well known in nation-states) a dual situation in which each person must choose sides. In every case, this new situation reveals both sides' political impotence: that of the violent group, because of its inability or impatience with using routine negotiation or representation procedures (in formally democratic societies); and that of the nation-state, because of its failure to institutionalise the conflict, and for having brought the contesting peripheral sectors to the conviction that violence is the only means left for meeting their demands (Braud, 1993).

In that sense, "state terrorism" and the responding "nationalist terrorism" are contrary elements which reinforce one another. Hacker (1976) explains that the state employs terror as an instrument of domination and intimidation, while terrorism is the imitation of terror methods by those who are not in power, by those who have been depreciated and who believe that terrorism is the only way they have to be taken seriously. The identity of the subversive group goes beyond simple interest and even reason; it separates and produces its own moral standard: that of the community, in whose name and in defence of which the group can and should be aggressive, without remorse. Membership in a group which decides everything, which imposes obligations and responsibilities, incites the individual to unselfish acts and heroic sacrifice. The group thus becomes the symbol on whose behalf the individual can satisfy his or her elemental desires for power and exteriorise repressed aggression. The violent group's relationship to the national community of the state is polarised, and their interaction becomes ruled by what Devereux (1970) calls "antagonistic acculturation", which produces a physical and social stress within the group in question.

The newly formed violent nationalist movements in the early 1970s were influenced by Third World liberation movements. Some years later, the evolution from an industrial society to a post-industrial society, which in the West led to the blossoming of new social movements, was an especially propitious moment for violent nationalist movements. In the social movements and in the violent nationalist or ideological groups, there was a strong emphasis on identity: membership in all these organisations was not based on class or state, but was marginal, often frequently recruited in local pockets of exclusion (by sectors in the case of the new social movements and ethno-territorially in the case of violent nationalist movements). In the case of the violent ideological groups, the evolution which moved them away from social movements and toward an internal war against the state could bring them into internal conflict; but in the case of the violent nationalist groups, the struggle against state structures, from the beginning, formed an essential part of their programme, assuming there was no deviation from their objectives. Successes in this sphere (even if they were symbolic and sporadic) brought them support and legitimation within the community.

But this sliding toward a "war against the state" gradually weakens the link between violent nationalist groups and new social movements, reinforcing the many differences that existed between them from the beginning. Alternative movements do not align themselves along the left-right axis or define themselves along class lines, while the violent nationalist movements do, at least in their discourse. The former, born out of the marginal sectors of advanced society, conceive of that society in distinctive categories -sex, age, etc.- or mobilise around the risks posed to the human race by specific threats. Violent nationalist movements, on the other hand, always strive toward totality. Values defended by the former, such as individual dignity and autonomy, come into conflict with the latter's strong leadership, collectivism and imitation of the state.

In the 1970s, some violent nationalist groups attempted to include social movements within an organised front of rejection with national content, but respectful of those movements' autonomy. However, these initiatives were soon extinguished and replaced by others tending toward submitting those social movements to their discipline, and if that were not possible, creating factions which would be faithful to them. Biltzar (Euskal Herriko Batzarre

Nazionala), led by Herri Batasuna at the end of the 1970s, was soon buried in order to incorporate social movements faithful to a coalition called ASK, affiliated with KAS.

Identity Temporal Dimension

In the temporal dimension, the violent nationalist groups identity is oriented toward utopia, toward a future in which -from their point of view- thanks to the heroism of the armed vanguard, their national group will be liberated from their current insufferable situation, and all their problems will be solved. A return to the past can be also effected, but not as a form of nativism: the reconstruction of an identity related to national origin selects a sequence of real or imaginary historical events that offer the image of a people in constant struggle against external invaders and internal enemies, a struggle whose unavoidable culmination and highest expression is the current guerrilla army. If the national ethnic group has had any political organisation in its history which was later interrupted, the radical nationalist groups establish historical continuity between their present and those episodes. In this way, long-term historical legitimation is duplicated with medium- and short-term political legitimation.

The *Green Book* of the Irish Republic Army (IRA), published for internal use in the late 1970s (Coogan, 1987), reconstructs that organisation's historical legitimation. Since the 9th century, Ireland had been the first nation north of the Alps, with its own literature and civilisation. The Norman invasion in 1169 set off eight centuries of constant Irish struggle against the occupiers, who subjected the whole island to economic exploitation, imposing its social and cultural domination in the south and turning the six northern counties into an old-style British colony. The struggle against imperialism intensified after the 1916 uprising; its high points have been the war against the Free State in the south and the six northern counties resulting from the partition in the 1920s, the campaign of explosives in England in 1939-40, the attacks on border posts between the two Irelands in 1959-62, and the current and most heroic campaign, as the book calls it, carried out by the (Provisional) IRA.

The IRA is described as the legal army of the Republic of Ireland formed by the Dáil (parliament) constituted after the Sinn Féin victory in the 1918 elections throughout the island, and the successor to the signers of the 1916 Proclamation. The Dáil of 1921 had resolved that if the British enemy decimated its ranks, its executive powers would be transferred to the Army of the Republic, which would constitute itself as a provisional government. When the Dáil approved the treaty of "surrender" which created partition, the IRA withdrew from it and recognised the minority deputies who had not signed the treaty as the custodians of the Republic. In 1938, the surviving deputies turned over their authority to the IRA, which converted itself into a provisional government heir to the Dáil. Thus, the IRA has the moral right to pass laws for the whole island and to lead the resistance against the occupying forces and its domestic collaborators- i.e., the British Army, the police force of Northern Ireland and the Army of the Free State, or Republic of Ireland (although the latter are not considered military targets).

Two documents, dated 1977 and 1990, contain the historical reconstruction of the Corsican resistance struggle against France, carried out by the Corsican National Liberation Front (FLNC). The focal point of this reconstruction is the historical figure of Pascal Paoli, "the father of the country". Corsica was freed from Genovese occupation from 1755, the year that Paoli was designated head of the Corsican nation in Casabianca, until 1768. Paoli's activity in those years, as an enlightened personal friend of Rousseau, is portrayed as a model to be imitated. Paoli wrote the Corsican Constitution, an antecedent of the US and French constitutions, and organised the government: the communes, administered by three elected magistrates in each community, were organised into *Pievi*, and these into provinces; the representatives of the *Pievi*, elected by the people, formed a National Assembly, the Consulta; this, in turn, elected the Supreme Council, which in turn appointed the *Giunta*, the High Court of Justice, and the General of the Nation. Paoli developed agriculture, commerce, port activity and industry, and created a currency, so that Corsica could live on its own resources; he laid the foundation of an educational system, establishing schools in towns and founding the University of Corti, which had free admission and where Corsican professors taught; he created a national defence system supplied by naval shipyards and manned by a peasant militia.

But in 1768 Genoa sold the island to France. According to the documents, war broke out between France and Corsica; the Corsican victory in Borgo was followed by the 1769 defeat in Ponte Novu. Paoli sailed to England; after Corsican resistance lasting until 1775, French settlers began to arrive, and they had the collaboration of local nobles such as Bonaparte. Paoli approved the French Revolution in 1789 and was once again appointed head of the island (*Babbu*) in 1793; but France outlawed his followers. When the French were defeated, they were replaced by a British protectorate; when France reconquered the island in 1796 it prevented the Corsican parliament from meeting anymore. After Corsican Napoleon Bonaparte took power in France, he put General Morand in 1803 in charge of a merciless repression of the rebellions which had continued on his native island, and at the same time he created two battalions of Corsicans, thereby combining repression with manipulation. From 1819 on, cultural alienation was imposed. The University of Corti was closed, obliging young Corsicans to study in the metropolis; pastors and peasants were looked down upon, and only expatriates enjoyed prestige. The Corsicans became an instrument used by French colonialism against other peoples, and in World War I, 35,000 of them died on battlefields. During the course of World War II, France publicised an alleged oath taken in Bastia stating that the Corsicans would be French forever; it did so that the hatred of the occupation was transferred to other Mediterranean peoples, such as the Italians, and it created an insidious amalgam of Corsican nationalism and fascism.

According to the documents cited, the FLNC is the response to this colonial oppression and the expression of the struggle against it: "The FLNC is the vanguard of the Corsican people, and the profound expression of its national aspirations. The Corsican nation has been forged in struggles against all invaders...two centuries of lies and alienation have strangled our people's national consciousness. That legitimates our struggle, because colonialism negates our national rights."

In the 1950s, at the beginning of Basque nationalism, several booklets published by the EKIN group, the antecedent of ETA, created a historical sequence. ETA subsequently simplified the temporal context of its legitimisation: there had been a single struggle among the Basque people against the Spanish invader, manifested in the Carlist Wars and in the war of 1936-1939, culminating in the war it currently wages. This concept is summarised in the slogan *akzoko eta gaurko gudariak* (fighters of yesterday and today), popularised by the Basque radical nationalist movement since the late 1970s.

Nonetheless, that movement has vacillated in building a bridge with the near past, avoiding the establishment of continuity with the Basque government of 1936 and 1937. That government emanated from the Spanish Republic, and its territory was limited to three Basque provinces, excluding Navarre; the parties which constituted that government -and which during the Franco years continued to form part of the Basque government in exile (PNV on the nationalist side, with socialists and republicans on the state side)- directly opposed ETA from the 1960s on. But that government had organised the Basques' anti-fascist struggle during the Civil War, and for that reason radical nationalist groups were interested in maintaining some kind of link with it. This was accomplished through the figure of Telesforo Monzón, the Basque government's counsellor of the interior in those times and co-founder of Herri Batasuna; and through the presence in Herri Batasuna, as a sort of historical relic, of Basque Nationalist Action, a small nationalist party at the time of the Republic which formed part of that Basque government during the years of war and exile.

Ideological Elements

Various types of ideology can be distinguished among the utilized by violent nationalist groups: subversive nationalists, class revolutionaries, those who justify violence, those who exalt the vanguard in struggle and those who tend to radicalise new social movements. All of them are fused together, in any case, in a single *Gestalt* of ideology and identity.

Distinct combinations of vanguard Leninism and the ideologies of Third World liberation struggles have been explicitly utilised by violent nationalist movements: Maoism, Guevarism, Fanon's concepts regarding the Algerian revolution, and the theorists of Latin American guerrillas Guillén and Marighela.

In 1915, Lenin had come to his own conclusions after reading Clausewitz's treaty on war. War, which for Lenin is a political instrument, should serve to fight against all imperialism and for socialist revolution. It is a test of force directed at breaking the enemy's will; defence should serve to re-establish the balance of forces, and the attacks should be daring.

Mao applied this theory to the case of a prolonged civil war. Revolutionaries should attain a progressive turnabout in the balance of forces, until the enemy is annihilated. In theory and practice, Mao verified the Clausewitz principle, according to which war does not halt political relations between belligerent forces. When war becomes civil and is prolonged for decades, "psycho-political" methods predominate, aimed at winning over the people's support. Thus, revolutionary war has a long defensive and strategic stage due to the inferiority of its forces, and must be combined with tactical offensives.

The influence of Mao's thought on violent nationalist groups does not end in the military terrain; the political equations he established in the course of the war against Japan would later conciliate those groups with Marxism in the 1960s. Mao asked in 1938: "Can a Communist, who is internationalist, be a patriot at the same time? We sustain that he not only can be, but must be." But while revolutionary war is a war of annihilation, in which the class enemy cannot survive and should therefore be destroyed, the national war of liberation does not propose to annihilate the foreign enemy, but rather to achieve that enemy's withdrawal from national territory. Thus, a decisive element in the liberation fighter's activities is influencing the enemy's will and the enemy country's internal public opinion, so that it will change its direction. (This distinction would inspire the peripheral nationalist movements' concept of negotiation with the state).

In his work *La guerra de guerrillas* (Guerrilla Warfare, 1968), Che Guevara defined the three changes that the Cuban revolution had introduced: 1. popular forces can win a war against a regular army; 2. one should not wait for all the conditions for making revolution to materialise, because the insurrectional struggle will cause them to emerge; 3. in underdeveloped America, the fundamental terrain for armed struggle should be the countryside. (The first two points would feed the conviction that the creation of an armed group can be the flame that unleashes a vast national liberation movement).

Paradoxically, the writings of a French expert in counterinsurgency, Delmas (1972), would be used in the opposite sense. According to Delmas, the difference between revolutionary warfare and guerrilla warfare lies in the psychological nature of the former. The revolutionary fighter is a militant and a soldier (embodied in the *gudari*-militant of the first ETA); likewise, revolutionary warfare assumes the defence of the poor: "Revolutionary warfare takes on a terrorist face because those who lead it carry the resentment of the poor and promise to avenge that resentment".

Delmas -who made his conclusions from the war in Algeria, in which major urban centres were sites of insurrection- gives great importance to what he calls "urban terrorism," in counterpoint to the works of Mao and Che Guevara. The main objective of this terrorism, according to him, is to "destroy administrative, social, religious... political hierarchies placed in command positions by the trust that had been given to the established powers. They are, therefore, the enemies of the revolutionaries, who strive to isolate them from the population, and who bring them down if they do not want to unite with them. At the same time, the revolutionary organisation creates 'parallel hierarchies'—i.e., an administration designed to replace the power against which it is rebelling." To achieve its purposes, urban terrorism employs the following methods: propaganda, "including the most utopian [consisting of nationalism, religious crusade...]," the infiltration of insurgent personnel within the population; and the commission of spectacular attacks aimed at stirring up the population. He describes the success of the first terrorist actions as essential for obtaining results. (The theory of "parallel hierarchies" was cause for reflection among some violent nationalist movements about how to organise their legitimating community).

Fanon (1968) had a new theory on the role of violence, having been strongly influenced by Sartre's *Critique of Dialectic Reason* (1960). The relationship between the colonist and colonisation, according to Fanon, is exclusively violence, without any mediation. For that reason, the liberation of the colonised can only be accomplished through force. It is this violence that the nation will engage in, because the fighter takes his homeland with him no matter where the guerrilla army goes: "Decolonisation is the encounter of two congenitally antagonistic forces.... Their first confrontation has been violent, and their cohabitation -the exploitation of the colonised by the coloniser- has proceeded with the help of bayonets and cannon.... The colonist has used, and continues to use, violence against the colonised..... In capitalist countries, between the exploited and the exploiter are a multitude of moral professors and counsellors. In colonial regions, on the other hand, the gendarme and the soldier... give the colonised person advice not to move with the rifle butt or napalm." Among the colonised, this situation generates the intuition that their liberation is only possible through force. In Fanon's mind, "violence is the universal mediator. Colonised people liberate themselves in and through violence." Violence is also the matrix of national reconstruction, because it is totalising and national, and entails the liquidation of regionalism and tribalism. "In guerrilla warfare," Fanon states, "it is not important where you are, but rather where you are going. Each fighter carries the homeland between his or her nude heels."

In his 1961 prologue to the first edition of Fanon's book, Sartre states: "This irrepressible violence is not an absurd tempest, nor the resurrection of savage instincts, nor even the result of resentment: it is human beings making themselves... The weapon of combat is their freedom. The survivors, for the first time, feel a national soil beneath their feet. At that moment the nation no longer is distant from them: they encounter it wherever they go, wherever they are." (Violent nationalist movements would take as their own this concept exalting violence as a factor for the reconstruction of the human personality of the coloniser and, above all, the nation's generating matrix).

While these authors fed the ideologies of violent nationalist movements in the 1960s, the thinking of the Latin American urban guerrilla theorists (Abraham Guillén, Brazilian Carlos Marighela) did the same in the 1970s. The point of departure was the works of Régis Debray, the father of the political-military concept of armed struggle and the "revolutionary focal point" theory. Effective leadership of a popular war can only be assumed by an executive group with sufficient technical skills, centralised and united on the basis of identical class interests—i.e., a revolutionary general command. In this general command there should not exist any difference between "political" and "military" leaders because, according to Debray, militants receive political training more rapidly in the guerrilla army than in the school of cadres.

Debray had theorised about rural guerrilla warfare. Abraham Guillén was the first to adapt these concepts to urban guerrilla warfare: "Strategically, in a country with a large urban population, the centre of revolutionary warfare should be in the city, based on quick and mobile surprise attacks, with superior fire power." Guillén criticised the historical left in this way: "The passivity of domesticated animals: the Communists that coexist are petty bourgeois; velvet socialists; radical bourgeoisie indifferent to parliamentary democracy, trade unions without class consciousness; students with the souls of seminarians; and the depoliticised popular masses, without revolutionary leaders who have a heroic view of life".

The *Mini manual* of Marighela (1970), the Brazilian guerrilla leader who died in 1969 in a confrontation with the police forces, is an authentic prescription for the machinery and forms of urban guerrilla warfare. He enumerates the forms of urban guerrilla action: "1. attacks; 2. incursions and invasions; 3. occupations; 4. ambushes; 5. street tactics; 6. work stoppages; 7. desertions and diversions, captures and expropriations of weapons, munitions and explosives; 8. prisoner rescues; 9. executions; 10. kidnappings; 11. sabotage; 12. terrorism; 13. armed propaganda; 14. war of nerves." The urban guerrilla army continues to be defined through political and military parameters, but in Marighela's concept the military predominates over the political: "Better trained men, more experienced and dedicated to urban guerrilla warfare and simultaneously to rural guerrilla warfare... This is the central nucleus, not of bureaucrats and opportunists hidden behind structures, not of empty lecturers, not of writers of resolutions which are valid only on paper, but of men in struggle." (Violent nationalist movements would incorporate into their set of theories these scornful judgements which would make Guillén and Marighela of the left).

In the 1970s, ideologies arising directly out of the new social movements were incorporated, this being the third contribution in chronological order. But these movements -ecology, antinuclear, feminist-are not violent; some, such as the pacifists and the antimilitarists, explicitly oppose the use of force. Thus, its use by violent nationalist movements occurs through a second-degree discourse which gives them a radicalised national significance: defence of the country regarding ecology, resistance to the repressive state structure in regard to pacifism and antimilitarism, and condemnation of the authoritarian patriarchal state with respect to feminism. The values of those movements are included in and subordinated to the state-armed vanguard central conflict; those social movements which prove incompatible with that orientation are finally discarded, or their exclusion is sought. Links are created with those sectors of post-industrial society that are especially marginalized: employees of low-paid services, long-term unemployed, etc.

The community that identifies with the armed group, which situates the nation-state at point zero on the scale of legitimisation, puts the armed group at the very top level of the scale. In a long conflict, state repression overflows the armed group, spreading into its social environment, punishing it for its "apology for terrorism" (even more so when this community contains points of contact between identification with the armed group and concrete support for its actions). Responses against the repression mould the nature of the nationalist grouping, the anti-repressive stance replacing ideology (because the community is fed by experiences of the militants and their families suffering).

Thus, the violent nationalist groups become impermeable and increasingly indifferent to the fact that today, contrary to the situation in the 1960s and 1970s, the justifications of violence which had certain intellectual authority have disappeared from the horizon of Western democracies—due to the evolution of Western Marxism and the obsolete nature of revolutionary discourses (Braud, 1993). The ideologies which originally fed them are therefore not banished from their pantheon. If to this it is added that the centre is incapable of perceiving the violence it channels toward the radicalised periphery, which generates sentiments against repression, and that it only perceives the response violence directed at the state, the radical group's solipsism increases, and its lack of communication with the outside world becomes total.

Identity in the Armed Group and in the Legitimating Community

The identity construction by the armed group and the legitimating community should be understood as a process mirroring the nation-state, but a warring state, characterised by a concentration of power. In this mirror-imaging, all "liberal" elements inherited from the Enlightenment disappear. Whereas according to the Enlightenment there was a supposed social contract which served as the basis of the political society, here the community members identify with the force of the group-state. Whereas there existed a group of individual-monads united in their consent of the state, here there emerges an indestructible "we" consisting of the anti-repressive community and the armed group fused together. Whereas it had been thought, in a hedonistic-utilitarian way, that the state should facilitate its citizens' individual happiness and constitute the welfare state, here a morality of individual self-sacrifice for the good of the community is implanted. Whereas there had been a belief in the linear and unending progress of society, here there appears a faith in the act of rupture which will produce *ex novo* a regenerated society and nation state. Whereas the state apparatus had been conceived as divided in three powers—the executive, the legislative and the judicial -which must oversee each other in order to diminish the potential for oppression-, here the maximum concentration of all the powers are put in the hands of the group-state.

Like all warring states, the group-state, like armies, feels the horror of "betrayal," and the more it is subject to a hostile external siege, the more it is inclined to detect and punish betrayals. The more violence to which it is subjected, the more an external violence against the enemy develops, and other internal violence to prevent the dissolution of the group. "This violence," affirmed Sartre, "aimed at avoiding the group's dissolution, creates a new reality: the behaviour of betrayal," and the traitor's elimination is based on the positive affirmation that he or she is or has been a member of the group.

Members of armed groups tend to be anti-intellectual. There are several causes for this attitude. One is that the experience of long years of struggle shows that the state must be fought with weapons and not with words. This conviction is reinforced by the memory of the early formation of the armed nucleus, when ideological effervescence generated or accompanied the emergence of divided factions. Once the anti-repressive stance fills up the ideological space of the legitimisation community, many intellectuals come to be seen as superfluous (because their functions can be carried out, with less risk of heterodoxy, by journalists, who work on a short-term basis, writing messages day by day), or, even worse, as factor eroding the group's unity. On the other hand, apologetic works, anti-repression literature, journalistic and legal writings are welcome, as well as, where there are linguistic differences, literary production in ethnic languages.

Another reason explaining this anti-intellectualism is that in long-term conflicts the armed group ends up being an exceptional case of elite in reverse, whose socio-economic and educational status is notably inferior to that of the average status of the members of its social milieu. It is true that its members are idealised by the legitimisation community, but that idealisation does not save them from exile, imprisonment, and on occasion, death. For this reason, the level of social strata from which those members come -in which the expected psychological gratification can encourage them to face up to probable future risks- tends to decline in the long run.[I] The ever greater presence among its members of workers or unemployed from large cities makes the group feel more and more instinctive antipathy for an intellectual world of which they ignore everything; the armed group only tolerates those intellectuals who recognise it as the leading vanguard.

It is the violence exerted by the armed group which maintains the cohesion of the ideological-identity *Gestalt*. The members of those armed groups that eventually disband -something which occurs frequently among those groups which see themselves as the "rearguard" of supposed revolutionary parties- either enter individually in the mother organisation which could give meaning to their armed struggle, or accept a "reinsertion" into civil life, characterised by the destruction (or the demystification) of the initial triple myth (national, political and social) which had given the armed struggle unity.[II] These former militants not only stop seeing their former group as the expression of a national centre-periphery conflict, but also eventually deny the very existence of that conflict, which explains their frequent entrance into state-wide parties.

The legitimisation community is structured as a "subculture of violence" (Ferracutti, 1968), characterised by its triple character as a community opposed to repression, a community of fear and a community at war. The state's national domination, considered illegitimate by that community, does not land any more upon an inert mass of isolated individuals: the consubstantial aspect of these communities is that they are cohesive and oppose state violence. Thus, oppression becomes repression, and the responses to repression (anti-repressive demonstration) are assumed and elevated to the rank of identity symbols by the legitimisation community. This opposition to repression assures the cohesion and continuity of the group, and contributes to making the community intolerant of all deviation.[III]

The legitimisation community is also a community of fear, characterised by an intense perception of aggression, a sensation of real personal or collective danger, a concept of society as victim, the putting on the same level of violence and self-defence, and a sentiment of terror (Ruiz de Olabuenaga *et al*, 1985). It also perceives itself as a community at war, along with the psychological effects which according to Bouthoul (1970) accompany war situations: moral Manicheism, the sharpening of the notions of friend and foe, the cult of dead heroes and the inversion of values seen in fighting groups, in which the most noble and most brutal qualities live side by side.

The mechanisms used by the community to legitimate the group-state are those of identification and sacralisation. Marcuse (1970) indicates that the concept explained by Freud in this work *Collective psychology and analysis of the ego*, according to which the superego is dominated by personal images (the father, the boss), belongs to a former world of social relations: in today's world, he affirms, the authority symbols are anonymous.

The members of the group-state are also anonymous, but their anonymity is that of persons who are forced to live clandestinely, so as not to be arrested by the police. The power the group-state gives itself is the greatest that can be conceived - and which defines state power as such: that of life and death. But except for a few exceptions, its members wind up arrested (researchers calculate the average period in which ETA members are on the streets as about three years), and suffer the life of someone who has been first arrested and then imprisoned. Thus, the identification of the counter-community's members with their group-state is more intense and more emotional than that of the citizens of nation-states with their governments; the sentiments that provoke that identification acquire a quasi-religious character, based on the legitimating mechanism of sacralisation. That character is fed, in Western Christian civilisation, by the symbols of the Sermon on the Mount, the death of Jesus on the cross, and his glorious resurrection from the dead. The Shiite tendency of the Islamic religion provides another type of messianic parameter, based on the memory of the legitimate Caliph's martyrdom and the return of the hidden Imam, whose significance is nevertheless coincident with that of the Christian culture.

In the West, the vanguard nature of violent nationalist movements tends to create organisations in the form of a waterfall of vanguards.

Radical Identity in Nationalist Systems

The factors in a system of nationalist movements that lead its radical wing toward violence occasionally affect the national ethnic group as a whole, as in Sri Lanka and the Sikh Punjab, where the Tamils and Sikhs, respectively, are discriminated against as groups.

In some composite states in the West, we see sometimes the spectacle, surprising at first, that - once a degree of territorial autonomy has been attained and the moderate wing of the nationalist movement has achieved a certain degree of power - the violence perpetrated by the radical wing, rather than diminishing, intensifies. Nevertheless, if we look carefully we can see that this reaction is an inexorable consequence of the logic of that violence.

Even though this violence is presented as an instrument of political action, its identity is actually fed by exclusion. But when the violent nationalist movement has consolidated, it is no longer a matter of the national ethnic group's exclusion, but of the exclusion suffered by its own community, which the armed group translates into violence. The institutional hegemony and the actions of moderate nationalist groups radicalise the anti-system grouping in its rejection of the political framework. The armed group now has an additional reason for intensifying its attacks: that of deepening the trenches that separate the two currents and thereby eliminate the existence of border zones between the radical community and moderate nationalists. The incorporation of systemic nationalism - contradictory as it may be - to the state consensus forces the armed cell to exert a higher degree of violence than before, in order to maintain the unity of the three elements of the initial myth created in the moment of totalization-fusion, and to maintain the cohesion of the anti-repressive community which legitimates it.[IV]

The radical wing's increased violence explains why, in some cases such as the Basque Country, the nationalist ethnic collective, when alarmed, gives majority support to the moderate wing (although this may not be the case where the violence of the armed group is of low intensity, as in Corsica) The moderate nationalists may feel obliged to make a common front with the state centre in defence of the imaginary trio of pluralism, humanism and pacifism, faced with the utilisation of violence by the armed group-state (whose logic, meanwhile, deprives legitimation to its institutional efforts); even if the maintenance of a discourse of conflict may permit it sporadically to remake the nationalist family's "we".

> The case of the Basque Country is paradigmatic in this sense. ETA-m consciously made the decision to increase the mortality of its attacks in late 1977, at the moment the Basque pre-autonomy regime went into effect, which coincided with the moderate Basque Nationalist Party (PNV) having first access to hegemony. A decade later, the PNV's consolidation of institutional power and the deployment of the autonomous government's police force (Ertzaintza) sparked generalised street violence led by the youth sectors of the legitimisation community, whose targets were the autonomous government's policemen and political parties, Basque or not.

In these peripheries, the leaders of the centre parties and the systemic nationalist parties formed what Dahl (1963) defines as "polyarchy," which consists - according to Goodwin (1988) - of "government by a series of minorities, some guided by self-interest, others by public interest, all in favour of accepting the established form of politics. Their political goals are within the framework of the limits of consensus".

But if a current of authors criticises Dahl because his construct ignores "political marginals," these radical marginals cannot be ignored: the cell fights the system with weapon in hand, while another wider circle supports those who fight. This new kind of Rejection Front - which harbours those alternative movements it can control - creates a new counterpoint between the institutional bloc and the radical social movements, which are identified by the institutions with the armed group and are stigmatised in the same way as it. The "pluralism" and "consensus" proposed by the institutional forces are counterpoised to many of these alternative movements' initiatives, the latter being considered to be part of the violent nationalist world. Thus, the armed group creates its adversary for the second time, and the legitimisation community-armed group whole is defined as the new "internal enemy." But this time it is not just the enemy of the nation-state, but also of the institutional forces of the periphery, nationalist of not.

Thus, the armed group's violence expresses a double exclusion: that of the national ethnic group from the state, and that of the anti-system nationalist movement from systemic nationalism with institutional responsibilities, a result of the lack of communication between the two. The latter, even if it enters into partial conflict with the centre, will develop a subsystem of alliance with it, based on their common opposition to violence (possibly utilising the existence of that violence as an element of pressure in its negotiations with the centre). Likewise, under these circumstances, it tends to favour - given the challenges of the international market - economically strong "pro-development coalitions" over pacts with the set of social forces in the territory.

Northern Ireland confirms the validity of the aforementioned in a contrary sense, since the pact between the two nationalist Catholic currents was one of the factors which contributed to de-activating political violence in 1994. With Sinn Féin having fallen into marginality between 1925 and the 1960s, the majority Catholic party in the North was the Nationalist Party, a party of traditional "notables" and professionals which only presented itself in electoral areas where gerrymandering had not prevented the existence of Catholic demographic majorities. Sinn Féin was reborn at the end of the 1960s, as was the moderate nationalist current, re-established in 1970 in the name of the Social and Democratic Labour Party (SDLP).

The SDLP, although a member of the Socialist International, is not a labour party, but rather a nationalist party. It was formed in 1970 on the basis of three currents: the remains of the Nationalist Party, bodies which emerged from the civil rights movement such as John Hume's Citizen's Action Committee, and Catholic trade unionists grouped since the 1950s into the Republican Labour Party, led by Gerry Fitt. The popularity of Fitt and Hume made it possible for the SDLP, in the 1973 regional elections, to win 22% of the votes in the six counties and 19 seats; from that time on it was recognised in London and Dublin as the *de facto* mouthpiece and mediator for the Catholic community. Its reform programme, structured into several stages and oriented toward "the promotion of the Irish people's majority cause, based on the agreement of the majority of the people in Northern Ireland", was aimed to convince the Protestant majority of the benefits of union, for which Hume had EEC support.

But except for the brief re-establishment of self-government in Northern Ireland during 1973-74, the six counties have been deprived of institutions since direct rule was introduced in 1972; as a result, the SDLP has not been able to gain access to hegemony, but even to an institutional presence. Hostility between the two currents could not intensify through the social and political exclusion of one of the currents at the hands of the other. This permitted some years later, in 1993, the SDLP-Sinn Féin political accord which unblocked the conflict and produced an IRA cease-fire in September 1994. What could bring Northern Ireland back to its previously violent situation would be the blocking of communication between the two currents, rather than the rupture of the truce announced by the IRA in February 1996 (Seiler, 1982).

In Corsica the two nationalist currents formed since 1977 by the National Liberation Front (FNLC) of Corsica and the Union of the Corsican People (UPC), despite their distinct ideologies (the FNLC is anti-imperialist, while the UPC is ethnicist and not socialist), share certain characteristics, because they were born in the same years - the 1960s and 1970s - as the product of a single movement opposed to French domination and the hegemony of the Corsican clans; the limited volume of the Corsican national electorate has maintained both forces tied to the same social base. This explains why ties have never been broken between the UPC and the successive political parties related to the FLNC. Fluid communication between the two currents has kept the FNLC's political violence at low levels: from the 1970s until the early 1990s, only 16 fatalities occurred. But this fluidity has also blocked that armed group's impermeability to the conflicts that emerged in Corsican civil society, leading at the end of the 1980s to civil and armed fractionalisation.

Access to territorial power by a nationalist current which was previously radical can invert the roles of the actors in a system of nationalist movements and radicalise previously moderate currents which have been excluded from that power. In Palestine, the replacement in 1987 of the PLO's armed anti-Zionist struggle by the civil resistance of the Intifada made it possible for the former to moderate its programme and accept the UN resolutions which recognised the state of Israel according to its 1967 borders. That allowed for the initiation of public talks, culminating in the Oslo accords in September 1993. But it was also in 1987 that the activity of Islamic groups - Islamic Jihad and Hamas above all, which was created that year- which had been tolerated until then as rivals of a revolutionary PLO, was radicalised. Islamic militants and leftists of the Rejection Front confront daily the Palestinian National Authority's police with almost the same intensity as they confront the police of the Jewish state.

Notes

[I] In the Basque Country, the profile of an ETA member changed during the 1980s. The percentage of women rose considerably; the society of unemployed which the country had become, with the consequent increase in marginal behaviour, augmented the number of unskilled workers in ETA's ranks. The typical ETA member in the 1980s was an unskilled worker from the outskirts of one of the four capitals of the Basque Country. The average age of its members also increased considerably, to above that of Herri Batasuna voters. A study by the Vasco-Press agency on the sociological extraction of ETA members, published in 1988 and based on data related to 220 persons arrested in the previous two years, came to the following conclusions: 1. The median age of ETA's social base has registered a growing tendency, while the number of young people has gone down significantly... 2. Large urban areas situated on the peripheries of Bilbao and San Sebastián are the main sources of ETA grass-roots recruits. The main cities continue to be important. 3. The rural environment is now secondary in terms of support for ETA. 4.Most of the support... comes from workers with little or no professional skills. The number of students is minimal".

[II] Both processes took place following the self-dissolution of ETA-pm (7th Assembly) in 1982. In these cases, reinsertion is accompanied by a radical mutation of identity. Given that the nation-state generates its own morality and logic, its greatest victory over an armed nucleus is that of convincing its members, individually or collectively, to exchange the former identity of the "we" that founded the armed group to identification with the state. The dynamic of "reinsertion" has much in common with the sacrament of confession: just as the sinner recognises the Church's moral authority, the "reinserted" individual renounces the morality of his or her former armed group in order to accept the morality of the state.

[III] According to Barth (1976), "in most of the political regimes where there is less security and people live under a greater threat of arbitrariness and violence outside of the primary community, that very insecurity acts to repress interethnic contacts. If an individual depends for his security on the voluntary and spontaneous support of his own community, self-identification as a member of that community will be explicitly expressed and confirmed; any behaviour deviating from the norm will be interpreted as a weakness in their identity and, therefore, in the bases of security".

[IV] H. Arendt (1961) explains in her paradoxical way that there is an inverse relationship between power and violence: the situation of everyone against one person is the extreme form of power, while the situation of one against everyone is the extreme form of violence. Therefore, violence is greater when the power being fought against is greater, and when the group that legitimates it is reduced in relative terms.

13 Peripheral Nationalisms and Violence (2): Organisation and Context

The Instrumental-Rational Dimension

The combined armed group-legitimating community whole aims to achieve certain goals of a national content, which are almost always socialist in nature, using a common strategy and tactics specific to each of the two elements of the whole. The resources mobilised to this end are ruled by the principle of rational choice, although they are limited by the fact that decisions are taken clandestinely (Crenshaw, 1983; Della Porta, 1996).

The limitations are not only due to the psychological effects arising from this situation, as the ongoing nature of organised political violence generates growing contradictions between the identity and instrumental dimensions of the whole. If the move toward instrumental rationality requires from all peripheral nationalisms that they surpass their ethnic boundaries in order to achieve their goals, such a step is all the more necessary for movements which need to accumulate the maximum degree of strength in order to counteract the response, hostile in its extreme, of the nation-state.

In terms of discourse, at least in the West, these nationalisms, which are almost always left-wing, have no difficulty in adopting a pluri-ethnic national programme. However, one of the consequences of prolonged violence is the polarisation of attitudes and solidarities which make neutrality difficult, if not impossible, and strengthens the boundaries between ethnic and ideological groups involved in the conflict (Crenshaw, 1983). That is apparent, not only in the case of inter-ethnic borders, but also in intra-ethnic ones: where systemic nationalisms and anti-systemic peripheral nationalisms co-exist, they increasingly come to see each other as enemies.

Excluding those situations of open national war - which has been the case in the Third World, but not in the West - the tendency toward reduction of the social bases supporting violent nationalism is accompanied by a gradual deterioration of the instrumental nature of violence. Almost without exception these nationalisms seek, not total victory over the nation-state, rather the attrition of the state with a view to forcing it into negotiation over a national programme. Put this way, violence leading to this aim requires the fulfilment of two conditions: to be visible, that is, having a fairly high media profile; and respect for certain boundaries (the breaking of which would be counterproductive for negotiation, because they would give rise to security issues which would dominate the questions at the heart of the negotiations, and because they launch an uncontrolled spiral of violence...) (Braud, 1983).

However, it is precisely the identity dimension of long-term violence that works against the requirements made by the instrumental dimension. In effect, violence is not only conceived as an instrument of negotiation, it is also a means for achieving the collective identity of the group and fusing the heterogeneous elements of the triple social, revolutionary and national myth. Inherent weariness which arises in long term conflicts, disaffection and growing hostility in previously neutral ethno-national sectors, forces the armed group to brake more and more thresholds. Its actions cannot take place without the moral support of its legitimating community; nevertheless, in as far as the rupture of thresholds generates a cordon of hostility and isolation around the community, then its members will strengthen their identification with the armed group as the only means of salvation, and close their eyes to indiscriminate acts of violence on its part.

The logic of negotiation is gradually replaced, more in fact than in the level of discourse, by that of "terror," that is, intimidation of the out-groups. It is of little relevance that this logic is justified instrumentally, claiming that what is important is that pressure is placed on the masses of the centre so that they, in their turn, force the state to agree to negotiations. The effect achieved is exactly the opposite, as it leads to direct mass support for strengthening anti-terrorist legislation and to tacit consent to illegal anti-terrorist tactics. In this way the circle closes, blocking the instrumental dimension of radical nationalism; and its violence and the violence of the centre feed off each other, forming two circles with very different diameters, one large, the other small, which endlessly revolve around themselves.

Strategic and Tactical Programmes

On the instrumental level, the strategic programme of violent nationalisms, their social and national goals, is split into two, one long term and the other short term; and it contains a tactical programme, the measures implemented to achieve strategic goals, which is divided for the two parts, civil and armed, of the whole. Armed groups set themselves an objective; the drive to create in the initial phase their legitimating community and to maintain its cohesion during the mature phase.

The long term strategic programme almost always involves attaining a separate and independent nation-state (which is presented as the result of national insurrection). When this goal is not explicitly formulated, the programme never sways from the demand for national self-determination and the right to exercise it by the entire population of the ethnic territory.

The existence of a short-term programme means, to some extent, that the idea of expelling the enemy from national territory by armed force has been rejected. Achievement of the goals set by this programme is presented as the consequence of negotiation between the armed group and the state, and it is this negotiation upon which the end of violence depends. For this reason, these programmes are very different, as each case is related to the specific nature of the centre-periphery conflict both are facing.

In the Basque Country, ETA's tactical programme is the result of an external historical fact, that is the transition from the Francoist dictatorship to Spanish parliamentary democracy, a process which the armed group interpreted in its own way. The new political set up did not satisfy the aspirations of radical Basque nationalism. The democratic Spanish opposition, for the sake of the reform agreed with liberal sectors of the Francoist regime, abandoned its posture in favour of the right to self-determination of the different nationalities within the state. ETA's tactical programme of rupture with Francoism, developed in 1975, the year of the dictator's death, has been adhered to since then with few modifications. It defines the Spanish system of government, formed during the transition period, as a continuation of the dictatorship, and argues that only acceptance of KAS's tactical alternative would be recognised as the end of Francoism.

The theoretical elements in ETA's political strategy, taken from the classics of revolutionary war, have undergone three phases. The first, which lasted until the years 1965-1968, crystallised in an imaginary "revolutionary war, which proposed the destruction of the enemy on Basque territory. The second phase, which stretched from this period until the first years of the transition, 1974-1977, referred to the "spiral of action-repression." The third strategic phase, the path of negotiation, represents to a certain extent an acceptance that the spiral has come to a halt; negotiation with the state is put forward as a goal once the armed group is convinced that popular insurrection is no longer viable.

The theoretical basis of negotiation was initially developed by ETA-political-military; in 1977 ETA-military, the only organisation of the two which still exists, made that basis its own. ETA-pm claimed in 1975, in Maoist terms, that the phase of bourgeois democracy, previous to popular war, corresponded to a programme of minimum principles which entailed a democratic rupture with the Francoist past. The central points of the programme were the territorial integrity of the southern Basque Country (the Basque provinces and Navarre) and recognition of the right to self-determination, principles which constitute the background to the KAS alternative. As a popular war based on insurrection did not seem feasible, a long 'war of attrition' was proposed, [...] "with the aim of forcing political negotiations, the terms of which would be decided by the correlation of forces." In 1978 ETA-m asserted that it would take full advantage of the left-wing *abertzale's* peaceful struggle and armed struggle" in order to achieve the goals set out in its programme, on the understanding that the intensification of the struggle would create an insoluble problem for the government, which would not be in a position to heighten its repression since this would broaden the "social base of support for armed struggle and undermine the base of Francoist reform." The "radicalisation of the reactionary wing," which ETA had anticipated, led to the failed *coup d'état* of February, 1981, which did not in any way induce the government to concede to the demands made by KAS, resulting in fact in increased repression against the radical sector as a whole. ETA, without modifying its programme, reaffirmed the need for a long war of attrition, arguing that such a war could only be favourable to ETA's and the Basque people's goals since "to resist is to win".

In its *Green Book*, written in the 1970s, the IRA sets out its long and short term strategic objectives, together with its tactics. The long-term objective is the creation of a democratic socialist republic throughout the island of Ireland. The republic will have a democratic and socialist government, be in favour of decentralisation, co-operativism and the revival of the Irish language; it will oppose social-imperialism and, in the international sphere, align itself with former British colonies. These objectives will be achieved through struggle against injustice and economic and social inequality, created by British imperialism in Ireland. Therefore, the short term objective is a British withdrawal (or "Brits Out!").

Tactics entail differentiating friends and enemies. Incidental enemies must not be confused with those who oppose short and long-term goals. Enemies, whose position is based on ignorance, need to be educated, although the first to be educated should be IRA members themselves; the IRA should not create enemies by needlessly harming people.

The *Free State* armed forces and police are regarded as illegal and treacherous, but they are not military targets, which is the case for the Royal Ulster Constabulary (RUC), the northern Irish police force. Until the events of 1969 British troops were not included among IRA targets, but have been regarded as such since then. SDLP members are perceived as "enemies of the people" because in 1973-1974 they participated in a northern Irish executive which "detained and tortured Irish people", but this only made them an object of ridicule.

Armed struggle should be a "war of flies" aiming to "strike British elephant' belly", an offensive war that causes sufficient deaths so as to lead for widespread calls for the withdrawal of the British army. It should have the following goals: make it impossible for the British government to rule in the six counties in a way that is not military-colonial; make financial investment unfeasible through a bombing campaign in Northern Ireland; support the war activities through national and international propaganda; and protect the armed organisation by punishing criminals, collaborators and informers (Coogan, 1987).

The FLNC defines itself in its 1977 and 1990 documents as "the vanguard of the Corsican people and the highest expression of its national aspirations". The Front should organise political and military activities as it is "the only organisation able to maintain political leadership that will not deviate from the ultimate goal." However, in its documents, the FLNC, which admits the unwillingness of Corsicans to cause great quantities of blood loss, distinguishes between "armed propaganda" (carried out in recent years) and the "armed struggle," the conditions of which it must be ready to create. For the FLNC, the armed struggle "does not mean wanting to defeat France; it is not about confronting the military and police apparatus which is superior to us in arms and resources: there will not be another Ponte Novu [...] Our dissuasive and repressive actions are determined by our people's perception of colonial repression".

The "armed propaganda" phase, which deliberately excludes human fatalities, has as its goal the consciousness-raising of the Corsican people in social, economic and cultural spheres. However, this is a transitional phase; political and military actions should aim to create the necessary conditions for the unleashing of an "armed struggle." This struggle will force the French state to recognise the national rights of the Corsican people; its stages will be determined by the evolution of the Front: " organisation of the masses and of the counter powers, importance of the material resources available, ability to resist heightened colonial repression.." This phase will lead to the achievement of the right to self-determination when the defeat of its colonial instruments compels France to negotiate. France must commit itself, under the supervision of the United Nations or another international organisation, to guaranteeing the modalities of applying this right. During the transition period, the government of the island will be held by the French state and fighters for national liberation".

The Palestinian case is proof that the tactical programme of violent nationalism changes when its strategic objectives are modified. The Palestine National Charter, approved in July 1968 by the first Palestinian National Council of a PLO controlled then by Al Fatah, defined Palestine in a way which was irreconcilable with the existence of any Israeli state. According to the Charter, Palestine is an indivisible territorial unit (Article 2); it will be the Palestinian Arab people who will decide their destiny once their homeland is liberated from Zionist occupation (Articles 3 and 4). All solutions which do not entail the total liberation of Palestine should be rejected (Article 21); the Palestinian Liberation Organisation, which represents all Palestinian forces, is responsible for the liberation of the homeland (Article 26). The armed struggle is defined as "the only route to the liberation of Palestine" (Article 9).

The Charter was also incompatible with Resolution 242, adopted in November 1967 by the UN Security Council, which made the restoration of peace in the Middle East dependent on the withdrawal of Israeli forces from the territories occupied after the Six Day War in 1967 (Gaza, West-Bank and the Golan Heights), but it likewise demanded respect for the sovereignty and territorial integrity of all states of the region (including Israel's).

After the start of the "Intifada" in 1987, the following year the PLO introduced the big tactical change consistent with handing over the initiative of the struggle to those involved in the popular uprising, thereby minimising the role of armed resistance. This made the vital tactical change possible which meant for the PLO approval of Resolution 242 in the UN, and therefore, accepting the idea of a Palestine state in occupied territories adjacent to Israel. The entire world is today witness to the hopes and uncertainties over the peace process which arose from this change.

The programme of armed groups is not twofold, but threefold; together with the strategic and tactical aspects there is the creation and consolidation of their legitimating community.

In the case of ETA this latter aspect of the programme has been reworked several times. During the first half of the 1960s ETA set itself the goal of creating utopian "parallel hierarchies." The "theoretical basis of revolutionary war" developed in 1965 the most realistic theory of the action-repression-action spiral prevailing for a decade, the practice of which generated an anti-repressive community. The spiral is described in the following way: 1. ETA, or the masses under ETA's direction, take part in a provocative action against the system; 2. the state apparatus of repression strikes a blow to the masses; 3. in so far as repression is concerned, the masses react in two opposite and complementary ways, panic and rebellion. This is the appropriate point at which ETA gives a counter blow that will reduce the former and increase the latter reaction.

When the spiral of action-repression has advanced sufficiently, the time will have arrived at which to play off the contradictions between reactionary capital, by provoking its use of repression, and liberal sectors, who will seek negotiation. In this way the system will be weakened and the final stages of revolutionary war can be completed.

In the debate which took place during the years 1973 and 1974 about what type of struggle should be pursued in, mass struggle or armed struggle, ETA-military opted for the autonomy of the latter. For this reason its members announced in its *Agiri* (Manifesto) of autumn 1974 their intention to withdraw from every type of mass struggle, and called for all nationalist grass roots organisations to self-organise. The emergence of the Socialist Nationalist Co-ordinating Body (KAS) in 1976 as the leadership bloc, and of *Herri Batasuna* in 1978 as a Popular Unity, was a response to this call. When this stage was complete, ETA used KAS at the beginning of the 1980s as a vehicle for structuring civil groups, which then became a faithful legitimating community subordinate to the armed group.

The FLNC developed this theme in 1981 in its theory on the creation of "counter powers," which proposed the integration of already existing groups, not set up by the Front, into its legitimating community. These included trade unions, cultural, ecological, defence and solidarity groups, parent-teacher associations, neighbourhood groups, national movements, legal movements, etc. These counter powers should be vehicles of self-organisation which "allow the Corsican people to assume responsibility for their own destiny at all levels of social, economic, political and cultural life..." The legitimating community had begun to organise itself with the foundation of the Nationalist Committees in 1976 which became the electoral channel of the group in the 1980s, a task which was subsequently taken on by the MCA and *Cuncolta*. In the same environment the *Donne Corse* (Corsican Women) was established, as was the prisoners' support group A Riscossa, the U Ribombu journal, the Corsican Workers' Trade Union, etc.

In the 1970s the IRA was up against the reverse kind of problem; since the 1920s a Catholic community existed in Northern Ireland radically politicised; however, at the time when the "Troubles" exploded, its structures were too open, making it vulnerable to unionist and British repressive measures. The Staff Report, an organisational internal publication, argued at the start of the "Troubles" that certain rules of clandestine life-style should be applied both to the armed group and to its legitimating community.

A return to secrecy and to strict discipline was proposed. A new IRA structure had to be constructed which would be made up of new recruits, unknowns, and they would be incorporated into a system of cells, instead of the previous neighbourhood battalions, more appropriate to a long-term struggle. The cells, under the leadership of brigades, would be composed of four people; the command's operation officers ranked above the brigades. The cells carried out executions, robberies and placed bombs, but they did not have control of arms or explosives. They were authorised to carry out operations outside their territories with the aim of confusing the British.

Civil organisations lost power in so far as they assumed a supporting role for IRA activities. The women's organisation *Cumman na Ban* dissolved; one part was absorbed into the IRA cellular structure and the other became involved in civil and administration tasks. The youth organisation *Na Fianna Eireann* became an underground organisation, without a public profile, which educated its members for active service in the IRA. The clubs, which were involved in youth and community work, pensioners, etc., became support bases and a means of collecting funds.

What is striking is the clarity with which this document sets out the subordinate role of Sinn Féin with respect to the IRA: " Sinn Féin must follow the orders of those with authority in the Army in every regard. It will provide full time organisers from the main republican areas. Sinn Féin must radicalise itself (under the leadership of the Army) and agitate against social and economic measures which are harmful to the welfare of the people. Sinn Féin should infiltrate other organisations with the aim of gaining support and sympathy for the movement. It should fully develop the role of the publicity and propaganda departments" (Coogan, 1987).

This means that in the 1980s and 1990s, especially after the IRA ceasefire, an opposite process has been in operation, which is essentially the growing autonomy of civil organisations of the republican movement in Northern Ireland; this process is the context in which the undeniable role of the Sinn Féin leadership in overcoming the conflict makes sense.

Mobilised Resources: Organisation of the Armed Group

The resources mobilised by an armed group in order to fulfil its objectives are its organisation, exclusivity or not in the use of armed struggle, its attacks, its social capital, comprising the social groups which legitimise it, its relation with its legitimating community and the political organisation of the community-armed group whole. Its structure of opportunity, always orientated, by definition, toward a system of conflict organises itself in a triple context; that of the ethno-national society, the state, and the international context.

Peripheral nationalist armed groups are rarely guerrilla bands. Leaving aside those exceptional cases in which the group exercises control over a section of territory, as was the case of the Tamils in the Jaffna peninsula until 1995, or the Kurds in Turkey or in Iraqi Kurdistan, absolute clandestinity is the case in the urban context. Their organisation, according to Grabowsky (1987) is hierarchically structured, with members divided into small groups of three to five people. Each of these groups comprises a cell in a structure which is based on non-horizontal communication among the different cells nor vertical hierarchic communication. Members of the different cells remain anonymous among themselves, and the identity of a superior is known only by one member of the subordinate unit. The urban structure allows fighters to live separately and fight together, thereby allowing them to disperse and reunite in agreed places without major problems. This system favours anonymity and secrecy surrounding the identity of leaders; those who become notorious often do so posthumously or when they have been arrested.

> In certain national conflicts in the West, such as in Northern Ireland, where suburban areas suffer blatant discrimination and the marginalisation of its resident ethnic group, - as is the case in West Belfast - the armed group may carry out parallel police functions with regard to combating drugs and delinquency. These functions, which resemble territorial control, were engaged in before and after the August 1994 IRA ceasefire.

Clandestine rules can be relaxed in "sanctuaries," territories situated in neighbouring states where a certain tolerance is shown toward members of the group (as long as this territory is used as a withdrawal base and not as a base for action). However, given that this tolerance depends on something as variable as the relation of forces between states, and that a sudden change of interests could transform members of the group into bargaining chips, no mature armed group will confidingly display its structures there.

> "Black September" put an end in 1970 to the PLO's sanctuary in Jordan; the Palestinian sanctuary in Beirut itself came to an end in 1982 with the Israeli invasion of Lebanon, which brought about the dissolution of numerous foreign armed groups which had benefited from the quasi non-existence of Lebanon as a state.
>
> ETA's sanctuary in the south of France was always under siege, but came to an end one year after GAL commenced its activities in Iparralde (French Basque Country) on account of the deportations of Basque refugees ordered by the French socialist government at the end of 1984, after it approved Spain's request to join the Common Market.

Armed groups are hierarchical in nature, which corresponds to their perception of themselves as armies, armies not subject to the rule of any civil power. This fact, together with their clandestine nature, makes co-option from the pinnacle the only means possible of becoming a member of the governing body. Leaving aside exceptional cases where permanent sanctuaries exist, the leadership is constantly renewed, on account of losses through detentions, expulsions, and the deaths of its members.

It is rare to find a document such as the IRA's *Green Book* which contains indications, even as general as the ones mentioned here, on organisational structure, as armed groups tend to keep them secret in order to prevent them falling into the hands of the police forces. The only source of external information available are press releases resulting from leaks by the police, which in their turn summarise statements made by activists who have been detained (being therefore of doubtful reliability).

> ETA's structure in the 1980s is probably similar to that of other armed groups. The organisation is structured around three levels of command: full-time "illegal" commandos, whose members have police records and who carry out the most important operations; "legal" commandos, made up of members unknown to the police and who include several sub-groups (contacts, messengers, and intelligence gatherers); and support commandos which provide whatever is necessary for full-time commandos.
>
> During this period, according to police sources, ETA's governing body was made up of between 25 and 40 members resident in the French Basque Country, acting under the authority of an executive committee of seven members. The organisational apparatus was built around propaganda, information, the political office, illegal commandos, legal commandos, and international relations. Finance and border operations, logistics, training, as well as operational commands were directly under the authority of the

executive committee. Members were given training on French territory on the use of arms, combat techniques, explosives, communications, etc. The courses lasted two weeks and included ideological training, but not information on the organisational structure of ETA (Clark, 1984).

Spanish intelligence services estimate that at that time there were about 500 ETA members. At the start of the 1980s, according to Lieutenant General Casinello, ETA's strength was around 200 armed men divided among twenty to thirty commands that were fully engaged in activities (illegal), while the rest, around 300, took part in occasional activities (legal).

Press releases indicate that in the 1990s the IRA is under the leadership of a seven member military executive (Army Council) that meets once a month. Under its authority is a governing body made up of fifteen individuals, the finance office (in charge of the "revolutionary tax" and raising funds abroad), the logistics officer (responsible for the supply of arms and explosives) and those whose work focuses on support (that is, the bases for withdrawal in the south of Ireland). The operations officer (who co-ordinates information and selects targets), under the authority of the Army Council, also co-ordinates the Northern Front (Ulster), the Central Front (Britain and British bases in Europe) and the Southern Front (the Republic of Ireland). The armed struggle was waged on these three fronts by between 200 and 500 active members, organised at a local level in battalions, the nucleus of which was brigades (cells made up of three or four members). The support work, in which thousands of sympathisers were involved, encompassed intelligence, surveillance, transport and storing arms.

Exclusivity of the Armed Struggle

The mobilisation of resources by an armed group is most favourable if it has achieved the exclusive use of political violence in the ethno-national territory (which is always sought after by the group in order to attain legitimation as a group-state). This is sometimes attained by political means, by broadening the limits of the legitimating community; and on other occasions by waging violence against rival armed groups.

In the case of armed cross-border ethnic groups operating in different states, exclusivity is impossible. In spite of the pan-nationalism put forward in their discourse, the mimetic nature of armed groups with respect to the states which they are combating forces them to fight on separately against each state, which leads to the emergence of as many groups as there are state territories. In these cases, the armed groups in question may mutually ignore each other, although they frequently fight among themselves, more politically than militarily, with the aim of attaining a hegemonic position in the core of the trans-state ethnic group.

Between 1975 and 1985 in the southern Basque Country there were three armed organisations, ETA-pm, ETA-m and the Autonomous Commandos, and from 1981 onwards a fourth was added to the list, *Iraultza*, a small group which engaged in low-intensity violence linked to the labour problematic; this group was the only one to survive in tandem with ETA-m after 1985. There were then, until 1985, three concepts of armed struggle: 1. that practised by ETA-pm, which conceived of its activities as a rearguard struggle guaranteeing popular gains; 2. that practised by ETA-m, for whom armed confrontation with the state should be carried out by a leadership bloc made up of political parties and mass organisations, that is KAS, whose vanguard was ETA itself; 3. the Autonomous Commandos, whose first operations go back to the end of 1978, and for whom armed actions should be the result of a "civil war waged between working class self-organisation assemblies and bourgeoisie."

In the Basque Country, the struggle for exclusivity did not translate into open confrontation. ETA-pm's rearguard war led to its disappearance when its supposed civil vanguard (known as EIA at the start and later as Euskadiko Ezkerra) ended up denying the existence of the national conflict, this led to its fusion with the Basque sector of Spanish socialism in the 1990s. The Autonomous Commandos dissolved due to the lack of a social environment which would provide them with legitimation. More than a few of those members of both groups that did not accept the disappearance of armed struggle are today in the ranks of ETA-m.

After the division of the IRA in 1970 into two sections: "provisional" and "official," the latter subsequently dissolved, without a violent confrontation between both organisations, in a development which has certain parallels with the ETA politico-military process. Nevertheless, a radical splinter group of the "officials" survived in Ireland which led to the creation of the Irish National Liberation Army (INLA) in 1975, which has close links with the Irish Republican Socialist Party (IRSP). Both these organisations expressed their disagreement with the August 1994 IRA ceasefire, although the INLA did not break it by carrying out any attacks.

In Corsica, conflicts dividing the nationalist community have affected the main armed group, the FLNC, which underwent several splits between 1988 and 1990 caused by French socialism's attempt to tighten the links with Corsican nationalism. The situation gave rise to three armed groups: *Resistanza*, the "historic" FLNC and the FLNC "habitual channel." During the period 1995 and 1996, these latter groups have been engaged in violent confrontation between themselves.

Between 1973 and 1981 several armed groups emerged in Sri Lanka. These were: in 1973, the Eelam Tamil Liberation Organisation (ETLO); in 1975 the Eelam Revolutionary Organisation; in 1981 the Eelam Revolutionary Front, pro-Soviet in its orientation. The Tamil Eelam Liberation Tigers (TELT), set up in 1975, ended up becoming a group-state through the elimination of rival groups.

The case of Palestine is unique in this respect. From 1964 onward there was a government in exile which had won certain international recognition, the Palestine Liberation Organisation (PLO), composed of several civil and armed Palestinian groups. Arafat' Al Fatah won control over the executive organs of the PLO, but he had to respect their plurality. In fact, armed organisations of the Palestinian resistance such as Habash's FDLP and Nayatmeh's FLP and Saika had been members of the national executive until 1995. Tensions among the different groups led to the withdrawal of the PLO from some of them: the Arab Liberation Front, Jibril's FDLP General Headquarters, the Palestinian Liberation Front, the Palestinian Popular Struggle Front, the Palestinian People's Party... The situation changed after Palestinian National Authority came into effect in Gaza and the West-Bank; those groups which do not recognise its legitimacy have to confront a Palestinian police force under its control.

Armed groups which emerge in the heart of cross-border ethnic groups have diversified in different states. One such group is *Iparretarrak* (IK), "those from the north", emerged in 1973 in Iparralde, (French Basque territory), small and carrying out fewer attacks than its "bigger brother," ETA. Until 1994, IK had not been looked upon by ETA with sympathy, since ETA needed calm in Iparralde, which it used as a withdrawal base. Therein laid the logic of ETA's discourse that the fate of the Basque Country in its entirety depended on ETA's success in the southern Pyrenees. The French government's change of attitude toward ETA members taking refuge in its territory from the mid 1980s onward probably encouraged ETA to accept from the mid 1990s the autonomy of IK.

The break up of armed groups belonging to the same ethnic collective reaches its highest point in Kurdistan, a territory divided among four states. Historically, the DPK, led by Barzani, active in Iraqi Mosul from 1945 onward, aimed since its foundation to control the pan-Kurd movement; from 1975 onward it has faced opposition from within Iraqi Kurdistan by Talabani's PUK. After 1980 it was the PKK, under the leadership of Ocalan "Apo," in Turkish Kurdistan, who claimed hegemony within the Kurdish movement. Rivalry among the different armed groups has recently become violent, with open confrontation taking place in Mosul, Iraq, between the DPK and the PUK.

Arms, Attacks, Thresholds of Violence

With regard to the political violence (or "armed struggle") waged by a group, it is necessary to analyse the instruments, or weaponry, the thresholds of violence, targets, and the symbolic meaning of attacks.

The degree of weaponry sophistication varies greatly among groups. According to Clark (1984), finding arms is not an obstacle in the West given that the international illegal arms market is an open market. (In the Basque case, routes go through Belgium, Switzerland, Germany and Holland, and the focus points are Brussels and Paris. The weapons are transported in false containers and are handed over to the ETA leadership in France. However,

Clark adds that what ETA does not obtain abroad, it acquires from Spanish arsenals.)

The contradiction pointed out by Della Porta (1996) in violent ideological groups, between attacks expressing the motives of social movements and defensive-aggressive attacks against the state, is not present in armed nationalist groups. War against the state does not mean deviating from initial goals, rather it constitutes the core of the programmatic objective. On the contrary, of continuing relevance is the definition between "external" attacks, giving publicity to their goals, and "internal" attacks directed toward guaranteeing the survival of the organisation. The latter refers to fund-raising via armed robbery, kidnapping, "revolutionary taxes," or the defence of organisational structures through punishments meted out to collaborators (informers or traitors) with the state.

Analysis of ETA-military's attacks from 1977 onward allows us to classify the acts of a stable armed group. In its documents at the end of the 1970s ETA defines the following as targets in its armed struggle: members of the security forces and armed forces, both of which accounted for the largest number of fatalities during that period, being the core of its war against the state; the nuclear power station of Lemoniz; and, from 1980, those engaged in drug-trafficking (the latter two are expressions of the motives of social movements).

Three further categories, which ETA does not list, should be added to these four; they are the defence of the organisation, the so-called "revolutionary tax,"[I] the kidnapping of individuals for a ransom to be paid in order to secure their release, and attacks which could be classified as "internal," that is, carried out against former members or collaborators whom ETA regards as "traitors." The selective nature of these attacks was to become degraded after 1985, the year in which indiscriminate violence such as car bombs and letter bombs began to be used. The number of potential targets increased and became un-discriminate: hyper-markets, residential barracks inhabited by civil guard families, political party leadership (attacks which ETA had itself forbidden for a long time because it considered politicians to be "adversaries" and not "enemies"). Armed confrontation between the *Ertzaintza*, police under the authority of the Basque Autonomous Government, during the 1990s, is a unique example of hostility between an armed group and the police force of a territorial government controlled by a systemic nationalist party.

In the Basque Country there is a type of violence which borders between instrumental and volcanic; it refers to that which is carried out by the most radical sectors of the legitimating community, almost always groups of young people, against the police forces (state or regional) or against public property such as transport or telephone boxes. Street violence, or *kale borroka* as it is known in the Basque Country, intensified in the 1990s in tandem with a notable decline in the number of ETA's attacks.[II] Although this violence did not involve the armed group, its degree of organisation and the fact that it formed part of the general strategy of the whole suggests that there is a clear mix of volcanic and rational violence.

The identity logic operates in long-term conflicts in favour of the armed group's progressive crossing of the thresholds of violence. Braud (1993) identifies three thresholds. The first is bloodshed; between damage to property and people there is a trench which requires a very different analytical approach by media. The second threshold refers to the class of victim; eliminating barriers which protected certain sectors of the population gradually spreads a sense of insecurity and makes expressions such as "indiscriminate attacks" and "innocent victims" commonplace. The third threshold is based on the scale of intensity, that is, the number of victims, the frequency of the attacks, the "choreography of the violence" (Jenkins, 1975).

The nature and number of fatalities depends, in part, on technical aspects such as the skills of the armed group and efficiency of police repression; it also depends, especially in long-term conflicts, on its "social acceptability" by the legitimating community. This explains the apparently surprising fact that for long periods of time, in ETA's case from 1959 to 1975 and in the case of the FLNC throughout the 1980s, the number of deaths was low, undergoing a sudden increase which cannot be explained by technical factors.

When nationalist conflict becomes a war between the state and an adversary that controls part of the territory, what happens in the Third World, the number of fatalities escalates. In Sri Lanka, for example, the number of violent deaths between 1983 and the present day is 50,000. In the West the figures, always lower in comparison, depend on the nature of the conflict. In Corsica they do not exceed 100. In Macedonian, between 1891 and 1902 ORIM was responsible for the deaths of 4,000 Turkish people. In Northern Ireland, between 1972 and 1992, there were 2,956 victims (2,041 civilians and 915 soldiers and RUC members; out of the total 69% were Catholic, while only 31% were Protestant).

In the Basque Country, fatalities caused by the different sectors of ETA from its foundation through to the 1980s was close to 800. After Franco's death the number of deaths were as follows: 9 in 1977, 67 in 1978, 72 in 1979, 88 in 1980 (the huge jump in the number of casualties took place in the early years of the transition period), 38 in 1981, 44 in 1982, 44 in 1983, 31 in 1984, 37 in 1985, 41 in 1986, 52 in 1987, 19 in 1988, 19 in 1989, 25 in 1990, and 45 in 1991. During the 1990s the number of victims has gone into decline, in contrast to street violence which has been on the increase. In 1996, only five fatalities were recorded, although all had far-reaching social consequences.

In long-standing social conflicts, acceptance of crossing thresholds of violence by the legitimating community tends to work mechanically. The armed group is convinced that an increase in the number and intensity of its attacks will wrest from the state what is impossible to attain through dialogue. But this "strategy of terror" reinforces even more the exclusion of its legitimating community, which heightens the growing violence of the armed group. The spiral of action-repression-action, which some groups had theorised at the outset, is replaced by another spiral, that of exclusion-violence-exclusion.

Organisation of the Legitimating Community

The legitimating community is formed by three concentric circles. The interior nucleus consists of the relatives and allies of those most repressed, prisoners, exiles, etc., as well as former prisoners. The second nucleus, which is broader, refers to those sectors of the population subjectively united to the group-state by the legitimising mechanisms of identification and sacralisation. Together with these two nuclei of the "faithful" there is, in those cases where a political party representing the group takes part in elections, a third external circle that is frequently larger than the other two put together. This circle, without approving, or not entirely approving of the use of political violence, adheres to the whole on the basis of the principle of rational choice, whether it be because of the radical nature of its national aspirations and goals, because of the left-wing nature of its social programme, or because of its rejection of other systemic parties of the centre or periphery. This third circle has to be won over by the whole in every occasion, which explains variations in its electoral representation (which is almost always lower than other parties due to the high degree of identified voters because of the whole's isolation).

The legitimating community can infiltrate already existing nationalist parties, radicalising them and winning them over to its logic. When it is the community itself which creates the political party, driven by the armed group or by the "leadership bloc," the state may respond by making the party illegal. In this case the consequences are usually not what the state had anticipated as the party may change its name or survive extra-legally, participating in elections as an electoral coalition of independent candidates.[iii]

The resulting dispute over whether it is right or not to participate in elections is settled usually in Solomonic style. The party puts forward candidates, but those elected do not take their seats; the abstention could be permanent or partial, in the latter case this means the occasional presence of its representatives in the chambers or Houses in order to give maximum publicity to the rejection of certain resolutions. Therefore, electoral activity is aimed at de-legitimising other parties, centralist or systemic nationalist, and the political framework created or respected by them. Participation in elections is, accordingly, about assessing the degree of social support which the movement can depend on. Likewise, these parties serve as support for mass mobilisations, always driven by the higher levels of the vanguard pyramid.

The movement organises itself on three levels which are hierarchically structured: the armed group (or group-state) as the undeniable vanguard; radicalised social movements under its control; and the political party (or parties). Sometimes, a second level, the "leadership bloc", acts as a link between the armed group and the civil echelons, disciplining the civil legitimating community, meaning that there are four levels. Contrary to what takes place in the sphere of theoretical production, which is almost always poor and schematic in nature, these wholes, necessarily, have every appearance of being highly intelligent and skilled organisers. They set up operational structures, which are flexible and built on the basis of watertight compartments, designed to protect external circles from legal repression.

> The structure of radical Basque nationalism as a whole is a good example of the aforementioned. From the beginning of the 1980s onwards, KAS and *Herri Batasuna* (HB) used the term Basque National Liberation Movement (MLNV) as a means for referring to all groups belonging to the whole, which were structured on four levels: ETA-m as KAS's vanguard, the organisations belonging to KAS as the "leadership bloc" (the party and mass organisations), HB (political parties and individual members) as a People's Unity coalition under the leadership of KAS, and an independent and sympathetic social base in the Basque Country, supportive of HB in elections. Basque Nationalist Action, one of the political parties belonging to *Herri Batasuna*, joined as an historic legitimating factor.
>
> A 1983 document entitled "KAS Leadership Bloc" sets out the structure of the organisations belonging to KAS. The text, which does not challenge ETA-m's vanguard role, allots the task of "globalising" mass and institutional activities to the political party HASI. ASK (Socialist Nationalist Committees) is responsible for the "globalising" of social movements "related" to the MLNV (amnesty organisations, anti-nuclear committees, neighbourhood associations and Basque language groups). LAB is a nationalist trade union, and *Jarrai* is the youth wing of KAS. The document attributes a monolithic character to all organisations belonging to KAS; it demands strong allegiance and homogenisation "without concessions to other activities or tendencies in its core." (During the 1990s the political party HASI dissolved, while KAS assumed its "globalising" tasks and its vanguard role. In 1996, sectorial organisations withdrew their membership from KAS.)

The complexity of the PLO owed, in part, to the fact that it was a government in exile; it was also due to its nature as a whole, like those mentioned above, encompassing armed groups and social movements within its core. The Palestine National Council was made up of 480 elected members, elected on the one hand by organisations such as Palestine Resistance (Fatah, FPLP, FDLP, Saika - after the withdrawal of the rejection forces -), by trade unions (the National Confederation of Palestine Workers, General Unions of journalists, women's groups, students, teachers, engineers, lawyers, doctors, artists and rural workers), and on the other hand by representatives of the Palestinian communities (including the Diaspora spread throughout the world, refugee camps, and independent individuals). From 1973 onwards the Council elected a central body made up of ninety members which appointed an executive committee composed of eighteen members, headed by a president (Arafat). Ministers belonging to this executive committee directed the different PLO departments (its ministries). Under the authority of the president was the military department, which encompassed the Palestine Liberation Army, the resistance forces and the militia (there was, therefore, not only one armed group, but several, which, at least in theory, were under the control of the state in exile, that is the PLO). The remaining departments were Political, Education, Community Organisations, Health, Occupied Homeland, Information and Palestine National Funds. The PLO has differentiated itself from other group-states in so far as it has been recognised by the vast majority of Palestinians from the interior as the legitimate representative of their homeland. All endeavours by the Israeli government to create alternative organisations of Palestinian representation have failed.

In a process which is the reverse of the disappearance of a political party press in the West, a phenomenon brought about by the transformation of parties into catch-all parties dependent on opinion-making press and audio-visual media, *Egin* in the Basque Country, *U Ribombu* in Corsica and *An Phoblacht* in Ireland play a key role for both the armed groups and the legitimating communities in question. Without publicity the activities of the former are less symbolically effective, which is why they require loyal media that will transmit their messages. The isolated nature of the legitimating community, hemmed in and subject to stereotyping and hostility, makes it vital for it to have a bridge which interprets the outside world in accordance with its ideological identity dimension. These publications are established usually as independent press funded by the community; it may take years for the armed group to assume control of them, although when this process is complete, it is total. If the media is a daily newspaper with a substantial circulation it has three categories of employees: those which develop the political line; those belonging to management, which rarely coincide with the first category; and the workers, linked to the media through labour relations and not necessarily coinciding with its editorial line. Nevertheless, the economic problems caused by the reduction, or elimination of financial sources, threats of closure and law suits make professional neutrality difficult. The need to ensure the survival of the media leads the whole to provide backing which often exceeds that offered to the electoral party (in the Basque Country over recent years, the MLNV has tended more toward the creation of support groups for *Egin* than building up *Herri Batasuna*). It is political journalists who establish the editorial line of the media who are the true organic intellectuals of the whole, a hegemonic position which they defend bitterly.

The Ethno-National Context

In the context of ethno-national territory, the exclusion of the periphery as a whole, or of the anti-system nationalism, in those cases where a system of nationalisms exists, feeds political violence. The global exclusion of the ethnic group is more frequent in Third World states based on vertical societies, as is the case of the Kurds, Tamils and Palestinians, than in the West. Nevertheless, they can emerge in authoritarian situations in the West, as in Spain under Franco. They have also occurred in a periphery of the periphery, as in Northern Ireland, where the Protestant/unionist majority discriminated on an ongoing basis against a substantial Catholic/nationalist minority, a situation out of which violence arose.

When it is the social hierarchy of the ethnic group which supports centralism (as is the case of the Corsican clans, organised on the island as branches of French political parties), the exclusion of certain sectors of the ethnic group promotes violence, although owing to its social minority nature it faces difficulties in the process of consolidating itself. That is why the intensity of FLNC armed actions is lower than in Northern Ireland or in the Basque Country.

The legitimation which parties of the centre enjoy on ethnic territory could be so overwhelming that it condemns the armed group (which is often born out of propagation) to a sporadic armed struggle. In these cases, instead of a legitimating community there is widespread sympathy within some sectors of radical nationalism which are a minority on their territory. Such would be the situation in France of the Breton Liberation Front (FLB) and, to a certain degree, of *Iparretarrak* in the Basque Country.

> The FLB's early actions, which defined itself as socialist and anti-capitalist, took place in the 1966-1967 period (that is ten years prior to the outbreak of armed struggle in Corsica). Its operations were aimed not to cause fatalities. The FLB has attacked French government offices, the Palace of Versailles, postal sorting offices, the Brennilis nuclear power station, and the multinational company Shell which was responsible for the *Amoco Cadiz* oil

slick on the Brittany coastline. At the end of the 1960s about sixty people were detained and then later freed by General De Gaulle. In the 1970s, the letters LNS (Socialist Nationalist Liberation) appeared next to the FLB, but this group disappeared shortly afterwards. From 1974-1975 ARB (Armed Revolutionary Breton) appeared written next to FLB. Since 1982 there have been a few sporadic actions carried out, and their documents are few.

The FLB-ARB strongly criticise the Breton Democratic Union (UDB) which it accuses of being the "Breton regionalist fringe, faithfully allied to the French repressive state." Only one small group does not condemn its attacks and continues publishing its statements in its press, this is EMGANN, which defines itself as the "Breton socialist movement for national liberation." Although Breton youth supports cultural expressions such as dance and music, the Breton language itself is in a situation of crisis. Out of a population of 2.8 million, only 250,000 people speak Breton, although 650,000 understand it. The *Diwan*, the equivalent of Basque *ikastolas* (Basque-speaking schools) have an enrolment of about one thousand children a year. In the first half of 1996 a renewal of violence took place; the FLB-ARB attacked a court house in Rennes, water and electricity companies and several buildings belonging to the French government.

The political hegemony of moderate nationalism in the institutions of ethno-national territory, when it occurs, does not exclude the presence of intense political violence, as the case of the Basque Country has proved again and again. In these situations, systemic nationalism demands from the centre the reinforcement of its power with a view to detaining the spread of anti-system nationalism; it also asks for greater control over the police in exchange for its commitment to the anti-terrorist struggle. The intelligence work of this regional police force in terms of penetrating the ethno-national society is more successful than that of the state police; however it reinforces the exclusion of the legitimating community and embitters its most radical sectors, bringing about the spread of civil violence led by youth. The spiral of violence-exclusion-violence once again takes off in a process of reverse feed which is, on this occasion, of an intra-nationalist nature.

The involvement of systemic nationalism in the anti-terrorist struggle is, nevertheless, counteracted by two factors; one is occasional, its rejection of the "dirty war" waged by state police forces; the other is structural, originating from its national objectives, which maintain within their discourse the flame of conflict with the centre. In fact, its defence of the right to self-determination coincides, on occasion, with anti-system nationalism. This situation, likely to generate tensions between the regional government and the moderate nationalist party, as well as within the ranks of the party, prevents an already intense conflict[IV] reaching its logical conclusion, which is a total breakdown of the links between the two nationalisms. This sets as a possible scene, in a future impossible to date, the alternative most feared by the centre, that of the re-composition of the instrumental-rational unity of nationalism under a common national programme supported by, and given electoral legitimation by, a nationalist majority.

The State Context

Public opinion of the state is overwhelmingly hostile to the armed group and its environment. The increasing political violence carried out by the armed group progressively polarises not just the centre but also, where they exist, the different ethnic peripheries that have suffered the political violence of the armed group. The latter interprets the absence of (or minimal) violence in those ethnic communities as a lack of national resistance, in which case it identifies them with the centre and operates in both in the same indiscriminate way. (ETA's indiscriminate bombings and car bombs in Catalonia from the mid 1980s onward should be understood from this perspective.) The legitimating community sees itself also as the vanguard of the struggle throughout state territory against the centre.[v]

If all governments need loyalty and widespread consensus in order to carry out repression (Dowse and Hughes, 1982), then support can be won relatively easily outside the ethno-national territory of the armed group. The widespread approval given by the majority of the population to anti-terrorist laws passed in Western states to combat revolutionary or nationalist enemies (and sometimes both) from the beginning of the 1970s onward is evidence of this.

Nevertheless, the illegal use of these laws, in terms of torture or the unleashing of a "dirty war," has generated anti-repressive reactions in ethno-national territories which systemic nationalisms have, with delays and reservations, ended up supporting.

Those conflicts emerging out of the decision-making process by those responsible for state repression have to be added to the aforementioned. In effect, the centre does not have unified ranks in the anti-terrorist struggle, although this is not apparent in the image projected for public consumption; in reality there are a range of political actors involved with numerous political interests and changing commitments. In the political sphere of the anti-terrorist Madrid Pact and, particularly in the case of its Basque counterpart the Ajuria Enea Pact, contradictions arise due to the insistence of nationalist parties that the centre-periphery conflict remains unresolved, and the refusal of centralist parties to see the parallels between state dirty war and ETA terrorism. Likewise, the existence of a plurality of police forces engaged in the offensive against the armed group accounts for the emergence of rivalries among the different branches of the security forces (in Spain, the Civil Guard, the National Police and the High Body of Police Forces).[VI]

When the conflict reaches the scale of an open national war between a powerful centre and a small resistant ethno-national group, the different postures adopted sometimes become instruments of the various factions in their struggle for power in the state. The events of the Russian-Chechen war cannot be understood outside of this perspective.

Chechnya-Ichkeria is inhabited by a Caucasian population of one million who are Sunni Muslims (730,000 of whom are ethnically Chechen). This region was an autonomous republic from 1936 in the heart of the Federal Republic of Russia until 1944, when Stalin removed its status and deported its people to Kazakhstan and Siberia. In November 1991 it declared itself an independent republic with Dudaiev as its president. In a process contrary to the experience of other autonomous republics, such as Tatarstan, the failure of the Moscow political talks in 1994 resulted in the invasion of Chechnya by Russia in December of that year. The decision to invade was taken not only for economic reasons (since Chechnya is rich in oil and two pipelines run through its territory taking oil to Azerbayan among other places), but also because the Russian "war party" had chosen Chechnya as an example where it could engage in a show of strength that would discourage other independence initiatives in the Caucasian powder keg. The invasion was accompanied by a propaganda campaign in the Russian centre promoting the image of the "Chechen mafia."

This war, which in one and a half years took the lives of 80,000 people in Chechnya, turned out to be disastrous for Russia. Its troops, which engaged in large scale destruction of buildings throughout the capital, Grozni, were unable to defeat the Chechen resistance, the leaders of which, Vasaiev, for example, carried out spectacular raids and mass kidnappings in the Russian Federation. The hand over of power in December 1995 to Zangaiev, the "man from Moscow," instead of resolving the problem, exacerbated it even more. With the death of Dudaiev in April 1996, the new man in power in Chechnya was the intellectual Bandariev. The results of the presidential elections in Russia, which forced a circumstantial agreement between president Yeltsin and General Lebed (the runner up with 15% of the votes) opened the door to negotiations between Lebed, appointed General Secretary of Security with Special Powers, and Masjarof, the Chechen Chief of Staff. The negotiations concluded on 31st August of that year, with an agreement to hold a ceasefire that included the withdrawal of Russian forces from Chechnya within one year and the announcement of a referendum in Chechnya in five years on the definitive status of the country. Chechens decided to hold presidential elections in January 1997.

However, the Russian "party of war," considering this agreement to be an obstacle to its goal of winning power in Moscow, did everything possible to boycott it. A front composed of the head of government, Kulnikov, the chief of the president's office, Chernomirdin, and the minister of the interior,

Chuvais, accused Lebed of having given in to the Chechens and of being involved in a plot to take power in Russia. President Yeltsin's dismissal of Lebed in October 1996, on the basis of his supposed "unacceptable and damaging errors" with regard to Russia gave rise to fears that the Russian-Chechen agreement would be revoked. Although it was saved at the last minute, indiscriminate attacks in Chechnya (against Red Cross personnel, among others), which the Chechen leadership attributed to Russian security services, have prevented the attainment of a stable political climate in the area.

The International Context

In the world context "international terrorism" as a condensed image of what is abominable has replaced in the pantheon of political demons all the old stereotypes such as anarchism and communism. This image, which amalgamates the violence of some peripheral nationalisms with very different phenomena (Third World guerrilla movements, ideological terrorisms, war by proxy carried out by certain states on foreign territory) emerged at the start of the 1960s in the heat of attacks carried out in the West by groups which were attempting to settle conflicts that had arisen in the Near East.

The process was controversial from the outset. Different concepts of the Palestinian question held by the West and Israel on the one hand, and by Third World States on the other created disagreements in the heart of the United Nations about whether certain groups would be considered liberation movements or terrorists. Two United Nations resolutions passed by the General Assembly in 1970 condemned international terrorism; however, the former Soviet Union had, with the support of newly independent ex-colonies, stipulated in March 1969 that "nothing could prevent the use of armed force in accordance with the United Nations Charter, including its use by dependent peoples, with regard to exercising their right to self-determination."[VII] From this point onward, Western states and Israel renounced to attain a general consensus in the UN and initiated three types of actions: practical government actions; police co-operation; and legal measures consisting in anti-terrorist legislation and extradition treaties (Freedman, 1976).

This perspective led the West, from the second half of the 1970s onward, to regard political terrorism either as a perverse and irrational criminal phenomenon, or as a consequence of an enormous sinister international conspiracy, promoted and backed by the Soviet Union together with some Third World states.[VIII]

In the West (the United States, Israel, Europe) the stereotype and the reality of international terrorism has led to closer co-operation among the various police forces whose members are often targets of attacks and may become personally involved in the pursuit of terrorists beyond their own borders. In the United States the authorities have backed sudden strike attacks against states which it defines as "terrorist" (Libya, for example). In western Europe, this same approach has facilitated the creation of a common police and judicial policy.

Some US authors, such as Jenkins (1975) and Alexander (1976), have theorised the need for concerted international intervention with regard to these developments. According to Jenkins, 'few nations are in agreement as to what constitutes international terrorism. It is a new type of war waged without territory and arms, in the way that we know them; there are no neutral forces and the innocence of civilians is not accepted'. The author asks whether, in the face of growing terrorism, governments will simply accept the new concept of war defined by terrorists or whether will they co-ordinate their war tactics between nations and wage a military offensive against terrorist enemies where they exist. Alexander describes terrorism as a new type of offensive in terms of technology, victims, threat and response. He writes that 'terrorism will be more than the sporadic destruction of law and order; it will threaten the survival of civilisation itself'.

With the exception of armed groups' legitimating communities, which tend to be very small, public opinion in nation-states unreservedly supports the anti-terrorist crusade. Political violence carried out by these groups is equivalent to a war; however, as Casamayor (1983) argues, it is a unilaterally declared war in which the combatants do not, even tacitly, acknowledge each other. This situation explains the anxiety of the population, an anxiety which has very little to do with those feelings related to a war. Unilaterally declared armed violence crystallises anguishes of the contemporary era and furnishes the psychological factor which explains the rapid spread within the population of the stereotype of international terrorism.

In some states, such as Spain and the United Kingdom, it is political violence of a peripheral nationalist nature which has given rise to this mood. In others, such as Germany and Italy, it is ideological violence. Third states have suffered the consequences of conflicts that have arisen beyond their borders.

France, which has experienced localised peripheral low intensity violence but rarely ideological violence, has been the target for attacks that have as their epicentre the Near East. During the 1960s and 1970s some authors here criticised the "centralised bureaucratic state" and argued in defence of the "right to difference" (Lefèbvre, 1968). They praised the values of indigenous cultures, contrasting them with individualist behaviour, banality and consumerism, characteristic of "fast food and concrete" cultures (Chesneaux, 1984).

Nevertheless, these perspectives, inherited from May 1968, rapidly disappeared from 1975 onwards, the year in which anti-colonial wars came to an end, when a world formed exclusively by nation-states emerged and when it became extremely apparent everywhere that the newly founded states were going to be equally as defective as the old ones. In this context publications appeared in France denouncing "regional nationalisms," which, according to this line of thinking, either served multinational interests or imitated Third World models. At the same time, ideologies claiming that western democracies would be the lesser of evils in a world full of totalitarian governments acquired increasing currency. The portrayal of international terrorism as one aspect of totalitarianism, the most dangerous, ensured that all armed struggles were tarred with the same brush.

In 1981, on assuming power, the French Socialist Party was caught between these two lines of thinking. The failure to convince many of those who were given an amnesty to return to the "democratic path" opened the door to a new concept of terrorism, according to which it was a phenomenon lacking in political content, motivated by criminal concerns. The turning point came with international terrorist actions which coincided with the Israeli operation "Peace in Galilee" and with the subsequent withdrawal of Palestinian organisations from Lebanon. In 1982, attacks such as those in the Malesherbes Boulevard, Marbeuf Street and particularly in Rosiers Street, which increased the toll of victims by the end of the year to 21 fatalities and 191 injuries, spread the image throughout France, and especially in Paris, of terrorism as a multinational dark plot, and contributed to public opinion rejecting all armed organisations of whatever origin. This explains why, in addition to pressure from the Spanish government, there was a change in posture within the French socialist government from 1984 onward, which was pursued and strengthened by subsequent right-wing administrations, with regard to Basque refugees on their territory (Quadruppani, 1989).

In this way the unity of all forces of formal democracy, both left and right-wing, within Europe began to take place under the anti-terrorist logic. Vervaele (1986) explains the meaning of this collaboration: "the anti-terrorist struggle served as a means whereby the existence of a type of central power was legitimised, a legitimation which commanded the attention of the population, distracting it from other problems, such as economic crisis...The anti-terrorist struggle, the defence of democracy, is fundamental to the extent that the government and opposition must commit themselves to it or be seen to be against democracy, everything else being secondary in comparison to this".

European police co-operation has emerged out of this context, and it became officially recognised as a strategy in 1977 when the Council of

Europe set up the Trevi Group, an organisation based in Rome and composed of high-ranking government officials from the Home Offices of the various member states. The main functions of the Trevi Group revolve around an exchange of anti-terrorist tactics, of different approaches adopted by the police and intelligence information on armed groups.

With regard to the judicial sphere, in January 1977 the Council of Europe approved an agreement for the Repression of Terrorism. According to Vervaele (1986), this agreement aimed to ensure that political crime, which has a history dating back to the French revolution and has been recognised as such in various countries in their anti-extradition stances, is eliminated as a concept. The French president Giscard d'Estaing proposed in 1978 the creation of a "European judicial sphere" entailing automatic extradition for both non political crimes and for terrorist activities, in tandem with European police co-operation.

> As Mattelart (1976) reveals, these big police spheres, both within Europe and beyond, requiring high level technology, represent a meeting point at its most sophisticated level between ideology based on "anti-terrorism" and the expansive logic of communications and information technology of multinationals, which in most cases have their headquarters in the United States. In the 1970s, with the gradual ending of "hot wars" in the world, such as those in Africa and Vietnam, the supposed transformation of these multinationals into peace-time industries began; it was then that their technology, whose cycle of military application was ending, was turned on to the civilian population. With a view to "civilising" this technology, an alliance grew up between these firms and the state apparatus. From 1975 onwards in western Europe, projects such as that in Wiesbaden in Germany where a central computer is engaged in the compilation of a "global information systems," in which business and state policy act in harmony, proliferate. From an economic and political perspective, this entails increased marginalisation for peripheral states, who are restricted to the role of consuming, albeit consuming in abundance, but never producing, electronic gadgets and information technology destined for use by the police.

Generally speaking, particularly in the West, the use of political violence by an armed nationalist group represents an additional obstacle for the ethnic community in whose name it acts to seek international support for its cause. The interest which a third state might have in a newly emerging nation-state, whether it be for geo-strategic motives or because of its ethnic linkage with this community, could be undermined or eliminated altogether by the fear of being seen in the international sphere as an accomplice of terrorism.

However, it may also be the case that certain states use armed nationalist foreign movements in order to carry out a war by proxy against certain rival states. In many cases this strategy is not acknowledged (it refers to terrorist activities instigated by the West against Third World states, as well as attacks carried out in the West by certain Arab "terrorist" states such as Libya, Iran and, to a lesser degree, Iraq, Syria and Sudan). On other occasions this strategy may be open and cynical in nature, taking place without any affinity between the state which instigates it and the armed group. The series of alliances and break-ups that took place during 1996 between Near East states and the different armed groups within the Kurdish people, who are entirely without true supporters, can only be explained by a cold (and somewhat sickening) theory of games based on the principle that the enemy of my enemy should be treated as my friend.

Finally, it may also be the case, as in the Near East, that individual terrorist agents end up becoming mercenaries acting at the behest of third states through a process whereby, according to Wiewiorka (1988), an armed group, in order to secure weapons, turns to "providers who will also be those who propose reciprocal exchanges and who, in the end, will make the terrorists into mercenaries for a cause which was not initially theirs".

Beyond the West, in those states organised on the basis of the exclusion of some of their vertical societies, ethno-national movements may find support when the identity markers which have caused the exclusion are shared by other states, neighbouring or otherwise. This is the case of the Palestinian movement, excluded from the construction of the Israeli state on account of it being Arabic and Muslim, and which has been supported in its struggle against Zionism, albeit relatively speaking, by Arabic and Islamic states. In other continents, such as Africa, the armed resistance waged by tribal groups or specific clans may be supported by states that are ethnically related to them in situations where the states which exclude them are their rivals.

Diasporas may also be a source of support for nationalist movements, depending on the circumstances behind the emigration and also the influence which the Diaspora has over the host state. The Irish-American community in the United States has inherited the memory of British colonial oppression, particularly so in the case of the Great Famine last century, which was responsible for the emigration of over a million and a half people; it is to this community that Sinn Féin owes its current popularity in the United States. This situation is in direct contrast to the lack of support which ETA has among the Basque Diaspora throughout Latin America, descendants of the old process of colonisation or work-related migration in the 19th century, but also of Spanish civil war refugees who identify with the PNV or the Spanish republican government of the time. The Corsican Diaspora promoted itself abroad thanks to the clans' clientele relations with the French administration, for which reason almost the entire Diaspora strongly supports French centralism. (A section of the *pied-noirs* who took up residence in Corsica after the independence of Algeria are descendants of those Corsicans who participated in the French colonisation of Maghrib a century ago.)

When the international community is entirely against them, which is more often than not the case, violent peripheral nationalisms are only able to diffuse this hostility in one way; by declaring a truce in their armed actions with a view to opening negotiations with the nation-state. Rejecting terrorism may, in these circumstances, result in a certain degree of calculated pressure being put on the nation-state in question by neighbouring states, or by the dominant bloc of states, to take into consideration the political demands of the armed movement, with the only goal being the preservation of stability within the region.

Notes

[I] A practice started in 1975, consisting in the extortion of payment to the organisation (ETA) of a quantity, one-off or periodical, made by small and medium-sized Basque businesses.

[II] In any case, this phenomenon should not be confused with processes such as the Palestinian "Intifada." From 1987 onwards this was the catalyst for the organisation of Palestinian society as a whole through the establishment of People's Committees, considered illegal by the Israeli government, and the resignation of Palestinians from their posts in local administration.. The *kale borroka* is more in line with youth violence inspired by Islamic groups and the opposition front in Gaza to Palestine National Authority and its police forces.

[III] The resulting political party, anti-system in nature, is an "object-group" as described by Guattari (1972). While subject-groups are groups which has established themselves, adopting their own internal law and becoming responsible for their own behaviour, object- groups are under the imposition of an external law, living in a hierarchically dominated situation with regard to subject-groups.

[IV] Since there is confrontation between the regional police force and the armed group, as well as increasingly hostile clashes between the two sectors.

[V] Herri Batasuna has put forward candidates in European elections in Catalonia and Galicia, in spite of the fact that there was a progressive nationalist movement with strong electoral support in the latter .For that reason, HB tried unsuccessfully to bring about a split in the Galician National Bloc prior to the 1987 European elections.

[VI] Since 1991, the year in which the Basque Autonomous Police (Ertzaintza) was deployed and became progressively involved in the fight against terrorism, the inter police force tensions have increased even more.

[VII] In spite of the international repercussions of the attack carried out by a Palestinian commando in the 1972 Munich Olympic Games, in which all Israeli athletes who were taken hostage died, the UN General Secretary was instructed that same year by the General Assembly to "study the underlying causes of terrorism and violent acts which have their roots in misery, frustration, injustice and desperation and which drive certain people to take human lives and risk their own to achieve radical change."

[VIII]   Claire Sterling's book "Le réseau de la terreur: enquête sur le terrorisme international" (1981), could be considered part of the ideological campaign aimed at constructing the aforementioned stereotype of international terrorism as a plot. According to Sterling, terrorists are "professional assassins responsible for death in 44 countries and four continents: they are methodically trained, heavily armed, immensely rich and have strong backing, they move cold-bloodedly from country to country and are able to capture the attention of the entire world". After the preliminary session of the Tri-continental Conference in Havana in January 1966, the Soviet Union would have become Cuba's principal source of support, forcing Fidel Castro to leave his international politics in the hands of the Kremlin and abandon his posture of indifference toward the Palestinians. Sterling argues that "many ethnic communities (Irish, Welsh, Scottish, Cornish, Flemish, Walloon, Serbs, Croats, and Kurds) are members in every respect of the international terrorist community, content and surprised to find within it arms, money, refuge and training". Until 1991, researchers of international terrorism were very wary, if not hostile, to the theory of a Soviet inspired plot. Wiewiorka (1988) states that the idea of terrorism directly inspired and co-ordinated by a central state does not stand up to scrutiny when applied to the Soviet Union (or to its Bulgarian, Cuban or other annexes) in comparison to certain countries of the Near East.

14 The Reaction of the Centre: Anti-Terrorist State Nationalisms

Long term peripheral violent nationalisms bring about two types of responses from the centre: government repressive reactions of an instrumental nature and identity reactions of a centralist nature which are likely to cause, in some cases, new state anti-terrorist nationalism. The projective image of peripheral nationalism as the internal enemy, the object of stereotyping and stigma, depends on the intensity of the centralist reaction.

In Third World vertical states or those which have emerged from a process of fragmentation of broader political set-ups, the hostility of the centre may lead to persecution, pogroms or ethnic massacres. Horizontally integrated states in the West tend to restrict both the repressive measures and group hostility by channeling them, in those cases where a system of nationalisms exists, toward the armed group and its social environment.

Anti-terrorist state nationalism stresses pluralist, pacifist and humanist values, directly counterpoising them to those of violent peripheral nationalism. Government response in social, media, police, justice and international spheres is presented as rational; its repressive measures are not defined as indiscriminate, rather as subject to the principle of anti-terrorist preciseness. It is claimed that a different approach could lead the state apparatus of coercion into a downward slide of uncontrolled violence toward a military coup, thereby endangering its democratic base.[1]

However, although the normative discourse insists on the combination of efficiency and legality, all the factors involved conspire to separate one from the other. At the core of the coercive apparatus, where its members are permanent targets for armed groups, corporate responses emerge which breach legal boundaries. The search for short-term efficiency drives those responsible for the anti-terrorist struggle in the same direction. The hostile polarisation of the nation-state society, which is itself under the threat of terrorism, conceals this rupture with second degree legitimation which is expressed, not through a direct and straight forward defence, but rather through a denial of the facts and in a silence bathed in tacit approval. For their part, the authorities tend to resolve the crisis of loyalty within the coercive apparatus by letting them free to act, if they are not coordinating their response. In those cases when there is a community confrontation, in Northern Ireland for example, group violence of the majority ethnic group benefits from the support of the coercive apparatus of the state.

This does not prevent the state from adhering to the democratic discourse which lends rational legitimation to anti-terrorist measures. In this way, when it occurs, the "dirty war" is invisible, except for its victims, and takes place entirely underground. If the armed peripheral group mirror-images the most ruthless aspects of state response, that is its power over life and death, in its turn the underground state dirty war mirror-images peripheral terrorism in its most cruel and inhuman actions. Contrary to the number of torture victims, fatalities tend to be relatively few in the Wests "dirty wars" on account of the compulsory invisible nature of this type of war; in fact, they are fewer than those who die at the hands of violent nationalisms. However, given that the activities of those engaged in the "dirty war" are secret, they are not subject to the "social approval" of any community, which is the case of the armed group. In this way, mirror-image violence reaches its climax, where bodies are twisted in silence, on the border of horror in its purest form.

Peripheral collectives who are victims of this war close themselves down to the discourse of the centre which labels them as violent. The double moral standards of the state, evident to them, predisposes them to accepting the armed group's breach of the thresholds of violence, however indiscriminate and hateful the attacks may be. The spiral of reciprocal hostility which this initiates broadens the gap which separates them from the outside world.

For this reason the factor which impedes the cogwheels driving the two mimetic circles of violence forward can only come from outside of both parts, that is peripheral violent nationalism and anti-terrorist state nationalism, emerging either as international pressure or as social movements located outside of the opposing parts.

Normative Anti-Terrorist Policies: Communicative, Penitentiary, Police, Judicial and International

The consolidation of violent nationalisms leads governments to explore multiple approaches in an attempt to determine the nature of the anti-terrorist struggle. The main questions revolve around the type of terrorism which the state is confronting and the response it requires. Should counter-insurgency be based on the application of repressive, police and judicial measures, or should it be a battle waged to win over public opinion in the society affected by terrorism? Should the mass media have a relationship with those leading the anti-terrorist campaign and, if so, what kind of relationship? What policy should be adopted to secure international support? Should the armed forces be involved and, if so, to what extent? Or should the campaign be entrusted to special police units? Finally, should repression be restricted to members of the armed groups, or should it include the civil society which is socially and ideologically related to it and, if so, to what degree? What policy should be applied, if any, to those members of the armed group who wish to dissociate themselves from the group?

Counter insurgency strategies developed in the 1950s and 1960s with a view to defeating anti-colonial liberation movements were not particularly successful in achieving their aims in the context of violent, ideological or nationalist groups in the West.[11]

At the beginning of the 1970s a line of thinking established by the British writer Moss (1972) and developed by Wilkinson (1976) highlights the specific nature of strategies for combating urban guerrilla organizations which have social support - in the West, those armed groups which are nationalist - strategies essentially based on psychological operations which aim to prevent the alienation of the population. Moss argues that terrorism's efficiency is rooted in the response of the community to its activities, which, in its turn, depends on what approach the government adopts with regard to the situation. He therefore underlines the importance of "not provoking public hostility in order to isolate the guerrillas and cut off their sources of support. It is crucial to restrict the number of people considered to be an enemy to an absolute minimum".

According to Wilkinson, it is risky for a government to announce its intention to "put an end" to the subversive movement. Such a stance would leave no room for political maneuvering or for the option of winning over the rebel elements through dialogue. He approves of government passing anti-terrorist laws, but rejects the absence of procedural guarantees, and categorically disapproves of the use of troops in anti-terrorist activities, that is, 'outsiders sent to the area to carry out an unenviable task, living in bases segregated from the population'. For this reason, he supports a "third front," that is anti-terrorist forces selected from special units within the army which are equipped with "generous powers of intelligence".

> In vertically structured societies of the Third World it is the army which carries out the bulk of the anti-terrorist struggle. In the past it was the Ottoman armed forces which confronted the ORIM in Macedonia. Thereafter, it was the army that entered into combat against Palestinian "sanctuaries" in Lebanon, who took by armed force the Sikh Golden Temple of Amritsar, who forced the Tamil Tigers out off the Jaffna peninsula. It was the Turkish and Iraqi armed forces who have carried out massacres of Kurdish people in both states, and it is the Russian military who have destroyed entire cities in Chechnya. However, it is not only in these countries where the army has been involved in this type of activity; in one of the most advanced states in Europe, in spite of expert advice, military intervention in a national conflict has taken place. This is the case of Northern Ireland, where in August 1969, on account of the pro-unionist bias of the Royal Ulster Constabulary (RUC) British soldiers were deployed to form a "neutral security barrier" between the two communities. The reality of the situation soon revealed the hypocrisy of this perspective; from the time of their arrival the British troops "fraternized" with unionist groups as both shared the same goal of defeating the IRA. It was British paratroopers who were responsible on 30th January, 1972 for the massacre of Catholics in an event that became known as "Bloody Sunday" and which further deepened the conflict. Since they became a key target for the IRA, members of the British army have been providing logistical support to loyalist paramilitaries.
>
> If in Spain control of the anti-terrorist campaign was not handed over to the army during the transition period, it was due to the historic memory of the role of the army during the civil war years and the Francoist dictatorship which followed. In March 1981, one month after the failed military coup, the government decided to assign the task of anti-terrorist surveillance along the Pyrenean border to the armed forces. However, its own apprehension, rooted in the aforementioned memory, restricted the nature of military involvement and led to the selection of the Civil Guard as a "third force." The socialist

government pursued this policy throughout its mandate from 1982 to 1996.[III] In any event, military intelligence of the CESID has continued to play a role in the anti-terrorist struggle up to the present day.

The normative response of western states which are confronted with this type of conflict is based on a controlled anti-terrorist policy that tends to exert moderate pressure on the population which supports the armed group. This policy must be put into practice by loyal and disciplined personnel. Police repression must run in tandem with a media strategy, with techniques aimed at undermining the unity of the group (through dispersion of political prisoners and their reintegration into civil society), with a coordinated approach in the judicial and police spheres and with the search for international support.[IV]

This anti-terrorist approach is based on no concessions or dialogue with the armed group. The authors cast doubt on the usefulness of governments putting forward a solution to the causes, whether they be social or national, of the political violence, even if the predominant line is negative. Redressing injustice, according to Reinares (1996), has only resulted in the reduction of violent options in those situations where there is a lack of popular support for them (as in Quebec at the beginning of the 1960s or in the United States in the 1970s). All dialogue between state representatives and armed groups, it is argued, is interpreted as a sign of weakness on the part of the government, providing legitimation to the group to the detriment of democratic legitimacy. This does not exclude exploratory contacts or the opening of channels of communication.

The strategy of the state with regard to the mass media aims to deprive the armed group of public support in the national, state and international spheres. Communication, it is said, must counterpoise democratic values defended by anti-terrorist measures to the anti-democratic posture of the armed group in question. Although censorship may be efficient in the short term, it is counter productive in the long term; more effective are informal agreements which respect the freedom of the press.[V]

According to these authors, reintegration and dispersion measures aim to undermine the internal cohesion of armed organizations. Reintegration, consisting in measures which alleviate the prison regime, must be followed up by the prisoner's reinsertion into society and separating him from his group. For this reason it is important to continue with, even though it contributes to the radicalization of the intransigents (hard-line prisoners), to campaigns aimed at bringing attention to the violation of prisoners' rights and the suffering experienced by their families, as well as extreme measures such as hunger strikes and attacks against prison officials and guards (Reinares, 1996). The policy of dispersion, which runs in tandem with reintegration, prevents the concentration of prisoners as autonomous groups within the jails; it aims to undermine the community network which is based on solidarity ties that unite the group.

Reintegration must bring about a change in the identity markers of members of the armed group and, therefore, "repentance" for his or her past activism, which brings about the substitution of the morality of the armed group for the morality of the state. Some states involved in a confrontation with armed groups demand that the individual condemn their former colleagues. (In Italy, the creation of the figure of "repentant terrorist" became during 1981-1982 a crucial means of defeating armed organizations: the "Law of the repentant" stipulated the means whereby sentences would be reduced in relation to the importance and sincerity of collaboration. In Spain, although the legal aspect of repentance-denunciation remained in force, the protest of members of organizations which opted to dissolve, such as ETA-pm, against this policy made it sufficient to attain the reduction of sentence to express the will of renouncing political violence).

Prison dispersion has affected Basque activists much more than their Irish counterparts. In 1994, of the 700 Irish prisoners in jail, 600 were in Northern Ireland, 35 in the Republic, and only 35 in Great Britain. On the other hand, in Spain at the beginning of the 1990s, of the 535 ETA prisoners, only 55 were in jails in the Basque country, the rest were dispersed throughout the state (including prisons in the Balearic and Canary Islands as well as Spanish dominions in Africa). In terms of the security status of prisoners, only 40 were classified as level three, the lowest level; more than a hundred were level two, although they were not allowed to spend the night outside of the prison; and 180, the "intransigents" were level one, the toughest regime, in which they were allowed no benefits.[VI]

Police activity, assigned to a "third force", - even though it is coordinated by bodies in which the armed forces and state security are involved -, must demonstrate its efficiency in terms of infiltrating groups, detention of activists, and particularly in preventing deaths, attacks and kidnappings. Its specific tool is intelligence: the dissolution of these forces, it is said, led in certain cases, Italy in 1976, Spain in 1978 to 1980, to an increase in attacks. Reinares claims that mass detentions which aim to compensate for policy inefficiency have counterproductive effects in this respect; the British and Israeli cases prove that prolonged detentions without legal back-up, as well as internment without trial, radicalize those detainees and their friends who were not members of violent groups to the extent that they may seriously consider joining the movement.

Judicial measures involve the promulgation of anti-terrorist legislation and special courts, which, according to those responsible for these politics, are necessary to prevent unjust acquittals provoked by the intimidation of witnesses, and necessary also to protect the administration of justice from violent groups.

From 1970, anti-terrorist laws and special courts became commonplace in western Europe (both are in force in Spain; in Italy anti-terrorist laws are applied by ordinary courts; France and Germany have a mixed regime). The United Kingdom is a pioneering state in putting the finishing touches to exceptional penal codes. Internment was introduced in 1971 in Northern Ireland by virtue of emergency legislation, a measure followed in 1973 by the introduction of Diplock Courts. In Italy exceptional measures have undermined constitutional guarantees; the *Reale* law of 1975, entitled the "Law for Safeguarding Public Order" introduced a police logic into the judicial and penitentiary spheres. This law was approved by parliament on account of the support of the Italian Communist Party, and indicated the beginning of support of the European Parliamentary Left for emergency laws.

From 1976 onward, West Germany joined in the anti-terrorist struggle and within a short space of time came to lead the field in creating a single European police framework. The first German emergency law, which goes back to 1976, stipulates automatic imprisonment for those on trial for terrorist crimes as well as tight control over contact between prisoners and their lawyer; since 1977 the principle of isolation of the prisoner from the outside world has been in force. In parallel with these developments, the BKA (vice-ministry of criminality) set up in Wiesbaden the largest data base in the world. This data base, which by 1977 already held the personal details of more than five million people, including their finger prints, photos, etc.,, has been used not only by Germany, but by all European states, according to Vervaele (1985).

The Anti-Terrorist Law, covering crimes committed by armed groups, passed by Spanish Congress in July 1978, which prolongs the period of detention to seven days, took its lead from the aforementioned models (criticised by lawyers for contravening fundamental civil rights and for its discriminatory treatment of terrorism in respect to other crimes).

Of necessity, the transnational consequences of terrorist acts bring about, according to this line of thinking, international collaboration, which, according to Crenshaw (1983), is determined by the extent of co-operation between governments, by the convergence or otherwise of their political and economic interests and by the degree of *entente* existing between states and judicial authorities. This translates into the signing of agreements, such as the Schengen Agreement, that allows for cross-border police activity, extradition pacts and in the search for third states willing to take in potential deportees.

The theory of the "two wars" on which this anti-terrorist approach is based is aimed to win over the support of the ethno-national collective, thereby isolating it from the armed group. The "psychological war" waged in the sphere of public opinion has, in this respect, a double meaning: if, for armed groups, "to resist is to win," then the counter insurgency must resist the will of those determined to bring about the attrition of the state; on the other hand, the "psychological war" is understood as one of the tools, the most important one, in the battle to win over the support of the population.

In fact, the "psychological war" deals in a different way with the different nationalist tendencies within the ethnic group. In those cases where a system of nationalisms exists, the principle of "precision" reserves for systemic nationalism the strict application of human rights while, at the same time, it tolerates the violation of these rights in an exclusion area, where, in addition to the armed group, its legitimating community is located.

Plan ZEN (Special Northern Zone), set up in 1983 by the Spanish government, established the principle of "suspicion" as a determining criteria for government activity both in the Basque Country and Navarre with regard to the armed group and its social environment. Plan ZEN (which included the Basque Country and Navarre) set out as its main political objective "the political isolation and the weakening of the terrorist organization". The chapter describing "legal measures and juridical actions" deals extensively with "collaborators", announcing possible sentence reductions and pardons, and claiming that the figure "introduces an element of suspicion and psychological distrust into the heart of the criminal organization possibly contributing to its destruction". The chapter dedicated to "information and intelligence" proposes, with a view to identifying terrorists and "controlling" the population, the use of infiltrators through workers, such as porters, waiters, wallpaper hangers, who on account of their job are in contact with people. The chapter focusing on "psychological action" describes the measures for dealing with the "terrorist organization, collaborators and sympathizers"; some of the measures described might fall into the figure of "coaction" typified in criminal law. It openly favours the use of "black propaganda" through news which "as long as they are credible will be exploited". It proposes, with respect to sympathizers "sending them warnings, threats, etc., and telephoning them at untimely hours to keep them in a constant state of anxiety and uncertainty". The eighth chapter, devoted to "security and self-protection" contains a description of the criteria which a police agent should refer to when considering whether an individual is suspect; the vagueness of the criteria means that most young people in the Basque Country would fall into this category. It refers to factors such as short-term residence in their respective abode, being young, not living in a nuclear family, attempting to be anonymous, not taking part in family events and, even, the fact that one of the family members leave the home without just cause.

Anti-Terrorist State Nationalism: Divergence Between Ideology And Identity

If the appearance of all peripheral nationalisms provokes a centralist reaction on the part of the nation-state, this is at its most extreme when nationalism uses violence. Whether the reaction is so intense that it is permissible to speak of a new anti-terrorist state nationalism depends on a number of factors: the threat which peripheral nationalism represents to the territorial unity and integrity of the state, the intensity of the violence and nature of the thresholds crossed by violent activities, the vulnerability of state structures, the relative economic, political and cultural importance of the periphery in question with regard to the centre, the circumstances of the peripheral territory in terms of its distance from the political centre of the state or its insularity, the space occupied by violent nationalism in the system of peripheral nationalisms, the degree of unity or fragmentation of the armed group, etc. And of course (although this has not been the case in the West) on whether the conflict has led to an open war between the state and the periphery.

In Western Europe the insularity of Corsica, its disproportion in relation to the centre and the low intensity of nationalist violence has restricted its repercussions on French nationalism. The high intensity of violence in Ulster was counteracted by its isolation and the solidity of the British state, due to which its repercussions on the latter have been containable. The vulnerability of Spanish democracy, outcome of a transition from a dictatorship whose state apparatus remained intact, together with the economic importance, although this is in decline, of the

Basque Country, and the intensity of ETA's activities, has meant that the repercussions have been significant for Spanish nationalism. It is for this reason that Spain is the best place to study a phenomenon which is, in reality, universal.

Ideological and identity transformations affect both the elites and the masses. The political, social and cultural elite of the nation-state, which prior to the appearance of nationalist violence appropriated the meaning of values to the detriment of the periphery, now assumes a range of new counter values. Where radical nationalism exalts violence as the birthplace of the nation, state elites proclaim the virtue of pacifism as a value opposed to destructive violence, and of humanism as the supreme value which demands respect for human life. Where the violent group-state brings about the fusion of all aspects of power into a single element, national elites exalt the values of tolerance, the democratic conventions on which civil rights are based and the moral superiority of pluralism. The obsolete nature of apologetic theories of revolutionary violence allows elites to proclaim to be the only lawful holders of reason, in contrast to the meaningless dementia of terrorism.

The stereotypes of the periphery, at work since the beginning of the construction of the nation-state, may undergo modification in two respects. On the one hand, the intellectual elite of the state accentuates the devaluation of all that is related to peripheral nationalism, the characteristics of which are ridiculed and identified with irrationality; on the other hand, systemic nationalism's rejection of the radical tendency toward violence facilitates its partial recovery through pacifism, humanism and pluralism, which establishes the bases of a united ideological front with the centre.[VII]

Anti-terrorist reaction results in ideology and identity positioning themselves in the centre in opposite directions, for which reason the two pieces of the *Gestalt* are no longer soldered together. While the elites focus their ideological discourse on the values of peace and democracy, the masses - and the elites - become increasingly and hostilely polarised against violent nationalism, which, becoming the internal enemy, comes to occupy a growing space in their identity dimension.

All these factors contribute to polarisation: the state population believing itself to be the victim of an undeclared war the impact of which may be felt at any moment; national stereotypes which emerge in conflict situations, now acquiring their most virulent character; the international image of terrorism as what is universally abhorrent. Furthermore, state violence is felt in a national conflict the more that this is distanced from the centre, but not in the "eye of the hurricane," which is the centre itself.

The receptiveness of the masses to the "psychological war" directed by those responsible for counter insurgency is at its highest point, and the mass media broaden its effect. Given that armed groups carry out activities defined as crimes in legislation throughout the world, for example, murder, robbery, kidnapping, etc., the hostility of the media with respect to the armed groups, shared by the population, simply forms part of what Wiewiorka (1988) calls "the language of common sense." However, the lay out is more complex, as counter insurgency does not only involve war against the armed nucleus: it also involves combating its legitimating community, including its political ideas (punished as "apology for terrorism"), electoral choice, socio-community structure and even its lifestyle. The media treatment meted out to this community corresponds to the laws of the stereotype. And, like all stereotypical messages, although one section of the information is consciously distorted and manipulated (on occasion, according to the logic of black propaganda), the other, the largest section, is unaware of its use of the stereotypes and conceives of itself as independent and objective.

The projective formation of the figure of the internal enemy (in the sense in which Horkheimer and Adorno use this concept (1974), that of false projection) facilitates the incorporation of the masses into the psychological war, making it on occasion into a factor as functional in terms of nation-state internal cohesion as the figure of the external enemy had been during the historical construction of the early nation-states.

However, the peripheral internal enemy is in part real, the armed group, and in part imaginary, the legitimating community. For the latter, the discourse of peace and tolerance often constitutes, as Marcuse (1969) explains with respect to the language of law and order in modern democracies, repression itself. He writes, "This language, far from restricting itself to defining the enemy and condemning it, constitutes it, and in this way the enemy created does not appear as it is in reality, but rather as it should be so that it may fulfill the function attributed to it by the established order".[VIII]

Anti-terrorist psychosis is, in part, a spontaneous reaction by the masses against the brutality and inhumanity of the armed group's attacks; but it is also a sought for effect by government which aims to resolve, in this way, the crisis of political integration caused by its failure in the economic, social, international, etc., spheres, or the crisis of state cohesion generated by its vulnerability, its history or by the struggle among political parties or different factions for power.[IX] It is in this way that anti-terrorism become a legitimating mechanism of the state; and Girard's (1972) anthropological theory of the sacrificial victim as the replacement victim assumes its full meaning in the political sphere.[X]

In this way the circle is closed; if terror is the paroxysm of the affirmation of oneself denying the existence of the other, from where the absence of dialogue and relation with death derive (Braud, 1993), the logic of the enemy dictated by the denial of the other is shared, on the one hand, by the state and its national society, and on the other hand, by the armed group and its legitimating community.[XI]

Hostile identity polarisation proper to anti-terrorist state nationalism leads to a situation which denies the proclaimed values of pluralism, humanism and pacifism. This double schizophrenic morality spreads through the elites and masses giving rise to a double discourse, that which is private-group in nature and that which is public-official. It is often in the hiatus existing between ideological discourse and the feeling of identity that the anti-terrorist dirty war is engendered. It initially emerges out of the corporate reaction of members of the coercive state apparatus to the attacks of which they are the target, the "dirty war" being consented to, and later, coordinated by the authorities, who make it into a second degree political instrument.

Anti-Terrorist Violence and its Thresholds

The dirty war, like peripheral nationalist violence, has its own thresholds. To cross them it is necessary that those responsible can rely on widespread legitimation, denied in the discourse but affectively approved, provided by the ideology dimension of state anti-terrorist nationalism. These thresholds are: illegal use of legal means (torture as a method of interrogation); permanent emergency legislation in force in the territory of the peripheral group; uncontrolled violence on the part of the state coercive apparatus; state terrorism, carried out by mercenaries or police; an open war against the ethno-national collective; and the military coup, accompanied by the imposition of an authoritarian regime throughout the state.

In the case of violent centre-periphery conflicts the latter threshold, the *coup d'état*, should be disregarded (unless the democratic regime is vulnerable and the leaders of the coup use this as justification). This arises in response to widespread revolutionary and social violence in which the state coercive apparatus confronts the masses, perceiving of them as the enemy.[XII] However, it is precisely in peripheral conflicts that an alliance emerges between the coercive state apparatus together with the population of the centre against violent peripheral nationalism. These conflicts can lead to open war in those vertical societies where the state has been constructed on the basis of discrimination (or total exclusion) against an ethno-national group, as is the case in the Kurdish, Palestinian, Tamil, Chechen, etc., conflicts.

In the west, the anti-nationalist dirty war always stops in the phases prior to the last two thresholds; it cannot become large scale at the risk of revealing the disjointed nature of its different elements, heavily eroding the democratic image of the state. The characteristics which guide it, therefore, are invisibility, quantitative contention and the possibility of attributing these actions to individual groups.

In some conflicts in the West where there is an ethnic confrontation between communities, or specific situations such as insularity, out of the community that identifies itself with the state armed "vigilante" groups may emerge which use violence against nationalist violence. At the beginning of the 1970s in Northern Ireland paramilitary groups emerged which acted against the Catholic community; these groups were the Ulster Defence Association, the Ulster Volunteer Force, the Red Hand Commandos. In Corsica from 1977 onwards, the *barbouzes* of different groups, such as the Action Front Against Independence and Autonomism (FRANCIA) have acted in response to FLNC activities with bombings and the kidnapping of Corsican activists, some of whom, as in the case of Orsoni, were killed.

However, the principal roots of the dirty war are in the corporate reactions of members of the state coercive apparatus. Many causes generate these hostile reactions in their ranks: the tendency of all police forces to align themselves with the centre in conflicts against the periphery; the feeling of isolation experienced in ethnic territories by those who are engaged in combat against armed groups; personal hatred provoked by the fact that the "war against the state" has led to attacks against them personally; and the high level of impunity enjoyed by these deviations from legality, since the authorities fear, and some say it openly, that the application of penalties to the armed and police forces adds the feeling of grievance to the real risks undertaken by them in the anti-terrorist struggle, resulting in low morale... If all this is taken into consideration together with the pressure of public opinion to obtain concrete results in the struggle against terrorism, it is easy to conclude that the first threshold crossed is that of torturing detainees. In long-term violent nationalist conflicts, both in the Third World and the West, police torture of detainees is the rule and not the exception (a fact confirmed by reports produced by neutral organizations such as Amnesty International); the prolongation of the period of detention under emergency legislation facilitates this type of treatment.[XIII]

Torture is defined as a crime in penal codes, for which reason its practice is a source of embarrassment and never publicly acknowledged (although in certain situations it is admitted: the High Court recently allowed the Israeli police to exercise "moderate physical and psychological pressure" on detainees). Nevertheless, there is a certain consensus on the usefulness of its application in some cases, which is expressed in some juridical spheres via a theory on the basis of which certain conditions, determined according to the whim of the torturers, would be exempting circumstances from penal responsibility (Ibarra, 1987).

The next threshold is "uncontrolled" police terrorism, in other words, that terrorism which is not desired by the government. This refers to violent activities, carried out by members of the police or armed forces, on occasion in collusion with extremist groups, which are directed against political adversaries or the civilian population. The frequency of these "uncontrolled" actions is directly related to the degree of psychological insecurity experienced by members of the security forces or extremist groups. For this reason it is particularly intense during periods of transition, when the political or personal future of the security forces hangs under a cloud of uncertainty; in these circumstances the hostility of the ethnic population within which they operate is perceived as particularly threatening.[XIV] "Uncontrolled" violent actions tend to be spontaneous, in response to concrete political and social events, and rarely do they result in fatalities, although these are not unknown.

Instrumental terrorism and strategic *coup d'état* terrorism cross other thresholds. These types of terrorism aim, by carrying out deliberately bloody and indiscriminate attacks, to create an atmosphere of disorder and chaos which incites, or justifies, the need for the state apparatus to engage in a coup in order to assume political power.[XV]

Instrumental terrorism, in contrast to the two previously mentioned types, is under the control of authorities. It feeds off corporate reactions of the apparatus of state violence; but these are not spontaneous and born out of the "base," rather they are coordinated and directed by the authorities.

This kind of terrorism has a double objective. The first consists in carrying out counter-insurgency operations (such as putting an end to the "sanctuary," real or presumed, which the armed group enjoys in a foreign state, or obtaining information by kidnapping, torture and murdering some of its activists). The second consists in maintaining the loyalty of these governmental apparatus, eroded by tensions to which they are subjected, relieving these tensions by giving them free reign in their activities (it is this objective, not the first one, which is the more functional of the two). In the West giving the green light to these operations requires widespread social acceptability in terms of public opinion, based on the vague notion that "something is going on" and a consensus not to reveal the details of these activities, since if they were made public there would be general condemnation of them. All of this makes it important to contain, in terms of time and the number of attacks, this kind of terrorism, without which tacit legitimation could not be maintained, particularly if the international repercussions of some of the activities make their immediate continuity impossible. The complete invisibility of these facts means that they acquire an even greater degree of cold bloodedness than that which is characteristic of armed groups. This factor, taken together with its instrumental nature, allows the term state terrorism to be rightfully used in this case.

This type of terrorism may be carried out by the coercive apparatus of the state itself, or by mercenaries. Mercenary terrorism is not autonomous, it is one more piece in global politics, and its authors, whether they are politically-minded or not (usually common criminals) work for a salary and under orders of the higher echelons of the state apparatus, who control the internal logic of their activities. It is particularly in this case where the law of informative functioning, as described by Baratta (1985), is evident: the image of terrorism, he states, "is a product of manipulating the flow of information related to it, information that is subject to mechanisms of selection [...] being both ignorance on right-wing terrorism and a relative abundance of information on "red" terrorism evidence of a single phenomenon: manipulation of the flow of information".

> In Spain, "mercenary" terrorism has had three phases, each of them driven by the Basque conflict. The first phase stretched from the beginning of the last state of emergency under Franco, in April 1975, until the first few months of 1977; the violent activities which took place during this period were claimed by an organization called Anti-Terrorism ETA (ATE). The second phase ran from the summer of 1978 until the end of 1980, ending slightly prior to the attempted coup on 23rd February 1981; the group which claimed responsibility for violent attacks during this phase was the Basque-Spanish Battalion. The third phase, which was dominated by Anti-Terrorist Liberation Groups (GAL), began, after ETA-pm kidnapped and murdered a captain of the Army, with the kidnapping and disappearance of two Basque refugees, Lasa and Zabala, from Bayonne in October 1983. This phase ended suddenly in summer 1986, in tandem with the French police's expulsion of Basque refugees to Spain. Of GAL's 26 fatalities, a fair number were French citizens who were not involved in the conflict in any way. This terrorist group set itself two instrumental objectives: the break-up of the refugee community,

which it did not achieve, and the creation of a climate favorable to the latter's expulsion from France (which, in fact, it achieved from 1984 onwards).

Information on the activities that took place during the first two phases (1975 to 1978 and 1978 to 1980) which began to be released by the media from 1985 onward, was based on inquiries carried out by the French and Italian judges. Those who were revealed to be in charge of mercenary plots were identified as persons linked to Spanish police and army intelligence services. The third stratum, the lowest, was composed of authentic mercenaries, that is, by criminals recruited from the underworld in European or North African countries close to Spain, particularly French thugs, who worked only for a pre-agreed amount based on the job to be done. This was the most vulnerable strata, whose members were most likely to be detained and put on trial, above all in France in 1979 and 1980.

The intermediate strata was composed of ultra right-wing elements, the cast offs of various failed political processes (soldiers of the last colonial wars of France and Portugal in Africa, right-wing activists linked to Latin American dictators or North African countries) who from the beginning of the 1970s had been flooding into a state whose Francoist regime suited them. Of particular note among these people were the Italian refugees who had led the wave of "black" terrorism in their own country. This stratum, which had to co-ordinate its activities with individuals from the underworld, appeared to be particularly protected by those in command; in reality, there were very few detentions of its members.

It was the absence of this second strata in the third phase of "mercenary" terrorism, that is, the activities of GAL, that forced the commanders, the highest level, to become directly involved in the recruitment of members of the underworld, and which thereby led to court cases, from 1987 onward, against them. Their late statements opened the door to proceedings against those responsible in the Ministry of the Interior in 1994 and thereafter.[XVI]

The Double Discourse of the Centre and the Entrenchment of Radical Nationalism[XVII]

The use of illegal anti-terrorist practices that are not officially acknowledged creates the conditions which facilitate the corruption of those responsible. The funds used to finance intelligence services, which must be secret due to their very nature, as they are destined for secret operations such as the creation of networks of informants and missions carried out abroad, become doubly secret when there is a dirty war involved. This situation favors and facilitates the illicit enrichment of those who administer the funds and their subordinates (who justify themselves in terms of the personal risk which their profession compels them to run), protecting them with a cloak of secrecy over the years.

If the corruption reaches high levels, it gives rise to activities which are no longer directly related to the anti-terrorist struggle, such as charging illegal commission on administrative contracts. Sooner or later the spread of corruption reaches a point when it can no longer be hidden. The state-national society, tolerant with regard to the dirty war, is nevertheless not prepared to accept the corruption of those responsible for counter-insurgency, which de-legitimises those authorities which have not persecuted, have tolerated, or benefited from, these activities.

Revelations of the dirty war and the corruption which flourishes around it end up becoming a weapon used by both those state-wide political parties in their struggle for power and by the different media in their attempts to increase the size of their public. In effect, the revelations, more than being an expression of a centre-periphery conflict, become an internal question proper to the centre. While a strong political and parliamentary government prevents these criticisms emerging into the light of day, or waters them down, they will explode in the face of a weak government. It is these questions, related to the struggle for power in the centre, that explain the prolonged silences and long periods of time between the acts being committed and their becoming public knowledge.

Nevertheless, once this has occurred the publicity forces political parties, the mass media, judicial figures, and the elites and masses to assume a position. The authorities' argument in favor of the presumed innocence of those involved is accepted or rejected depending on political party alliances or rivalries. Some media (which are in turn attacked by their rivals) take pains to reveal political responsibilities through testimonies made by secondary players in the plot or former activists who have moved on and are now abandoned. The dividing line is not clearly drawn between support for and rejection of the governing party, since there are non-civilians implicated at the highest levels of counter insurgency activities. If systemic nationalism has maintained a tenuous critical position in relation to the dirty war, it now intensifies it. Judges who instigate proceedings are subjected to continual pressure by the authorities and certain media, who accuse them of wanting to become "star judges". In any case, the entire community in the periphery is aware of the differential treatment meted out to members of armed groups and those charged with involvement in the dirty war, who are rarely placed in custody.

These developments cause division among intellectual elites of the centre; some back the critics, while others look the other way. In any case, almost all stick to the Manichean discourse which places the violence of terrorists on one side and the peace of democrats, supported by mass uproar that surfaces every time the armed group carries out a bloody attack, on the other.

Society witnesses the performance sickened by the revelations of corruption, angered - but not terribly - by what is made public about the "other terrorism", and astounded (and occasionally amused) by the vicissitudes of some individuals involved in this tragic-comedy. In any case, the inquiries opened, although they hesitate to cross certain thresholds of the hierarchy of responsibilities, create a catharsis within the state-national society, which is no longer inhibited by the feeling of being an accomplice on account of its shock at certain aspects, the most macabre, of the dirty war.

The revelations in the media and political party condemnation may diminish if there is a change of government. The new government, in control of the counter insurgency apparatus, begins to favour, if not in its discourse, then in practice, a policy of shelving responsibility for investigated dirty war. The task of denouncing the facts is restricted to a small section of the media which are increasingly isolated, a few judges, who come under growing criticism, and by those political parties which have no chance of being in government. With this situation, the identity and ideological elements of the state anti-terrorist *Gestalt* once again close ranks in a distorted way. (One piece, nevertheless, remains outside of this set up: pacifist movements which arose in the heat of the anti-terrorist pacts, previously polarized against violent nationalism, that now take sides against past, present and future expressions of the dirty war. These movements, on converging with symmetrical movements that surfaced in the heart of radical nationalism, lead to a new indirect perspective on how to tackle the centre-periphery conflict).

The normative anti-terrorist policy had previously entrenched the armed group and its legitimating community. (The policy of dispersion of prisoners, if it does not violate legislation, diverges from the principle that the prisoner should be in a jail near his or her place of origin. This policy, leaving aside exceptional cases, does not undermine the armed group, but it does intensify the destitution and loneliness of the prisoners and increase their family and loved one's suffering. It also leads to a deep-rooted resentment against those responsible for this policy and the political parties which approved it).

The hypocritical language and morality inherent in the dirty war, once made public, reinforce the hostile polarisation around identity markers. The accusations made at the time by the radical sector, acquainted with all of this from the outset, being the principal victim, were mainly ignored or dismissed as policy of false condemnations. Now that the activities, although not all of them, have been made public, an official discourse is reformulated according to which only "democrats", and not "supporters of violence," that is, those who suffered from these activities, are in a position to condemn them. The frivolity with which these events are dealt with by many sectors of the media and the cynicism that makes them into political footballs infuriates radical nationalists, a feeling which runs in tandem with humiliation and suffering.

The divide separating radical nationalism from systemic nationalism widens, the former reproaching the latter for its initial inhibition and its current lack of steadfastness with respect to the dirty war. This prevents for the present (perhaps not for the future), any instrumental rapprochement to radical nationalism, which the common stand in terms of rejecting anti-terrorist state nationalism would otherwise have made possible.

Paradoxically, the image which anti-system nationalism has of itself improves. It considers the fact of openly acknowledging the need for violence, in its opinion the inevitable result of national oppression, to be morally superior to the violence which the state has engaged in, sometimes with greater intensity, while hypocritically denying its existence. Radical nationalism alleges that kidnappings and revolutionary taxes serve the financial requirements of the cause, not the personal interests of armed activists, whose Spartan lifestyle is counterpoised to the corruption of those involved in anti-terrorist activities. Above all, radical nationalism closes itself down to the state's discourse on peace and tolerance, the sound of which it mentally situates in the cells where torture victims and detainees who have disappeared are held.

The divide widens, the figure of the evil and hypocritical external enemy occupies the identity sphere in its entirety; and the legitimating community comes to accept any breach of the threshold of violence which the armed group engages in. As a consequence this community becomes increasingly abhorrent from the population of the centre, which convinces itself that the dirty war is finally over, and points out that the number of fatalities is much lower than that which the armed group is responsible for. In this way, the game of mirrors multiplies *ad infinitum* the masks of violence.

Notes

I   Contemporary violent nationalist conflicts in the West have not led to this situation. It has arisen only in exceptional cases of transition from authoritarian regimes where the coercive apparatus has been maintained intact, which it has tried on occasion to utilize the pretext of the terrorist threat in an attempt to return to the previous regime (as was the case in Spain in 1981), but unsuccessfully.

II   Counter-insurgency theories developed from the 1950s onwards by British, French and U.S. theoreticians, experts in the anti-guerrilla campaign in the Third World, were based on the following principles: i) removal and resettlement of local populations to ensure greater control over them; ii) development of specialized anti-guerrilla units; and iii) updating intelligence strategies, including torture, in order to undermine the insurgents' political structures. The U.S. expert Blaufarb (1977), a former CIA agent, notes the disappearance of this approach in the 1970s on account of its failure in Indochina.

III   The line of thought which inspired this policy was outlined by Lieutenant Colonel Jáudenes in a Congress on "International Terrorism" held in the Institute on International Questions (ICI) in 1982. He stated, "From the range of possible threats, it seems that the only occasions when the deployment of the armed forces may be justified are those involving terrorist groups of a nationalist character or in situations of widespread social unrest [...] In any case, the existence of paramilitary security forces, known as the 'Third Force', allows for the maximum possible delay in military intervention in the anti-terrorist struggle, or at least creates space for this intervention to be gradual".

IV   For Crenshaw (1983) the conjunction of efficiency with legality is a prerequisite for the success of government actions; to act otherwise would not only be morally wrong , but also politically disastrous. Uncontrolled responses, according to this writer, provoke a destructive cycle of terrorism and anti-terrorism, hard-line measures in combination with popular pressure produce martyrs and feed into terrorism. According to Braud (1993), it is vital that the deployment of force should be in proportion to the seriousness of attacks, for which reason governments must impose rigorous discipline on the specialized units who are assigned the role exercising this force. This discipline is attained through technical and psychological training aimed at achieving control over emotional responses. Dowse and Hughes (1982) claim that police and army loyalty may be achieved by recruiting their members from the ranks of the dominant classes, through a policy of neutralization or through concessions. However, this is not necessary in those western states confronting peripheral violent nationalisms; in these cases members of the coercive apparatus always unconditionally support the centre.

V   Communicative action forms an essential part of counter-insurgency tactics, which define it as "psychological war." Workshops held in Madrid in November 1980 in the Higher Centre of Studies of National Defence (CESEDEN) aimed to link the media of social communication with the anti-terrorist struggle; for this reason the directors of the main newspapers, except for *Egin*, as well as a number of journals and also audio-visual media personnel were invited. The following recommendations were put to journalists during the workshop by military experts: i) separate information from propaganda (that is, not publishing statements made by armed groups); ii) avoid semantic terms which favour terrorists; iii) reject any attempt at justifying terrorist crime (the victim must always be seen to be innocent). A CESEDEN spokesperson announced in 1982 that the agreement reached during the workshop had been adhered to by more than fifty directors of print, radio and television media. Lieutenant Colonel Manuel Fernández Monzón stated in the aforementioned ICI congress in 1982 that the need for an ongoing psychological war consisted in an information policy, "All states need to engage in permanent psychological activity with regard to their citizens as a means whereby what were previously called beliefs are maintained and complemented".

VI   In the mid 1970s, in tandem with advances in anti-terrorist thinking, a modification in prison strategies took place based on the theory of social reintegration which was applied in tandem with increasingly harsh prison conditions imposed on specific groups of prisoners. Baratta (1985) describes it in the following way: "Until 1975, all prison reform laws, such as those in Italy and Germany, were linked to the ideology of 'treatment.': prison must be the framework within which the reintegration of the internee should place. High security prison regime openly denied this [...] This had the effect of revealing that prisons were an end in themselves, the main aim of which was custody and maximum security and not the re-integration of the prisoner". However, those prisoners who, by definition, are not willing to be re-integrated into the type of state and society which gave rise to the prison regime, and who therefore are locked away in high security prisons, are precisely "political terrorists." This prison theory became common currency in the Spanish state during 1978, and led to the mass dispersion of Basque prisoners to new "maximum security" jails. The policy of winning over and reintegration (analyzed, among other authors by the Italian Ferracutti, 1979), proposed a double policy: dividing prisoners into different security levels with a view to splitting the group into "soft" and "hard-line", and dispersing them throughout different prisons (a policy put into practice in Spain from 1987 onward).

VII   Nevertheless, the parallel of equality drawn by elites between "reason," "democracy," "humanism" and "pacifism" on the one hand, and the state on the other, places organic intellectuals in an uncomfortable position with respect to current thinking in social sciences, which is clearly opposed to terrorism but sceptical with regard to the apology of the nation-state as a factor of democracy. That confers a certain provincial and obsolete character on their theoretical productions, although for different reasons to those related to violent nationalisms.

VIII Marcuse (1969) establishes a relation between the formation of the figure of the internal enemy and its identification with minority groups in advanced societies. According to his argument, "there is a flagrant contradiction, on the one hand, between the technological transformation of the world, which makes freedom and the attainment of a happy and free existence a possibility, and on the other hand, the intensification of the struggle for existence; this contradiction engenders, among the oppressed, widespread aggressiveness which, if it is not focused on a so-called national enemy who may be fought and hated, will attack anyone, white or black, indigenous or foreigner, Jew or Christian, rich or poor ... Never since the Middle Ages had been seen such a form of organized aggression, on a global scale, against those who are external to the world of repression, the 'inside and outside marginals'".

IX   Anti-terrorist psychosis took off in Spain in 1978, the year in which the Constitution was enacted, as a result of the increased death rate in ETA's attacks. According to polls published at the time, the fear of terrorism, which in 1975 was the biggest worry for 22% of the population, had increased to such an extent that this figure rose to 53% by 1978. In 1979 some print media published decalogues on self-protection which reflected and fed into this psychosis. Among the "ten security commandments" were recommendations such as "change travel routes frequently and use different means of transport," "do not accept calls from strangers who have not been introduced by a third part," "avoid driving," "do not open the door directly," "get to know the people around you, your children's friends, your daughter's boyfriend," and "try to carry a short-arm." This mind-set inspired the Plan ZEN (Special Northern Zone) in early 1983.

^X According to Girard, "a single victim can substitute all potential victims and all enemies [...] In those cases where there had previously been a thousand particular conflicts [...] once again a community emerges, united in the hate inspired by one of its members, the expiatory victim."

^{XI} For some authors (Hacker, 1976; Dispot, 1978) terrorism and state terror are complementary terms reflecting the same mimetic reality. Hacker explains that in certain conflicts, such as those in the Near East, state propaganda serves as a means for paranoid explanations aimed at maintaining the struggle against a dangerous enemy who, it is claimed, has powerful and secret allies and who represents only the tip of an enormous iceberg. The terrorist act, according to Dispot, is the "purest expression of a frenzy of justice and law" both on the part of those in power and of those in opposition. Dispot concludes that the expression "state terrorism" is a pleonasm, since all states have their origins and resources in terror, terrorism referring only to the state.

^{XII} It is in these situations where the dirty war assumes massive proportions, for example in Chile and Argentina, where tens of thousands of people have been murdered or "disappeared" under dictatorships; there is no comparison in the West with the dirty war waged in national conflicts against the population. Counter-insurgency theorists have described the techniques used in the dirty war in the Third World and Latin America. Grabowsky and Mickolus (1978) mention black propaganda, the deliberate spread of false information on terrorists with the goal of changing the public perception of them, and death squads, claiming that they are common to both dictatorships and democracies.

^{XIII} *Tortura en Euskadi* (Torture in Euskadi), written in the Basque Country in 1986, claims that torture aims, from the instrumental point of view, to force the detainee to collaborate, to depersonalize him and change his values. Official response to torture tends to embrace the following phases: claims to have investigated the allegations; denouncing them as false; passing them to the prosecutor to open proceedings against the accusers; and including the denunciation of torture in the list of "terrorist tactics." The torturers, according to this publication, are not sadists, rather they are professionals in their field; for these people "us" and "them" duality is a major factor in their group psychology. Ideological conditioning functions as a motive for torture: in Argentina, according to the "Sábato Report," this motivation consisted in the "defense of God and Christian values;" in the Basque Country other elements surface almost systematically: the ideology of law and order, the concept of the indivisibility of the Spanish homeland, anti-Basque sentiments, etc.

^{XIV} In fact, in Spain "uncontrolled" terrorism did not exist in any form during the 1950s and 1960s, the decades in which Franco's regime seemed to be strongest. It did, however, begin to emerge immediately after the "Burgos trials" in 1970, and became widespread during the period which stretched from the last state of emergency under Francoism, April to June 1975, until the first elections of June 1977.

^{XV} This was the case in Western Europe, particularly in Italy, at the end of the 1960s and beginning of the 1970s (Reinares, 1996); sections of the Italian police were forced to resign in 1978 on account of their involvement in the promotion of neo-fascist activities which resulted, on occasion, in civilian massacres. When these police networks were dismantled extreme right-wing terrorism disappeared in Italy.

^{XVI} These activities are in direct contradiction to the anti-terrorist instrumental policy which democratic states define as correct. Reinares (1996) indicates that the illegal use of public resources to complete the repression of terrorism by terrorist tactics may pay off in the short term; however it is very counter productive with regard to undermining the bases of the insurgent violence. In his opinion, although GAL (which the author defines as agreed to or directed by political and police authorities) put an end to ETA's sanctuary in France, it nevertheless eroded the anti-terrorist political consensus, discredited the police apparatus and provided arguments for ETA and its followers.

^{XVII} This section, although it obviously takes into account the particular characteristics of the Spanish case, theorizes a more general process.

15 Negotiations and the End of Violence

In centre-periphery violent conflicts each side perceives of peace as the triumph of its goals. For the nation-state peace will come out of the eradication of terrorist violence; for the armed group and its legitimating community, from their success in attaining national liberation. However, when the conflict is long-term the opposing sides end up internalising the impossibility of a total victory, which paves the way for negotiation (on the part of the armed group) and agreement through dialogue (on the part of the state authorities), sub discourses which co-exist in any event with the initial maximal discourse.

Stages Of Negotiation: Cease-fires And Their Breakdown

The state quickly becomes aware of the impossibility of eradicating the armed group. The latter's war is imaginary: it is not organised in battalions, and neither are there battle fields making a final defeat possible. The way in which the armed group wins battles consists in the mere fact of survival, it being enough to carry out sporadic attacks in order to be recognised as a group-state by its legitimating community. The detention and imprisonment of its members, far from weakening the ties which unify the whole, strengthen them; sacralisation functions even more in the core of a victimized community the more the group-state resembles a suffering collective.

So in this way, after a certain period of time - which could be several decades - the state's strategy moves down a scale, and its goal of defeating the violent group changes to that of agreement through dialogue and the hand over of weapons in exchange for a solution to the individual situations of prisoners and exiled members of the group (personal solutions measured on the basis of whether bloody attacks have been carried out). In this intermediate scale the state does not recognize the existence of the centre-periphery conflict (or even the link between this conflict and the violence of the armed group). It is therefore about an agreement without political content which spokespersons for the state explicitly differentiate from negotiations on the political status of the ethno-national group in question.[1]

Once the first phase of identity agitation is complete, for its part the armed group quickly realises the minority character of its legitimating community in the core of the ethno-national group, and takes note of the imbalance of power between the community on whose behalf it is acting and the rest of the state. This translates into an intermediate scale in which the strategic programme for achieving independence through a military victory over the foreign occupying forces opens the door to political negotiations with the state on a short-term national programme (with or without the prior declaration of a cease-fire). The armed group demands that the state recognises it as a legitimate interlocutor; political negotiation must be an agreement between sides who accord mutual recognition to each other as opposing forces. Given that the armed group conceives of itself (and it is also seen this way by its community) as a group-state, it grants itself exclusivity in the negotiations, which it considers it has a right to on account of its commitment to the struggle, thereby denying the role of representative to the other political forces within the ethno-national collective, even including those cases where the radical sectors which recognise it are in a minority within the system of nationalisms.

The third scale is that of negotiation between all forces within the periphery and the centre (brought about by the declaration of a ceasefire by the armed group, but endowed with the legitimation granted by majority' support and respect for democratic conventions), with the aim of having the periphery its right to decide its political status and the relation (or lack of relation) which it wishes to have with the state.

However, a number of obstacles make access to this scale difficult, in addition, obviously, to the refusal of the state to accept it. Although violent conflict usually emerges from situations where nationalism is strong, the nationalist forces do not always have the support of majority in ethno-national territory or the ability fully to convince non-nationalist forces to back it. Finally, systemic nationalism would look favorably on a negotiating alliance with radical nationalism (in fact some of its ideologues often theorise this possibility), but its main aim will be to maintain its hegemony within territorial institutions, in those cases where it has in fact attained this position. For this reason, systemic nationalism's alternative projects do not often include the intention to share power with radical nationalism to end the exclusion to which it had been subject. In this way, although the fact of being the two opposing sides, the armed group and the state, in the intermediate scale which does not exclude contact does create the illusion of possible agreement, the false bases upon which it relies make it impossible. This explains why in this type of conflict, the start of bilateral negotiations, preceded by a ceasefire, and their breakdown, followed by a renewal of violence, is so often the case.

The mechanics of these negotiations begin to develop when one of the two sides sounds out the other on the willingness to participate in them, obtaining a positive response; which almost always comes with the demand for a public declaration of a ceasefire by the armed group. The moment chosen to make this approach and the willingness to respond affirmatively to the approach depends more on internal dynamics of the interlocutors than on the interests of the ethno-national group or the state. These refer to, in terms of the armed group, the strength of the leadership, the rivalry, if they exist, of other armed groups, the degree of control over the legitimating community, the relative strength of this community within the core of the system of nationalisms, the internal or international vulnerability of the state, etc. The dynamics which determine the decision of the government are related to the assessment of the political leadership as to whether a solution to the conflict will favour its continuation in power.[II]

Once contact has been made the continuing insistence on exclusivity on the part of the armed group and the refusal of the government to negotiate any political change prevent the development of a common language which would make agreement feasible, which results in failure in the short or long-term. In this way, the violence that had been put on hold on the level of instrumental rationality breaks out, since it had not been defused at the level of the identity dimension on both sides. The dialectics of reciprocal hate lead, in effect, to both the state-national society and the armed group's legitimating community supporting the continuation of violence between the two contenders. Moreover, this dynamic is reinforced by, with regard to the state, those carrying out the dirty war and those who benefit from it. On the side of violent nationalism this refers to members of the leadership bloc, when this exists, who have not undertaken as great a risk as the members of the barmed group; since their authority over the legitimating community derives from their vicarious position in relation to the armed group, they do not have an interest in an outcome which would make them dispensable.

The Indo-Lanka Agreement in 1987, in which the Sri-Lanka government committed itself to authorizing a certain degree of self-government to the Tamils through the establishment of Provincial Councils throughout the island, was respected by the Tamil Tigers, who adhered to a ceasefire. The passing of an amendment to the Sri-Lanka Constitution which, in fact, annulled the Councils' powers, brought about the collapse of the Tamil Tigers' ceasefire. The Singhalese Popular Front, which had been triumphant in the 1995 elections, anticipated the renewal of negotiations with the Tigers, who controlled the Jaffna peninsula. However, the armistice was broken within four months on account of the government demand for a prior handover of weapons by the Tigers.

The history of Iraqi Kurdistan resistance is full of armistices and renewal of hostilities, in contrast to Turkish Kurdistan which has been blocked by the intransigence of official pan-Turkism. After the Gulf war, and under the protection of Western powers, Saddam Hussein was forced to agree to the formation of a Kurdish parliament in Mosul. Fratricidal struggles over recent years among the Kurdish people have allowed the Iraqi government to intervene in favor of one faction, the DPK, and against another, the PUK.

The PLO's change of agenda in 1987 in which it accepted the idea of a Palestinian state existing adjacent to the Israeli state opened the way to the 1991 peace talks in Madrid, under the sponsorship of the Western powers, which culminated in the 1993 Oslo Accords that led to the establishment of limited autonomy in Gaza and the West-Bank presided by a Palestinian National Authority originating from the PLO. The replacement of Labour Jewish government by Likud's meant, in practice, the stagnation of the Oslo Accords.

The French socialist proposal to grant Special Status to Corsica in 1982 led to a ceasefire on the part of the FLNC with the goal of a possible amnesty for its imprisoned members. Nevertheless, the ceasefire broke down when the right-wing in the recently established Assembly blocked all resolutions favorable to Corsican identity. The new historic FLNC ceasefire at the start of

1996 ended in the face of French government refusals to accept any of the nationalist demands.

For its part, ETA-military had stated that it would establish a ceasefire after the KAS alternative aims had been met. In 1988 it agreed to declaring a ceasefire prior to the start of negotiations; the Spanish government, in collaboration with its French counterpart, grouped together previously deported ETA members in Algeria with the aim of making dialogue possible. The talks failed because the Spanish government did not anticipate the negotiation of any political agenda and because ETA claimed to represent exclusively the Basque people. After the failure in Algeria, ETA's sophisticated theoretical constructions on hypothetical representation have come up against the refusal of successive Spanish governments to commit themselves again to public negotiations with the armed group.

In Northern Ireland the Provisional IRA announced its first ceasefire in June 1972; the detention of a number of republican leaders, including Gerry Adams and Martin McGuinness, who had risen to the fore of the movement at that time, generated a great reluctance within the IRA to hold new ceasefires. After the 1993 Downing Street Agreement which recognised the right to self-determination at two levels, within the island of Ireland as a whole and within the north of Ireland, the IRA announced a cessation of armed operations that appeared to be indefinite. However, the weakness of Major's government meant that he was at the behest of the unionists and led him to introduce pre-conditions which the Agreement did not contain, for example, the handover of weapons by the IRA at the start of negotiations. Consequently, the IRA broke its ceasefire in February 1996.

The End of Violences: International Pressure and a New Political Culture

In the scale of instrumental rationality it is only on the third level, that on which the armed organisation admits that it has not the entire representation of its ethno-national group and on which the nation-state acknowledges that a centre-periphery conflict should be resolved through the appropriate politico-legal modifications, where political agreement is feasible. There is, however, nothing in the processes of hostile identity polarisation between the two communities that predisposes them to putting pressure on the leadership cores to move from the second to the third level.

This pressure can only come from outside, or from the international context, as was the case with Palestine, Sri Lanka, Iraqi Kurdistan and Northern Ireland, or from (if international pressure is weak or non-existent) the emergence of a current of opinion generated by social movements and individuals who create a new political culture on the fringes of both sides.

The content of this political culture must also be based on pacifism, humanism and pluralism, rescuing the meaning of these values from the double discourse into which they have been embedded by the nation-state elites in defense of the centre. The radical claim for life and for the human being, on which pacifism and humanism are based, should be made from an emancipating perspective, a perspective that confronts the different political, economic and cultural mechanisms of domination and struggles to overcome them; which is incompatible with the defense of the nation-state, a centre for co-ordination of the various dominations and instruments that protect the nation-state from the response of those who are dominated. This political culture of emancipation is therefore incompatible with the logic of armed groups which, even though they adopt the protest of the dominated, contain within themselves the vices of the enemy they are engaged in combat with.

Neither can there be a pluralism which is not based on the willingness to understand the identity markers of the other (person or collective), on the full recognition of difference and, therefore, of the right to "autonomy" (the ability to set for oneself moral and legal laws, according to the Kantian meaning of this word). There is no pluralism unless there is respect for the right to a group's (whether it be social, political or ethnic) right to self-determination; the discourse ridiculing this right in which state-national elites engage is nothing more than a mask for imposition and the expression of the will to maintain the hetero-determination on which the different dominations are based.

The creation of this new political culture has to endeavor to change the identity markers of the radical group. The transformation of a complex which has been formed as a counter-state and counter-society into a group which will pressurise, pressure shared by other forces, the state itself in order to attain a change in society, and the consequent end of its self-perception as a whole in itself becoming a self-perception as part of a whole, affects components as passionately rooted in human beings as their identity markers and their affective identification. For this reason, it is only through a process of self-transformation of the armed group in civil organisation decided consciously by it that a radical change will take place in the whole. However, the state-national community must also modify its identity markers, deconsecrating something as changeable in the historic long-term as the subsistence of the nation-state in present territorial boundaries, without which agreement based on mutual recognition will be impossible.

The catalyst for this culture cannot be political parties; it must come from beyond them, with the aim of influencing them.[III] It would, nevertheless, be deceptive to call this culture a "third space." Systemic nationalism, which sees itself as a third force between the opposing sides, may assume exclusive rights over this space when, in fact, its claim to hegemony and institutional monopoly has played an important role in the creation of pockets of exclusion which have generated, or maintained, the violence of anti-system nationalism. The acceptance of pluralism by this nationalism must be expressed through its willingness to put an end to its hegemony and recognition of the principle of shared responsibilities. For this reason, more than a "third space" a culture of "no-man's land" should be spoken of, which, after being sown, should end up becoming, with the support of pacifists from the centre and periphery, a "common ground" for nationalist and non-nationalist forces of the periphery, and for peripheral and state-wide organisations and political parties.

The end of violent confrontations supposes, as has happened so often in history, the end of stigmas and stereotypes. According to Rodinson (1975), given that both are products of disputes and wars, the achievement of peace, a peace that is not based on the triumph of one side over the other, but rather on agreement between both, will lead to the disappearance of both. 'When peace is achieved between warring sides', Rodinson writes, 'an ideological demobilization takes place, and ethnocentric or racist war slogans are forgotten. Sometimes even, if the peace is lasting, if an alliance is concluded, an entirely opposite type of propaganda campaign may be launched in which the qualities of the enemy, which were categorically rejected shortly before, are suddenly brought to the fore'.

In any event, nobody is unaware of the obstacles making it difficult to reach this stage. In effect, the end of stereotyping and reconciliation between opposing sides use to take place some decades ago after the ethno-national collective has been granted the status of independent state; but in reality this scenario is present in very few contemporary centre-periphery conflicts. When the solution consists in greater self-government within the composite state, the authorities of the ethnic territory will always be dependent on the political fluctuations taking place within the government of the centre. If the latter strengthens its position at the cost of an already institutionalized periphery a new system of nationalisms may recur in which the old radical nationalism becomes systemic, facing now the response of new violent nationalisms. Contemporary Palestine demonstrates to its cost that both processes are possible.

Notes

[I]   Although the violence practiced by states of the Third World (for example, in Latin America) against insurgent movements with a community base is often much more intense than that used against peripheral nationalisms in the West, paradoxically, political negotiation there is much easier there than here, once it is started. Where there is no centre-periphery conflict, agreement does not demand outright the modification of the political framework of the state; the outcome may be the end of the socio-political exclusion which caused the response violence, it being enough, on occasion, to incorporate the leaders of the insurgent group into the process of state-national building. On the other hand, a stable political solution to national conflicts not only requires an end to exclusion, but also a change in the territorial status of the ethno-national group, which the state is much less willing to accede to since this change usually demands the modification of its own constitutional framework.

[II] Therein lies the conclusion that all the mediation work not expressly requested by the opposing sides is superfluous, not only because it is very likely to be ignored, but also because mediators take into consideration questions related to the circumstances in which the ethno-national group and the state find themselves; when what is of greater importance are questions related to the interests of the leadership core of both sides, by definition known only to themselves.

[III] Those who promote this culture will have to accept that the price is their lack of acknowledgement. The nation-state elite will subject their efforts to the same degree of stigma and stereotyping that violent nationalism is subjected to, thereby facilitating the confusion of one for the other; the hostility may even be greater as the elite becomes aware that their ideas, contrary to those of violent nationalism, will endure. The core of violent nationalism will refer to them as weak, or worse, as an internal enemy; they must therefore accept being offended by those who have been offended and humiliated by those who have been humiliated. Neither will they enjoy recognition should their project be successful; it will be the opposing sides, civilian or armed, and the political parties, who will take all merits in these circumstances. But it cannot be any other way.

Bibliography

Abou, S. (1981), *L'identité culturelle: relations inter-ethniques et problèmes d'acculturation*, Anthropos, Paris.

Acosta, J. (1979), *Historia y cultura del pueblo andaluz*, Editorial Anagrama.

Adler, A. (1975), *Conocimiento del hombre*, Colección Austral, Madrid.

Aguilera de Prat, C.R. and Vilanova, P. (1987), *Temas de Ciencia Política*, P.P.U.

Alberoni, F. (1984), *Movimiento e Institución*, Editora Nacional, Madrid.

Aleman, J. et al. (1978), *Ensayo sobre historia de Canarias*, Biblioteca Popular Canaria, Las Palmas de Gran Canaria.

Alexander, J. (1976), *International terrorism: National, Regional and global Perspectives*, Praeger Publishers, New York.

Algar, H. (1991), 'Sources et figures de la révolution islamique en Iran', in *Islam et politique au proche-Orient aujourd'hui*, Gallimard, Paris.

Almond, G. and Verba, S. (1970), *La Cultura Cívica. Estudio sobre la participación política democrática en cinco naciones*, Euramérica, Madrid.

Anderson, M. (1993), 'Frontiers: Changing Agendas', in *Peripheral Regions and European Integration*, Research seminar, Belfast.

Anderson, B. (1983), *Imagined communities, reflections on the origin and spread of nationalism*, Verso Edition, London.

Anderson, J. and Goodman, J. (1993), 'Regionalism in the EC: Changing Geopolitics in Ireland' in *Peripheral regions and european integration*, Research seminar, Belfast.

Ansart, P. (1975), *Les idéologies politiques*, PUF, Paris.

Arendt, H. (1961), *La condition de l'homme moderne*, Calmann-Lévy, Paris.

Arendt, H. (1972), *Du mensonge a la violence: essais de politique contemporaine*, Calmann Lévy, Paris.

Arendt, H. (1972), *Le systéme totalitaire*, Ed. du Seuil, Paris.

Arendt, H. (1982), *Los orígenes del totalitarismo*, 3 vols, Alianza, Madrid.

Arendt, H. (1982), *L'Imperialisme*, Fayard, Paris.

Arieh, J. (1978), *Le défi national: les théories marxistes sur la question nationale à l'épreuve de l'histoire*, Anthropos, Paris.

Aron, R. (1976), *Penser la guerre, Clausewitz*, Gallimard, Paris.

Arrighi, P.and Pomponi, F.(1984), *Histoire de la Corse*, P.U.F., Paris.

Badie, B. (1985), 'Formes et transformations des communautés politiques', in Grawitz, M. and Leca, J. (eds.), *Traité de Science Politique. Vol. 1: La Science politique, Science sociale. L'ordre politique*, P.U.F, Paris.

Baker, S. (1993), 'Borders and environmental questions', in *Peripheral Regions and European Integration*, Research seminar, Belfast.

Bakunin, M. (1976), *Etatisme et anarchie, tome IV, Oeuvres complètes*, Editions Champs Libres, Paris.

Balandier, G. (1981), *Anthropologie politique*, P.U.F, Paris.

Bandura, A. (1973), *A social learning analysis*, Englewood Cliffs, Prentice Hall.

Baratta, A. (1985), 'Violencia social y legislación de emergencia en Europa: situación en Italia' in *Democracia y Leyes Terroristas en Europa*, Imprenta Berekintza, Ipes –Bizkaia, Bilbao.

Bardon, J. (1992), *A history of Ulster*, The Black Staff Press, Belfast.

Barth, F. (1976), *Los grupos étnicos y sus fronteras*, Fondo de Cultura Económica, Madrid.

Bastian, S. (1994), 'Sri-Lanka. Ethnic conflict and the experiment of devolution', in *Nation-building and sub-cultures*, Journal of behavioral and social sciences, vol. 1994, 3. Kanagawa, Japan.

Bastide, R. (1975), *Le sacré sauvage*, Payot, Paris.

Bauer, O. (1979), *La cuestión de las nacionalidades y la socialdemocracia*, Siglo XXI, México.

Beiras, J. M. (1995), *O atraso económico de Galiza*, Edicions Laiovento, Santiago.

Bell, D. (1964), *El fin de las ideologías*, Tecnos, Madrid.

Bentham, J. (1973), *Fragmentos sobre el gobierno*, Aguilar, Madrid.

Berberoglu, B. (ed.) (1995), *Nationalism, Ethnic question and Self-determination in the 20th century*, Temple University Press, Philadelphia.

Berque, J. (1978), 'L'identité collective et sujet de l'histoire', in *Identités collectives et relations interculturelles*, Edition Complexe, Paris.

Bertaux, P. (1984), 'Africa', in *Historia Universal siglo XXI*, vol. 32, Ed. Siglo XXI, Madrid.

Beyhaut, G. y H. (1986), 'América Latina: III. De la independencia a la segunda guerra mundial', in *Historia Universal siglo XXI*, vol. 23, Ed. Siglo XXI, Madrid.

Bianco, L. (1984), 'Asia contemporánea' in *Historia Universal siglo XXI, vol. 33*, Madrid.

Birnbaum, P. (1985), 'L'action politique. L'action de l'Etat, différenciation et dédifférenciation', in Grawitz, M. and Leca, J. (eds.), *Traité de Science Politique. Vol. 1: La Science politique, Science sociale. L'ordre politique*, P.U.F, Paris.

Blaufarb, D. B. (1977), *The counter-insurgency Era: U.S. Doctrine and performance: 1950 to the Present*, The Free Press, MacMillan, London and New York.

Bloom, S. F. (1941), *The world of nations: A study of the national implications in the work of Karl Marx*, Columbia University Press, New York.

Bon, F. (1985), 'Communication et action politique: Langage et politique', in Grawitz, M. and Leca, J. (eds.), *Traité de Science Politique. Vol. III: La Science politique, Science sociale. L'ordre politique*, P.U.F, Paris.

Borjas, A. (1996), 'Chiapas: crónica de una reivindicación pendiente', Curso de Doctorado *Gobierno y análisis comparado 1996-1997*, UPV, Leioa.

Bourdieu, P. (1972), *Esquisse d'une théorie de la pratique, précédé de trois études d'ethnologie Kabyle*, Librairie Droz, Genève, París.

Bourdieu, P. (1977), *La reproducción. Elementos para una teoría del sistema de enseñanza*, Editorial Laya.

Bourdieu, P. (1980), 'L'identité et la réprésentation, éléments d'une réflexion critique dans l'idée de région'. in *Actes de la recherche en Sciences sociales*, n° 37, Paris.

Bourque, G.(1995), 'Quebec nationalism and the struggle for sovereignty in French Canada', in Berberoglu, B. (Ed.), *The National Question,* Temple University press, Philadelphia.

Bouthoul, G. (1971), *La guerra,* Oikos-tau Ediciones.

Braud, P. (1985), 'Du pouvoir en géneral au pouvoir politique', in Grawitz, M. and Leca, J. (eds.), *Traité de Science Politique. Vol. 1: La Science politique, Science sociale. L'ordre politique,* P.U.F, Paris.

Braud, P. (1993), 'La violence politique: repères et problèmes', in *La violence politique dans les démocraties européenes occidentales,* L'Harmattan, Paris.

Breuilly, J. (1982), *Nationalism and State,* Manchester University Press, Manchester.

Butt, P. A. (1993), 'Border Controls in the EC, post 1992. The case of Western Europe', in *Peripheral regions and european integration,* Research seminar, Belfast.

Calogeropoulos-Stratis, S. (1973), *Le droit des peuples a disposer d'eux-mêmes,* Bruylant, Brussels.

Caminal, M. (1996), 'El nacionalismo', in *Manual de Ciencia política,* Tecnos, Madrid.

Carrere D'encausse, H. (1991), *El triunfo de las nacionalidades: el fin del imperio soviético,* RIALP, Madrid.

Carrere D'encausse, H. (1985), 'Les régimes politiques contemporains. L'URSS ou le totalitarisme exemplaire', in Grawitz, M. and Leca, J. (eds.), *Traité de Science Politique. Vol. III: La Science politique, Science sociale. L'ordre politique,* P.U.F, Paris.

Casamayor, F. (1983), *Et pour finir, le terrorisme,* Gallimar, Paris.

Castellan, G. (1991), *Histoire des Balkans,* Fayard, Paris.

Castoriadis, C. (1975), *L'Institution imaginaire de la société,* Ed. du Seuil, Paris.

Castro, J. L. (1992), *La emergente participación política de las regiones en el proceso de construcción de la unión europea,* UPV-EHU, doctoral thesis.

Cazorla, J. (Ed.) (1983), *Fundamentos Sociales del Estado y la Constitución,* Universidad de Granada, Granada.

Chaliand, G. (1982), *Guerrilla stratégies: A historical anthology from the long March to Afganistan,* University of California Press, Berkeley.

Charlot, J. and Charlot, M. (1985), 'L'interaction des groupes politiques', in Grawitz, M. and Leca, J. (eds.), *Traité de Science Politique. Vol. III: La Science politique, Science sociale. L'ordre politique,* P.U.F, Paris.

Charnay, J. P. (1981), *Terrorisme et culture: (pour une anthropologie stratégique),* Centre d'Etudes et de Recherches sur les stratégies et les conflits. Les cahiers de la Fondation pour les études de défense nationale. Supplement to n.º 11 of "Strategique".

Chatelet, F. and Pisier-Kouchner, E. (1986), *Las concepciones políticas del siglo XX,* Espasa-Calpe, Madrid.

Chaussier, J. D. (1994), 'Pouvoir du territoire contre territoires du pouvoir, Société politique et notables politiques au pays basque', Colloque IEP Bordeaux, *Scènes du débat politique au pays basque: récomposition ou simulacre.*

Chaussier, J. D. (1995), *Pays Basque sud et Pays Basque nord-le comparatisme est-il possible?* , Colloque Aquitaine-Euskadi , IEP-Bordeaux.

Chesneaux, J. (1984), *Langues opprimées et identité nationale*, Colloque, Departement des langues et cultures oprimées et minorisées, Université de Paris 8, Saint Denis.

Chorbajian, L. (1995), 'The nationalities question in the former Soviet Union: Transcaucasia, the Baltics, and Central Asia', in Berberoglu, B. (Ed.), *The National Question*, Temple University Press, Philadelphia.

Clark, C. (1995), 'The possibilities (and difficulties) in stablishing a European-wide voice for Gypsies an other travellers in an age of post-modernity: a view from Britain', in *The criminalisation and victimisation of Gypsies in the new Europe*, Oñati International Institute for the sociology of law, Oñati.

Clark, R. (1984), *The Basque insurgents: ETA 1952-1980*, The University of Wisconsin Press, Wisconsin.

Clarke, M. (1974), *On the Concept of Sub-Culture*, British Journal of Sociology, December.

Clausewitz, C. V. (1955), *De la guerre*, Les editions de Minuit, Paris.

Colomer, J. M. (ed.) (1991), *Lecturas de Teoría Política Positiva*, Instituto de Estudios Fiscales, Madrid.

Connor, W. (1993), 'Elites and ethnonationalism: The case of Western Europe', Congreso internacional: *Os nacionalismos en Europa: pasado e presente*, Santiago de Compostela.

Converse, P. (1976), 'The stability of belief systems over time', in Niemi R. and Weisberg H. F, *Controversies in American Voting Behaviour*, Freeman, New York.

Coogan, T. P. (1987), *The IRA*, Fontana, London.

Costard, M. (1977), 'Las instituciones regionales y la regionalización de Europa', in *Documentación Administrativa*, n.º 175, revista de la Secretaría General Técnica, Presidencia del Gobierno, Madrid.

Cotarelo, R. (1988), 'Objeto, Método, Teoría', in Pastor, M. (ed.): *Ciencia Política*, McGraw-Hill, Madrid.

Crenshaw, M. (1986), *Terrorism, legitimacy and power. The consequences of political violence*, Ct. Wesleyan University Press, Middletown.

Culioli, G. X. (1990), *Le complexe corse*, Gallimard, Paris.

Dahl, R. A. (1963), *Who governs?. Democracy and power in an American city*, N. H. Yale University Press, New Haven, Conneticut.

Davies, J. (1993), *A history of Wales*, The Penguin Press, London.

Davies, J.(1970), *When men revolt and why*, Free Press, New York.

De Blas, A. (1984), 'En torno a la génesis tardía del nacionalismo español', *Revista de Política Comparada*, Universidad Internacional Menéndez Pelayo, n° 10 -11, Santander.

De Blas, A. (1984), *Nacionalismo e Ideologías Políticas Contemporáneas*, Espasa-Calpe, Madrid.

De Blas, A. (1988), 'Estado, Nación y Gobierno' in Pastor, M. (ed.), *Ciencia Política*, McGraw-Hill, Madrid , pp. 49-76.

De Blas, A. (1994), *Nacionalismos y naciones en Europa*, Alianza Universidad, Madrid.

Debray, R. (1972), *Révolution dans la révolution?*, Maspero, Paris.

Della Porta, D. and Rucht, D. (1994), *Left-libertarian Movementes in context: a comparison of Italy and West Germany, 1965-1990*, WZP Paper, Berlin.

Della Porta, D. (1996), 'Lógica de las organizaciones clandestinas: un análisis comparado en Italia y Alemania', *Sistema*, 132-133, Madrid.

Delli Zotti, G. (1993), 'The Alpe Adria Working Community and the Central European Initiative', *Peripheral regions and european integration*, Research seminar, Belfast.

Delmas, C. (1972), *La Guerre Révolutionnaire*, P .U .F, Paris.

Deloye, Y. (1994), 'La nation entre identité et alterité' in *L' identité politique*, PUF, Paris.

Descartes, R. (1989), *Discurso del método*, Tecnos, Madrid.

Deutsch, K. W. (1961), 'Social mobilization and political development', in *American Political Science Review*, 55.

Devereux, G. (1970), *Ethno-psychanalyse complémentariste*, Flammarion, Paris.

Dhondt, J. (1987), 'La alta edad media', in *Historia Universal siglo XXI*, vol. 10, Ed. Siglo XXI, Madrid.

Diani, M. (1992), 'The concept of social movement', *The sociological Review*, n° 38.

Diaz-Andreu, M.(1995), 'The past in the present: the search for roots in cultural nationalisms. The Spanish case', in the Congreso internacional: *Os nacionalismos en Europa: pasado e presente*, Santiago de Compostela.

Diner, D. (1981*)*, 'Israel: el problema del Estado nacional y el conflicto del Oriente próximo', in *Historia Universal siglo XXI*, vol. 36, Ed. Siglo XXI, Madrid.

Dispot, L. (1978), *La machine à terreur*, Grasset, Paris.

Djait, H. (1991): 'Culture et politique dans le monde arabe', in *Islam et politique au proche-Orient aujourd'huy*, Gallimard, Paris.

Documentos (*of ETA*), 18 tomes, (1979), Editorial Lur, Grupo Hordago, San Sebastián.

Dowse, R. E. and Hughes, J. A. (1982), *Sociología Política*, Alianza, Madrid.

Dressler-Holohan, W. (1981), *Dévéloppement économique et autonomisme. Le cas de la Corse: 1960-1980*, Comité d'organisation des recherches sur le développement économique et social. n° 6/7. Grenoble.

Dressler-Holohan, W. (1993), 'French Forms of Regionalization: a Specific Response to European Integration and to Claims for Regional Autonomy', in *Peripheral regions and european integration*, Research seminar, Belfast.

Droz, J. (1963), *Le romantisme politique en Allemagne*, Armand Colin, Paris.

Durkheim, E. (1975), *Textes*, tome II, Les Editions de Minuit, Paris.

Easton, D. (1953), *The Political System*, Knopf, New York.

Easton, D. (1965), *A System Analysis of Political Life*, Wiley, New York.

Eccleshall, R. et alt. (1993), *Ideologías Políticas*, Tecnos, Madrid.

Eccleshall, R., Vincent, G., Richard, J., Rick, W. (1984), *Politics and Ideologies*, Hutchinson, London.

Eder, K. (1983), *A new social movements*, Telos 5 –30.

Eidheim, H. (1976), 'Cuando la identidad étnica es un estigma social', in *Los grupos étnicos y sus fronteras*, Fondo de Cultura Económica.

Eliade, M. (1965), *Le Sacré et le Profane*, Gallimard, Paris.

Engels, F. (1969), *Ludwig Feuerbach y el fin de la filosofía alemana*, Península, Barcelona.

Engels, F. (1978), *Anti-Duhring,* Ayuso, Madrid.

Erikson, E. (1972), *Adolescence et crise: la quête de l'identité,* Flammarion, Paris.

Fanon, F. (1968), *Les damnés de la terre,* Petite Collection Maspero, Paris.

Fernández de Casadevante, C. (1985), *La frontera hispano-francesa y las relaciones de vecindad: especial referencia al sector fronterizo del País Vasco,* Servicio Editorial de la UPV/EHU, Leioa.

Ferracutti, F. (1979), *La coordination des recherches et l'application de leurs résultats dans le domaine de la politique criminelle,* Comité Européen pour les problémes criminels. Conseil de l'Europe. Affaires juridiques, Strassbourg.

Ferry, L. and Pisier-Kouchner, E. (1985), 'Théorie du totalitarisme', in Grawitz, M. and Leca, J. (Eds.) (1985), *Traité de Science Politique. Vol. 2, Les régimes politiques contemporains,* P.U.F., Paris.

FLNC (1977), *Manifeste du 5 mai au Peuple Corse,* Corti.

FLNC (1990), *Cunferenza di Stampa,* U Borgu, November 1990.

Foucault, M. (1975), *Surveiller et punir: naissance de la prison,* Gallimard, Paris.

Freedman, R. (1976), 'Soviet Policy toward International Terrorism', in Alexander, Y.: *International terrorism,* Praeger Publishers, New York.

Freres du monde, vol. 78 (1972), *Irlande,* Bordeaux.

Freud, S. (1921), 'Psychologie de groupe et analyse du moi', in *Essais de psychanalyse,* Payot, Paris.

Fromm, E. (1975), *Anatomía de la destructividad humana,* Siglo XXI Editores, Madrid.

Gallissot, R. (1976), 'Je ne crois pas en un seul peuple', *Pluriel-Débat* n° 6, Paris.

Gallissot, R. (1976), 'A propos de l'idéologie nationale de Jean-Yves Guiomar. Les différentes conceptions de la nation', *Pluriel-Débat,* n.° 5, Paris.

Gallissot, R. (1980), *Problématique de la Nation: mouvement ouvrier, nation et nationalité,* Notes dactylographiées.

Gallissot, R. (1982), 'Nationalisme', in *Dictionnaire critique du marxisme,* P.U.F., Paris.

Gamson, W, A., and Meyer, D. S. (1992), *The framing of political oportunity,* Paper prepared for the Conference "European/American Perspectives on Social Movements", Washington D. C.

Gari Hayek, D. (1995), 'Los nacionalismos periféricos ante la construcción política europea: el caso del archipiélago canario', Congreso internacional *Os nacionalismos en Europa: pasado e presente,* Santiago de Compostela.

Garvin, T. (1995), 'Nationalism and Separatism in Ireland (1760-1993): a comparative perspective', Congreso internacional *Os nacionalismos en Europa: pasado e presente,* Santiago de Compostela.

Gaxie, D. (1977), 'Economie des partis et rétributions du militantisme', *Revue française de Science politique* 27, n° 1.

Geertz, C. (1964), 'Ideology as a cultural system', in Apter D. (ed.), *Ideology and discontent,* Free Press, New York.

Geertz, C. (1973), *The interpretation of Cultures,* Basic Books, New York.

Geiss, I. (1981), 'Condiciones históricas previas de los conflictos contemporáneos', in *Historia Universal siglo XXI,* vol. 36, Ed. Siglo XXI, Madrid.

Gellner, E. (1988), *Naciones y Nacionalismos,* Alianza, Madrid.

Giner, S. (1982), *Historia del Pensamiento Social,* Editorial Ariel, Madrid.

Girard, R. (1972), *La violence et le sacré*, Ed. Bernard Grasset, Paris.

Goehrke, C. et al. (1984), *Rusia*, Historia Universal siglo XXI, vol. 31, Madrid.

Gonen, A. (1996), *Diccionario de los pueblos del mundo*, Anaya & Mario Muchnik, Madrid.

Goodwin, B. (1988), *El uso de las ideas políticas*, Ediciones Península.

Grabowsky, M. (1978), 'The urban context of Political Terrorism', in Sthol, M. *The politics of terrorism*, Marcel Dekker Inc., New York.

Graham Jones, J. (1990), *The history of Wales*, University of Wales Press, Cardiff.

Gramsci, A. (1978, 1983), *Cahiers de prison*, Editions Gallimard, Paris.

Grawitz, M. and Leca, J. (eds.) (1985), *Traité de science politique*, 4 vol., P.U.F., Paris.

Grawitz, M. (1985), 'Psycologie et politique', in Grawitz, M. and Leca, J. (eds.) (1985), *Traité de Science Politique. Vol. 3: L'action politique*, P.U.F., Paris.

Gray, A. (1988), 'The Americans of South America', *The minority Rights Group*, Report no. 15, London.

Guattari, F. (1972), *Pyschanalyse et transversalité*, Maspero.

Guevara, Che (1968), 'La guerre des guérrilles', Oeuvres, tome I, in *Textes militaires*, Petite Collection Maspero, Paris.

Guezennec, G. (1991), *La yugoeslavie autogestionnaire: bilan critique d'une époque prestigieuse*, Editions Créer, Paris.

Guilhaudis. J . F . (1976), *Le droit des peuples à disposer d'eux-mêmes*, Presses Universitaires de Grenoble.

Guillaumin, C. (1972), *L 'idéologie raciste*, Mouton, Paris.

Guillén, A. (1966), *Estrategia de la guerrilla urbana*, Manuales del Pueblo, Montevideo.

Gupta, D. (1995), 'Ethnicity, religion and national politics in India', in Berberoglu, B. (ed.), *The National Question*, Temple University press, Philadelphia.

Gurr, R. (1974), *Why men rebel*, Princeton University Press, Princeton, New Jersey.

Gurvitch, G. (1967), *Traité de Sociologie*, P.U.F., Paris.

Habermas, J. (1978), *Raison et légitimité: problemes de légitimation dans le capitalisme avancé*, Payot, Editions Gallimard, Paris.

Habermas, J. (1973), *La tecnhique et la science comme ideologie*, Gallimard, Paris.

Hacker, F. (1976), *Terreur et terrorisme*, Flammarion, Paris.

Hainsworth, P., Duncan, M. (1993), 'Northern Ireland: A Region in Comparative Perspective' *Peripheral regions and european integration*, Research seminar, Belfast.

Haupt, G ., Lowy, M. and Weill, C. (1974), *Les marxistes et la question nationale 1848 –1914*, Maspéro, Paris.

Hayek, F.A. (1960), *The Constitution of Liberty*, Routledge & Kegan Paul, London.

Hayes, C. (1966), *Essays on Nationalism*, Russel, New York.

Hegel, G.W.F. (1980), *Lecciones sobre la filosofia de la Historia Universal*, Alianza Editorial, Madrid.

Heraud, G. (1963), *L'Europe des ethnies*, Presses d'Europe.

Hetcher, M. (1975), *Internal Colonialism*, Routledge and Keagan Paul, London.

Hetcher, M. (1989), 'El nacionalismo como solidaridad de grupo', in *Sociología del nacionalismo*, Servicio Editorial UPV-EHU, Leioa.

Hirsch, F., Goldthorpe, J. (1978), *The political economy of inflation*, Martin Robertson, London.

Hobsbawn, E. and Ranger, T. (eds) (1983), *The Invention of Tradition*, Cambridge University Press, Cambridge.

Hobsbawn, E. (1972), *Les bandits*, Petite collection Maspero, Paris.

Hobsbawn, E. (1959), *Primitive rebels: studies in archaic forms of social movement in the 19th and 20th centuries*, University of Manchester Press, Manchester.

Hofmann, S. (1985), 'L'ordre international', in Grawitz, M. and Leca, J.(eds.) *Traité de Science Politique. Vol. 1: La Science politique, Science sociale. L'ordre politique*. P.U.F., Paris.

Horkheimer, M. and Adorno, T. (1974), 'Eléments d'antisémitisme', in *Dialectique de la raison*, Gallimard, Paris.

Hroch, M. (1985), *Social Preconditions of National Revival in Europe*, Cambridge University Press, Cambridge.

Hroch, M. (1993), 'The integrating and desintegrating circomstances in the Czech and Slovak national movements', Congreso internacional *Os nacionalismos en Europa: pasado e presente*, Santiago de Compostela.

Huntington, S.P. (1971), *The Politics of Modenization*, University of Chicago Press, Chicago.

Ibarra, P. (1987), *La evolución estratégica de ETA (1963-1987)*, Kriselu, San Sebastián.

Ibarra, P. (1995), 'Nuevas formas de comportamiento político', in *Inguruak* 13, Leioa.

Ibrahim, F. (1995), 'The Kurdish national movement and the struggle for national autonomy', in Berberoglu, B. (ed.), *The National Question*, Temple University press, Philadelphia.

Inglehart, R. (1991*), El cambio cultural en las sociedades industriales avanzadas*, C.I.S., Madrid.

Instituto de Ciencias Internacionales, I .C .I .(1984), *Coloquio sobre Terrorismo Internacional*, M . Huerta, Madrid.

Jenkins, B, (1975), *International Terrorism: a new mode of conflict*, Rand Corporation.

Kant, I. (1987), *Teoría y práctica*, Tecnos, Madrid.

Keating, M. (1996), *Naciones contra el Estado. El nacionalismo de Cataluña, Quebec y Escocia*, Ariel Ciencia Política, Barcelona.

Keating, M. (1995), 'Scotland, Nationalism and the UK State', Congreso internacional *Os nacionalismos en Europa: pasado e presente*, Santiago de Compostela.

Khakimov, R. (1995), *Federalization and stability; a path forward for the Russian Federation*, Bulletin, Harvard University.

Knippenberg, H. (1992), *The nationalities question in Soviet Union*, La Haye.

Kohn, H. (1949), *Historia del Nacionalismo*, F. C. E., México.

Konetzke, R. (1984), 'América latina. II. La época colonial', *Historia Universal siglo XXI*, vol. 22, Ed, Siglo XXI, Madrid.

Kriegel, A. (1983), *La race perdue*, P.U.F., Paris.

Krumeich, G. (1995), 'Juana de Arco y el nacionalismo francés en el siglo XIX', Congreso Internacional *Os nacionalismos en Europa: pasado e presente*,

Santiago de Compostela.

Laffont, R. (1971), *Décoloniser en France: les Régions face a l'Europe*, Gallimard, Paris.

Lagroye, J. (1985), 'La légitimation', in Grawitz, M. and Leca, J. (eds.) *Traité de Science Politique. Vol. 1: La Science politique, Science sociale. L'ordre politique*, P.U.F., Paris.

Laitin, D. (1996), *'Resurgimientos nacionalistas y violencia'*, in *Sistema*, Madrid.

Laplantine, F. (1974), *Les trois voix de l'imaginaire*, Ed. Universitaires, Paris.

Laqueur, W. (1979), *Terrorisme*, P.U.F., París.

Lavau, G. and DuhameL, O. (1985), 'La démocratie', en Grawitz, M. and Leca, J. (eds.): *Traité de Science Politique. Vol. 2: Les régimes politiques contemporains*, P.U.F., Paris.

Le Goff, J. (1984), 'La baja edad media', *Historia Universal siglo XXI*, vol. 11, Ed. Siglo XXI, Madrid.

Leca, J. (1985), 'La théorie politique', in Grawitz, M. and Leca, J. (eds.), *Traité de Science Politique. Vol. 1: La Science politique, Science sociale, L'ordre politique*, P.U.F., Paris.

Lee, J. J. (1992), *Ireland. 1912-1985. Politics and society*, Cambridge University Press, Cambridge.

Lefebvre, H. (1968), *La vie quotidienne dans le monde moderne*, Gallimard, Paris.

Lefebvre, H. (1976), *L'Etat dans le monde moderne*, 4 tomes, Union Générale d'Editions, Paris.

Lenin (1948), *Sobre el derecho de las naciones a la autodeterminación*, Ediciones en Lengua Extranjera, tome V, Moscow.

Lenin (1975), 'Qué hacer?'. 'El imperialismo, fase superior del capitalismo', *Obras escogidas*, 3 Vol, Akal, Barcelona.

Letamendia, F. (1994), *Historia del nacionalismo vasco y de ETA*, 3 vol., R&B Ediciones, San Sebastián.

Letamendia, F. (1994), 'Basque nationalisms and their strategies', in *Nation-building and sub-cultures. Journal of behavioral and social sciences*, vol. 1994, 3. Kanagawa, Japan.

Letamendia, F., Castro, J. L. and Borja, A. (1994), *Cooperación transfronteriza Euskadi-Aquitania (aspectos políticos, económicos y de Relaciones Internacionales)*, Servicio Editorial de la UPV-EHU, Leioa.

Letamendia, F. (1995), 'Basque nationalism and the struggle for self-determination in the Basque Country', in Berberoglu, B. (ed.), *The National Question*, Temple University Press, Philadelphia.

Letamendia, F. (1995), 'On nationalism in situations of conflict (reflections from the basque case)'. Congreso Internacional *Os nacionalismos en Europa, pasado e presente*, Santiago de Compostela.

Letamendia, F. (1995), 'European construction: Regional, State and supra-State levels', in *Ethnicity. National movements. Social Praxis*, Economica, Saint Petersburg.

Letamendia, F. (1996), *Identity and instrumental violence in peripheral nationalisms as social movement*, II European Conference on Social Movements, Vitoria-Gasteiz.

Letamendia, F., Gomez Uranga, M., Etxebarria, G. (1996), 'Astride two States:

cross-border cooperation in the Basque Country', in O'Dowd, L., Wilson, T. (Eds.), *Borders, Nations and States,* Avebury, Aldershot.

Letamendia, F. (1997), 'Basque nationalism and cross-border cooperation between the Southern and Northern Basque Countries', *in Regional &Federal Studies,* Frank Cass Journals, London.

Letamendia, F. (1997), *Les identités éthniques et nationales dans les nationalismes anciens et nouveaux,* Revue Internationale de Politique Comparée, Louvaine.

Levasseur, G. (1977), 'Les aspects repressifs du terrorisme international' in Guillaume, G. y Levasseur, G, (eds), *Terrorisme International,* Imprimerie Grandville.

Levi-Strauss, C. (1958), *Anthropologie structurale,* Plon, Paris.

Levi-Strauss, C. (1961), *Race et histoire,* Editions Gouthier, UNESCO.

Levi-Strauss, C. (1962), *La pensée sauvage,* Plon, Paris.

Lewis, B. (1991), 'Europe, Islam et societé civile'. 'L'identité chiite', in *Islam et politique au proche-Orient aujourd'hui,* Gallimard, Paris.

Linz, J. J. (1975), 'Authoritarian and Totalitarian Regimes', in Greenstein and Polsby (eds.), *Handbook of Political Science. vol. III. Macropolitical Theory,* Reading Mass, Addison Wesley.

Lipset, S. M. and Rokkan, S. (1976), *Party Systems and Voter Alignments: Cross National Perspectives,* N. Y. Free Press, New York.

Lipset, S. M. (1987), *El hombre político. Las bases sociales de la política,* Tecnos, Madrid.

Llera, F. J. (1994), *Los vascos y la política,* Servicio Editorial UPV/EHU, Leioa.

Locke, J. (1955), *Ensayo sobre el Gobierno civil,* Aguilar.

Loughlin, J. (1987), *Regionalism and Ethnic Nationalism in France: a case study of Corsica,* European University Institute, Florence.

Loughlin, J. (1993), 'Regions in the New European State', in *Peripheral regions and european integration,* Research seminar, Belfast.

Loughlin, J. (1995), *Nouvelles formes de l'Etat dans l'Europe contemporaine,* Colloque Aquitaine-Euskadi, Institut d'Etudes Politiques, Bordeaux.

Mabileau, A. (1985), 'Les régimes politiques contemporains. Les institutions locales et les relations centre-périphérie', in Grawitz, M. and Leca, J. (eds.), *Traité de science politique,* P.U.F., Paris.

Macridis, R.C. (1986), *Contemporary Politic Ideologies: Movements and Regimes,* Little Brown, Boston.

Magré F. J., Martinez Herrera, E. (1996), 'La cultura política', in *Manual de Ciencia política,* Tecnos. Madrid.

Maisonneuve, J. (1973), *Introduction à la psychosociologie,* Presses Universitaires de France.

Maiz, R. (1995), 'La construcción de las identidades políticas', in *Inguruak* 13, Leioa.

Maiz, R. (1995), 'The open-ended construction of a nation: the Galitian case in Spain'. Congreso Internacional *Os nacionalismos en Europa: pasado e presente,* Santiago de Compostela.

Mandrou, R. (1985), *Histoire sociale, sensibilités collectives et mentalités,* P.U.F., Paris.

Mannheim, K. (1956), *Ideologie et Utopie,* M. Riviére, Paris.

Mansvelt, B. J. (1993), 'Basque and Catalan nationalisms in comparative perspective', Congreso Internacional *Os nacionalismos en Europa: pasado e presente*, Santiago de Compostela.

Mao-tse-tung, (1971), *Obras escogidas*, Ediciones en lenguas extranjeras, Pekin.

Marcuse, H. (1968), *L'homme unidimensionnel*, Les Editions de Minuit, Paris.

Marcuse, H. (1968), *La fin de l'utopie*, Editions du Seuil, Paris.

Marcuse, H. (1969), *Vers la libération*, Les Editions de Minuit, Paris.

Marcuse, H. (1970), *Culture et société (1933-1968)*, Les Editions de Minuit, Paris.

Marcuse, H. (1970), 'Le vieillissement de la psychanalyse', in *Culture et Société*, Les Editions de Minuit, Paris.

Marcuse, H. (1982), *Eros et civilisation*, Les Editions de Minuit, Paris.

Marighela, C. (1970), *Acción libertadora, (Mini-manual del guerrillero urbano)*, Ediciones latinoamericanas, François Maspero, Paris.

Markale, J. (1985), *Identité de Bretagne*, Editions Entente, Paris.

Martinet, A. (1970), *Eléments de linguistique générale*, Librairie Armand Collin, París.

Martinez Sospedra, (1996), *Introducción a los partidos políticos*, Ariel, Barcelona.

Marx, K. (1956), *La ideología alemana*, Grijalbo, Barcelona.

Marx, K. (1956), *Tesis sobre Feuerbach*, Grijalbo, Barcelona.

Marx, K. (1970), *Contribución a la crítica de la economía política*, A. Corazón Editor, Madrid.

Marx, K. (1975-76), *El Capital. Crítica de la economía política*, 3 vol., FCE., Madrid.

Marx, K. (1968), *Manuscritos. Economía y Filosofía*, Alianza, Madrid.

Mattelart, A. (1976), *Multinationales et Systèmes de Communication. Les Appareils Idéologiques de l'imperialisme*, Editions Anthropos, Paris.

Mayes, D (1993), 'Peripherality and the Single Market', in *Peripheral regions and european integration*, Research seminar, Belfast.

McCarthy J. and Zald, M. N. (eds) (1979), *The dinamics of social movements*, CA. MA. Wintrop.

Melucci, A. (1981), 'Ten Hypotheses for the Analysis of New Movements' in D. Pinto (ed.), *Contemporary Italian Sociology*, Cambridge University Press, Cambridge.

Melucci, A. (1985), 'The symbolic challange of contemporary movements', in *Social Research*, 52, 4.

Meny, Y. And Thoening, J. C. (1992), *Las Políticas Públicas*, Ariel, Barcelona.

Michaud, G. (1978), 'Ethnotype comme systeme de signification', in *Identités collectives et relations inter-culturelles*, Edition Complexe.

Mickolus, E. (1979), 'International terrorism', in Stohl, M., *The politics of terrorism*, Marcel Dekker Inc, New York.

Milbrath, L. W. (1965), *Political Participation*, Rand McNally, Chicago.

Mill, J.S. (1965), *De la libertad. Del Gobierno representativo. La esclavitud femenina*, Tecnos, Madrid.

Moreno, L. (1995), *Escocia, nación y razón*, CISC, Madrid.

Mota, F. (1996), *El capital social en el estudio de los movimientos nacionalistas*, II Congreso de la AECPA, Santiago.

Muhlmann, W. E. (1968), *Messianismes révolutionnaires du Tiers Monde*, Ed.

Gallimard, Paris.

Nairn, T. (1979), *Los nuevos nacionalismos en Europa*, Península, Barcelona.

Nieburg, H. (1969), *Political violence: the behavioral process*, St. Martin's Press, New York.

Nielsson, G. P. (1989), 'Sobre los conceptos de etnicidad, nación y Estado', in *Sociología del nacionalismo*, Servicio Editorial UPV-EHU, Leioa.

Nigoul, C. (1981), *L'Autonomie: les Régions d'Europe, en quête d'un Statut*, Institut Européen des Hautes Etudes Internationales, Presses d'Europe.

Noiriel, G. (1994), 'L' "identité nationale" dans l'historiographie française', in *L' identité politique*, PUF., Paris.

Nozick, R. (1974), *Anarcy State and utopia*, Blackwell, Oxford.

O'Connor, J. (1973), *The fiscal crisis of the State*, St. Martin's Press, New York.

O'Dowd, L. (1993), *Reestructuración económica y desigualdad étnica en la periferia europea: el caso de Irlanda del Norte*, Conference, UPV., Leioa.

O'Dowd, L., Corrigan, J., Moore, T.(1993), 'Strengthening the Irish Border on the Road to Maastricht', in *Peripheral regions and european integration*, Research seminar, Belfast.

Oberschall, A., (1973), *Social conflicts and Social Movements*, Englewoods Cliffs, NJ, Prentice-Hall.

Obieta, J. A. (1980), *El Derecho de Autodeterminación de los Pueblos: un Estudio Interdisciplinar de Derechos Humanos*, Publicaciones de la Universidad de Deusto, Bilbao.

Offe, C. (1988), *Partidos políticos y nuevos movimientos sociales*, Sistema, Madrid.

Palau, J. (1993), 'El mito anti-serbio: claves, objeciones y comentarios', Congreso Internacional *Os nacionalismos en Europa: pasado e presente*, Santiago de Compostela.

Panebianco, A. (1988), 'La dimensión internacional de los procesos políticos', in Pasquino, G., *Manual de Ciencia Política*, Alianza, Madrid.

Parsons, T. (1968), *La Estructura de la Acción Social*, 2 vols, Guadarrama, Madrid.

Pasquino, G. (1991), 'Naturaleza y evolución de la disciplina', in Pasquino, G. (1988), *Manual de Ciencia Política*, Alianza, Madrid.

Pasquino, G. et al. (1988), *Manual de Ciencia Política*, Alianza, Madrid.

Pastor, M. (1994), 'Las Ideologías Políticas. Apéndice: La Ciencia Política en España', in M. Pastor, *Fundamentos de Ciencia Política*, McGraw-Hill, Madrid.

Pastor, M. (1988), *Ciencia Política*, McGraw-Hill, Madrid.

Pearson, R. (1995), 'The nationalist wave in Eastern Europe'. Congreso Internacional *Os nacionalismos en Europa: pasado e presente*, Santiago de Compostela.

Penrose, J. (1995), 'Reification in the name of change: the impact of nationalism on social construction of nation, people and place in Scotland and the United Kingdom'. Congreso Internacional *Os nacionalismos en Europa: pasado e presente*, Santiago de Compostela.

Percheron, A. (1985), 'La socialisation politique: défense et illustration', in Grawitz, M. and Leca, J. (1985), *Traité de Science Politique. Vol. 3: L'action politique*, P.U.F., Paris.

Perlmutter. A. (1977), *The military and politics in Modern Times: On professionnal, pretorians and revolutionary soldiers*, Yale University Press, New Haven, and

London.

Petrosino, D. (1995): 'Is it possible to invent ethnic identity? Some reflections on ethnic and territorial politics in Italy'. Internacional Congress *Os nacionalismos en Europa: pasado e presente*, Santiago de Compostela.

Pizzorno, A. (1978), 'Political exchange and collective identity in industrial conflict,' in Crouch and Pizzorno (eds.), *The resurgence of class conflict in Western Europe since 1960*, Macmillan, London.

Plamenatz, J. (1970), *Ideology*, Macmillan, London.

Postiglione, G. (1995), 'National minorities and nationalities policy in China', in Berberoglu, B. (ed.), *The National Question*, Temple University Press, Philadelphia.

Poulantzas, N. (1978), *L'Etat, le pouvoir, le socialisme*, P.V.F.

Pourcher, I. (1992), *La dynamique des territoires en France: l'exemple des politiques régionales*, Colloque du centre europeen de recherche sur les pratiques politiques.

Proudhon (1868), 'Du principe féderatif', *Oeuvres Completes*, tome VIII, Librairie Internationale, París.

Quadruppani, S. (1989), *L'anti-terrorisme en France ou la terreur integrée, 1981-89*, La Découverte, Enquêtes, Paris.

Rapoport, D. C. and Yonah, A. (eds.) (1982), *The morality of terrorism: Religious and Secular Justification*, Pergamon, New York.

Rawls, J. (1970), *A Theory of Justice*, Oxford University Press, Oxford.

Reberioux, M (1975), 'Maoismo'', in *Diccionario del Saber Moderno. La historia de 1871 a 1971, las ideas y los problemas*, tome 8, Ediciones Mensajero, Bilbao.

Reich, W. (1972), *La psychologie de masse du fascisme*, Payot, Paris.

Reich, W. (1971), *L'analyse caracterielle*, Payot, Paris.

Reinares, F. (1996), 'Fundamentos para una política antigubernamental antiterrorista en el contexto de regímenes democráticos', in *Sistema*, 132-133, Madrid.

Renan, E. (1934), *Qu'est-ce qu'une nation?*, Helleu éditeurs, Paris.

Ribó, R. and Pastor, J. (1996), 'La estructura territorial del Estado', in *Manual de Ciencia política*, Tecnos, Madrid.

Rodinson, M. (1975), 'Ethnisme et racism', in *Pluriel-Debat*, n.º 3, Paris.

Roheim, G. (1967), *Psychanalyse et Anthropologie*, Gallimard, Paris.

Rokkan, S. (1970), *Citizens, elections, parties*, Universitetsforlaget, Oslo.

Rokkan, S. and Urwin (1982), *The Politics of Territorial Identity: Studies in European Regionalism*, Sage, London.

Rokkan, S. & Urwin, D. W. (1983), *Economy, Territory, Identity: Politics of West European Peripheries*, Sage, London.

Rousseau, J.J. (1987), *Discurso sobre el origen y los fundamentos de la desigualdad entre los hombres, y otros escritos*, Tecnos, Madrid.

Rousseau, J. J. (1977), *Le contrat social*, Editions du Seuil, Paris.

Rubiralta, F. (1988), *Origens i desenvolupament del PSAN (1969-1974)*, Edicions de la Magrana.

Ruiz De Olabuenaga, J. I., Fernandez Sobrado, J. M., and Morales F. (1985), *Violencia y ansiedad en el País Vasco*, Ediciones Ttartalo.

Sabine, G. (1995), *Historia de la teoría política*, Fondo de Cultura Económica, Madrid.

Samary, C. (1993), 'Nationalismes: la tragédie yugoeslave'. *Le Monde Diplomatique,* 17, Paris.

San Román, T. (1995), 'The Gorgio law and the calé Gipsies' position in the Spanish State, in *The criminalisation and victimisation of gipsies in the new Europe,* Oñati International Institute for the sociology of law, Oñati London.

Sanchez, J. (1996), 'El Estado de Bienestar', in *Manual de Ciencia política,* Tecnos, Madrid.

Sartori, G. (1969), 'Politics, Ideology and Belief Systems', *American political Science Revie,* June.

Sartori, G. (1980), *Partidos y sistemas de partidos,* Alianza, Madrid.

Sartori, G. (1992), *Elementos de Teoría Política,* Alianza, Madrid.

Sartre, J. P. (1960), *Critique de la raison dialectique,* Gallimard, Paris.

Sartre, J. P. (1961), Préface a Fanon F., *Les damnés de la terre,* Maspero, Paris.

Schemeil, Y. (1985), 'Les cultures politiques', in Grawitz, M. and Leca, J. (eds.), *Traité de Science Politique. Vol. 3: L'action politique,* P.U.F., Paris.

Schmitt, C. (1972), *La notion de politique,* Calmann-Lévy, Paris.

Schmitter, P. and Lehmbruch, G. (1979), *Trends Towards Corporatist Intermediation,* Ca. Sage, Beverly Hills.

Seiler, D. L. (1980), *Partis et familles politiques,* Presses Universitaires de France, Col. Thémis, Paris.

Seiler, D. L. (1982), *Les partis autonomistes,* P. U. F., Paris.

Seiler, D. L. (1989), 'Peripheral Nationalism between Pluralism and Monism', *International Political Science Review,* vol.10, n° 3.

Seiler, D. L. (1990), *Sur les partis autonomistes dans la CEE,* ICPS, Barcelona.

Seiler, D. L. (1993), 'Inter-ethnic relations in East Central Europe', in *Communist and post-communist studies,* volume 26, California.

Seiler, D. L. (1994), 'L'Etat autonomique et la science politique: centre, periphérie et territorialité', in *L'Etat autonomique: forme nouvelle ou transitoire en Europe?,* Ed. Economica.

Seiler, D. L. (1996), *L'exemplarité du nationalisme basque: une perspective comparée,* Fundación Sabino Arana. Iruña.

Shafer, B. (1964), *Le Nationalisme. Mythe et réalité,* Payot, Paris.

Simon, P. J. (1976), 'Propositions pour une léxique de mots-clé dans le domain des études relationnelles', in *Pluriel-Débat,* n° 7, Paris.

Smith, A. (1976), *Las Teorías del Nacionalismo,* Península, Barcelona.

Smith, A. D. (1979), *Nationalism in the twentieth century,* Martin Robertsons, London.

Soldevila, F. (1978), *Síntesis de historia de Cataluña,* Destino Libro, Barcelona.

Soucy, P. Y. and Mascotto, J. (1979), *Sociologie politique de la question nationale,* Editions Coopératives Albert St. Martin, Montreal.

Soulier, G. (1985), 'Les institutions judiciaires et répressives', in Grawitz, M. and Leca, J. (eds.), *Traité de Science Politique. Vol. 2: Les régimes politiques contemporains,* P.U.F., Paris.

Stalin, J. (1945), *Le matérialisme dialectique et le matérialisme historique,* Ed. Sociales, Paris.

Sterling, C. (1981), *La reseau de la terreur: enquête sur le terrorisme international,* J.C. Lattés, Paris.

Strassoldo, R. (1996), 'Ethnic regionalism versus the State: the case of Italy's Northern Leagues', in O'Dowd, L., Wilson, T. (Eds.), *Borders, Nations and States*, Avebury, Aldershot.

Tafani, P. (1986), *Géopolitique de la Corse*, Fayard-La Marge, Paris.

Tap, P. (1979), 'L'identification est-elle une aliénation de l'identite?', in *Identité individuelle et personnalisations*, Presses Universitaires de Toulouse, Toulouse.

Targ, H. (1979), 'Societal Structures' in Stohl, M., *The politics of terrorism*, Marcel Dekker Inc., New York.

Tarrow, S. (1991), 'Ciclos de protesta'. *Zona Abierta*, n° 56, Madrid.

Tertsch, H. (1993), 'El mito serbio', in *La venganza de la historia*, El País.

Thoenig, J.C. (1985), 'Les politiques publiques. Présentation', in Grawitz M. and Leca, J.(eds.), *Traité de science politique*, P.U.F., Paris.

Tilly, C. (1975), 'Revolutions and Collective Violence', in Greenstein, F. and Polsby, N. (eds.), *Handbook of Political Science. Vol. 3. Macropolitical Theory*, Reading, Addison-Wesley, Massachussets.

Tilly, C. (1985), 'Models and realities on popular collective action', in *Social Research*, 52, 4.

Tilly, C. (1975), *The formation of National States in Western Europe*, Princeton, New Jersey.

Titarenko, M. L. (1994), 'Eurasianism; a paradigm of Russia', in *Nation-building and sub-cultures. Journal of behavioral and social sciences*, vol. 1994, 3, Kanagawa, Japan.

Tonnies, F. (1944), *Communauté et société*, P.U.F., Paris.

Tortura en Euskadi (La) (1986), *Revista "Eliza-Herria 2000"*, Editorial Revolución.

Touraine, A. (1985), 'An Introduction to the study of social movements', in *Social Research*, 52, 4.

Touraine, A and Dubet, F. (1981), *Le pays contre l'Etat. Luttes occitanes*, Editions du Seuil, Paris.

Touraine, A. (1978), *La voix et le regard*, Editions du Seuil, Paris.

Tyriakian, E. and Rogowski, R. (1985), *New Nationalisms of the Developed West*, Allen & Unwin, Boston.

Vallespin, F. (1990), 'Aspectos metodológicos de la Teoría Política', in Vallespin, F. (ed.), *Historia de la Teoría Política (1)*, Alianza Editorial, Madrid.

Vervaele, J. (1986), 'El Espacio Judicial y Policial Europeo elaborado por las Instituciones Políticas de Europa (Consejo de Europa y Mercado Común) versus La Protección de los Derechos del Hombre', in *Repercusiones de la entrada en la CEE: perspectivas desde Euskadi. La Europa Económica*, tome I, Ipes-Vizcaya, Imprenta Berekintza, Bilbao.

Vervaele, J. (1985), 'Perspectiva europea de la situación de los Derechos Humanos', in *Democracia y Leyes Terroristas en Europa*, Ipes-Vizcaya, Imprenta Berekintza, Bilbao.

Vilar, P. (1980), 'Pueblos, naciones, Estados', in *Iniciación al vocabulario del análisis histórico*, Crítica, Barcelona.

Vilar, P. (1982), 'Estado, nación y patria en las conciencias españolas: historia y actualidad', in *Hidalgos, amotinados y guerrilleros: pueblos y poderes en la historia de España*, Crítica, Grupo Editorial Grijalbo, Barcelona.

Vincent, A. (1992), *Modern political ideologies*, Blackwell Publishers, Oxford.

Vodopivec, P. (1995), 'Central Europe, ex-Yugoeslavia and the Balkans: new or old nationalisms?', Internacional Congress *Os nacionalismos en Europa: pasado e presente"*, Santiago de Compostela.

Vovelle, M. (1985), *Ideologías y mentalidades,* Ariel, Barcelona.

Wakabayashi, H.(1994), 'Dual process of Belgian federalisation', in *Nation-building and sub-cultures. Journal of behavioral and social sciences,* vol. 1994, 3, Kanagawa, Japan.

Waldmann, P. (1996), 'Sociedades en guerra civil', *Sistema,* 132-133. Madrid.

Weber, M. (1971), *Economie et societé,* Librairie Plon, París.

Welty, G. (1995), 'Palestinian Nationalism and the struggle for national self-determination', in Berberoglu, B., (Ed.), *The National Question,* Temple University press, Philadelphia.

Wiewiorka, M. (1988), *Societés et Terrorisme,* Librairie Arthème Fayard.

Wiewiorka, M and Touraine, A. (1980), *La prophétie antinucléaire,* Editions du Seuil, Paris.

Wilkinson, P. (1976), *Terrorismo político,* Ediciones Felmar, Madrid.

Wolfgang, M. E. and Ferracutti, F. (1967), *The subculture of violence: Metaphor and sacrament,* Tavistock, London.

Ziatdinov, V. (1994), 'The ethnic mobilization and Regional developement: the case of Tartars', International Congress on *Restructuring identities and political space in Europe,* Sieldce, Poland,

Zolberg, A. R. (1985), 'L'influence des facteurs "externes" sur l'ordre politique interne', in Grawitz, M. and Leca, J. (eds.), *Traité de Science Politique. Vol. 1: La Science politique, Science sociale. L'ordre politique,* P.U.F., Paris.

Zulaika, J. (1988), *Basque violence. Metaphor and Sacrament,* University Press, Reno, Nevada.

Printed and bound by CPI Group (UK) Ltd, Croydon, CR0 4YY

22/10/2024

01777620-0010